Microfluidics for Biomedical Applications (2nd Edition)

Microfluidics for Biomedical Applications (2nd Edition)

Guest Editors

Nan Xiang
Zhonghua Ni

MDPI

Basel • Beijing • Wuhan • Barcelona • Belgrade • Novi Sad • Cluj • Manchester

Guest Editors
Nan Xiang
Southeast University
Nanjing
China

Zhonghua Ni
Southeast University
Nanjing
China

Editorial Office
MDPI AG
Grosspeteranlage 5
4052 Basel, Switzerland

This is a reprint of the Special Issue, published open access by the journal *Biosensors* (ISSN 2079-6374), freely accessible at: https://www.mdpi.com/journal/biosensors/special_issues/0N8AZT6GC8.

For citation purposes, cite each article independently as indicated on the article page online and as indicated below:

Lastname, A.A.; Lastname, B.B. Article Title. *Journal Name* **Year**, *Volume Number*, Page Range.

ISBN 978-3-7258-2821-0 (Hbk)
ISBN 978-3-7258-2822-7 (PDF)
https://doi.org/10.3390/books978-3-7258-2822-7

Cover image courtesy of Nan Xiang

Contents

About the Editors

Nan Xiang

Nan Xiang is a Professor in the School of Mechanical Engineering at Southeast University, Nanjing, China. He is now the PI of the Microfluidics and Soft Robotics Group in Jiangsu Key Laboratory for Design and Manufacture of Micro-Nano Biomedical Instruments. He received his B.E. degree in Biosystems Engineering from Zhejiang University in 2009 and his Ph.D. degree in Mechanical Engineering from Southeast University in 2014. His research interests include inertial microfluidics, soft robotics, microfluidic cell separation, viscoelastic microfluidics, point-of-care testing devices, microflow cytometers, and dielectrophoresis. He has published over 130 peer-reviewed articles and applied for over 100 patents. His research has gained increasing recognition through important awards and honors, such as First Prize of Natural Science of the Ministry of Education, Top Cited Paper Award from the Institute of Physics, Jiangsu Science Fund for Excellent Young Scholars, Provincial Excellent Doctoral Dissertation of Jiangsu, the Six Talent Peaks Project of Jiangsu Province, 333 Young Academic Leader of Jiangsu Province, and the Zhishan Young Scholar Fellowship of Southeast University (first class).

Zhonghua Ni

Zhonghua Ni is a Professor in the School of Mechanical Engineering and the Director of the Jiangsu Key Laboratory for Design and Manufacture of Micro-Nano Biomedical Instruments, Southeast University, Nanjing, China. He received his B.E. degree from the China University of Petroleum in 1990 and his Ph.D. degree from Southeast University in 2001. His research interests include microfluidics, biomedical microdevices, and biosensors. He has published over 300 peer-reviewed articles in well-known SCI journals such as *Lab on a Chip, Analytical Chemistry,* and *Applied Physics Letters.*

biosensors MDPI

Editorial

Innovations in Microfluidics to Enable Novel Biomedical Applications

Nan Xiang * and Zhonghua Ni

School of Mechanical Engineering, Jiangsu Key Laboratory for Design and Manufacture of Micro-Nano Biomedical Instruments, Southeast University, Nanjing 211189, China; nzh2003@seu.edu.cn
* Correspondence: nan.xiang@seu.edu.cn

As a new technique for precisely controlling micro-/nanoparticles and fluids at the microscale, microfluidics has been attracting increased interest in the fields of material science, medical diagnosis, biological research, and even soft robotics. For use in biomedical applications, microfluidics has been widely employed in areas including biosensing, sample preparation, cell culture, and drug discovery. Some microfluidic products have already been successfully commercialized and have reformed traditional techniques.

Our first Special Issue, "Microfluidics for Biomedical Applications", was devoted to the most exciting technical innovations in the area of microfluidics, particularly in relation to biomedical applications. We successfully launched the second volume of this Special Issue following the triumph of the first volume. A total of twelve outstanding papers (including seven research articles and five reviews) are included in this Special Issue. We will now briefly introduce these twelve papers.

Cell co-culture is a powerful tool for studying the communications and interactions between cells involved in various cellular activities. Li et al. (contribution 1) review the recent advances in three-dimensional (3D) cell co-culture using microfluidics. They offer new insights into the design of 3D co-culture microfluidic devices and the detection techniques utilized in 3D co-culture microfluidic devices. Finally, they introduce interesting applications for these 3D co-culture microfluidic devices to address real biomedical demands.

The traditional soft lithography technique used for fabricating microfluidic devices is complex and time-consuming and requires a cleanroom environment. With 3D printing, we can rapidly create microstructures or microchannels in bulk materials, and this represents an important alternative for fabricating microfluidic devices. In their review, Li et al. (contribution 2) present a comprehensive summary of the use of light-driven 3D printing techniques to manufacture advanced microfluidic devices. More importantly, they summarize three light-driven 3D printing strategies for creating microfluidic devices for use in cell culture and tissue engineering.

The enrichment of rare circulating tumor cells (CTCs) from peripheral blood is a challenge due to the rarity of CTCs (typically less than 50 CTCs in 1 mL blood) and their large heterogeneity. Sen-Dogan et al. (contribution 3) designed a spiral inertial microfluidic device with hydrofoil-shaped pillars for the high-throughput and label-free separation of CTCs. Their results prove that the new spiral design performs significantly better than the conventional spiral design in terms of the recovery ratio of tumor cells and the depletion ratio of white blood cells. They also analytically validated the device performances using three different tumor cell lines (A549, SKOV-3, and BT-474).

The accurate detection of micro-/nanomaterials (e.g., biomolecules, exosomes, and cells) using resistive pulse sensors represents a promising enabling technology in microfluidics. However, the throughputs of most existing resistive pulse sensors are low. Xu et al. (contribution 4) developed a bipolar pulse-width, multiplexing-based resistive pulse sensor for the high-throughput detection of microparticles. Low errors—2.6% and 6.1%—were

Citation: Xiang, N.; Ni, Z. Innovations in Microfluidics to Enable Novel Biomedical Applications. *Biosensors* **2024**, *14*, 507. https://doi.org/10.3390/bios14100507

Received: 11 October 2024
Accepted: 15 October 2024
Published: 17 October 2024

achieved for particle sizing and counting, respectively. The practicability of their device was further demonstrated through the detection of HeLa cells.

Diabetes mellitus is an endocrine disease that is common globally and seriously affects human health. Glutamic acid decarboxylase antibody (GADAb) is regarded as a biomarker for the clinical diagnosis of type 1 diabetes. Tao et al. (contribution 5) developed a glass capillary solid-state nanopore for the detection of glutamic acid decarboxylase (GAD65), GADAb, and their antigen–antibody complexes. This glass capillary solid-state nanopore could be employed without any modifications and thus was cost-effective and easy to operate. Based on this nanopore system, the authors successfully achieved the differentiation of GAD65, GADAb, and GADAb-GAD65 complexes.

Magnetic micro-/nanoparticles (MMPs or MNPs) are widely employed in many biomedical applications, as they offer the advantages of high biocompatibility and diverse expansibility. Ger et al. (contribution 6) developed a microfluidic device embedded with a giant magnetoresistance sensor and successfully detected low-concentration MNPs at a velocity of 20 mm/s. A high detection sensitivity of 10 μg/μL for MNPs was achieved using a vertical magnetic field of 100 Oe and a horizontal magnetic field of 2 Oe.

Lunelli et al. (contribution 7) propose a microfluidic scheme using specifically functionalized MMPs inserted in polymeric microchambers. MMPs were functionalized with aptamers, antibodies, or small functional groups for coupling with specific antibiotics. These three different functionalization strategies were carefully compared. The authors found that the functionalization with aptamers allowed them to capture and release almost all tetracycline and to deliver an enriched and simplified antibiotic solution.

Stable and uniform droplet generation at a high throughput is critical in the accurate and efficient detection of digital nucleic acid. Luo et al. (contribution 8) developed a step emulsification microfluidic device with the advantages of flexible droplet generation capability, a small footprint, simple fabrication, low contamination, and high robustness. To increase the uniformity of the generated droplets, a tree-shaped droplet generation nozzle was designed by equating flow rates. Finally, the stable and uniform droplets generated by the step emulsification microfluidic device were employed for nucleic acid amplification and detection.

Intracellular delivery refers to the transportation of substances into cells and is a crucial process in various cellular applications (e.g., drug delivery and gene editing). Among the reported intracellular delivery technologies is mechanoporation, which employs mechanical forces to create temporary pores on cell membranes for delivering substances into cells. Wang et al. (contribution 9) review recent advances in mechanoporation. First, the authors review different mechanoporation technologies and highlight the applications of mechanoporation in stem cell research. They then discuss the integration of mechanoporation into microfluidics for high-throughput intracellular delivery with enhanced transfection efficiency.

Paper microfluidics represents a promising tool for rapid diagnostics and on-site analysis in resource-limited settings, offering the advantages of biodegradability and affordability. Kumar et al. (contribution 10) present a concise overview of paper microfluidics. Sustainable sensing applications of paper microfluidics in healthcare, environmental monitoring, and food safety are explored. Fabrication techniques, principles, and applications in paper microfluidics are also discussed.

The fluidically loaded bi-material cantilever (B-MaC) is a key component of microfluidic paper-based analytical devices. Kumar et al. (contribution 11) studied the dynamics of a B-MaC constructed from Scotch Tape and Whatman Grade 41 filter paper strips. They explored the stress–strain relationship to estimate the modulus of the B-MaC under different saturation levels.

Traditional drug development based on animal experiments is expensive and time-consuming. The organ-on-a-chip (OOC) simulates complex human organ microenvironments and physiological responses in a microfluidic device and offers a promising tool for cost-effective and efficient drug development. Yuan et al. (contribution 12) review the recent advances in OOC systems for drug discovery. Their review focuses on the design, fabrication, sensing capabilities, and applications of OOC systems. Technical challenges in this field are also discussed.

We strongly believe that microfluidics will lead to a revolution in biological research and disease diagnosis in the near future.

Funding: This research was funded by the National Natural Science Foundation of China (52375562, 51875103, and 81727801) and the Natural Science Foundation of Jiangsu Province (BK20190064).

Acknowledgments: The authors are grateful for the valuable opportunity to serve as the Guest Editors of this Special Issue, "Microfluidics for Biomedical Applications II". We thank all the authors for their contributions to this Special Issue. We also thank the editorial and publishing staff of Biosensors for their support.

Conflicts of Interest: The authors declare no conflicts of interest.

List of Contributions

1. Li, C.; He, W.; Song, Y.; Zhang, X.; Sun, J.; Zhou, Z. Advances of 3D Cell Co-Culture Technology Based on Microfluidic Chips. *Biosensors* **2024**, *14*, 336.
2. Li, X.; Wang, M.; Davis, T.P.; Zhang, L.; Qiao, R. Advancing Tissue Culture with Light-Driven 3D-Printed Microfluidic Devices. *Biosensors* **2024**, *14*, 301.
3. Sen-Dogan, B.; Demir, M.A.; Sahin, B.; Yildirim, E.; Karayalcin, G.; Sahin, S.; Mutlu, E.; Toral, T.B.; Ozgur, E.; Zorlu, O.; et al. Analytical Validation of a Spiral Microfluidic Chip with Hydrofoil-Shaped Pillars for the Enrichment of Circulating Tumor Cells. *Biosensors* **2023**, *13*, 938.
4. Xu, R.; Ouyang, L.; Shaik, R.; Chen, H.; Zhang, G.; Zhe, J. Rapid Detection of Microparticles Using a Microfluidic Resistive Pulse Sensor Based on Bipolar Pulse-Width Multiplexing. *Biosensors* **2023**, *13*, 721.
5. Tao, C.; Bai, Y.; Chen, J.; Lu, J.; Bi, Y.; Li, J. Detection of Glutamate Decarboxylase Antibodies and Simultaneous Multi-Molecular Translocation Exploration by Glass Nanopores. *Biosensors* **2024**, 14, 255.
6. Ger, T.-R.; Wu, P.-S.; Wang, W.-J.; Chen, C.-A.; Abu, P.A.R.; Chen, S.-L. Development of a Microfluidic Chip System with Giant Magnetoresistance Sensor for High-Sensitivity Detection of Magnetic Nanoparticles in Biomedical Applications. *Biosensors* **2023**, *13*, 807.
7. Lunelli, L.; Germanis, M.; Vanzetti, L.; Potrich, C. Different Strategies for the Microfluidic Purification of Antibiotics from Food: A Comparative Study. *Biosensors* **2023**, *13*, 325.
8. Luo, G.; Zhang, Y.; Wang, S.; Lv, X.; Yang, T.; Wang, J. Establishment and Validation of an Integrated Microfluidic Step Emulsification Chip Supporting Droplet Digital Nucleic Acid Analysis. *Biosensors* **2023**, *13*, 888.
9. Wang, R.; Wang, Z.; Tong, L.; Wang, R.; Yao, S.; Chen, D.; Hu, H. Microfluidic Mechanoporation: Current Progress and Applications in Stem Cells. *Biosensors* **2024**, *14*, 256.
10. Kumar, S.; Kaushal, J.B.; Lee, H.P. Sustainable Sensing with Paper Microfluidics: Applications in Health, Environment, and Food Safety. *Biosensors* **2024**, *14*, 300.
11. Kumar, A.; Heidari-Bafroui, H.; Rahmani, N.; Anagnostopoulos, C.; Faghri, M. Modeling of Paper-Based Bi-Material Cantilever Actuator for Microfluidic Biosensors. *Biosensors* **2023**, *13*, 580.
12. Yuan, S.; Yuan, H.; Hay, D.C.; Hu, H.; Wang, C. Revolutionizing Drug Discovery: The Impact of Distinct Designs and Biosensor Integration in Microfluidics-Based Organ-on-a-Chip Technology. *Biosensors* **2024**, *14*, 425.

biosensors

Review

Advances of 3D Cell Co-Culture Technology Based on Microfluidic Chips

Can Li [1,†], Wei He [2,†], Yihua Song [1], Xia Zhang [1], Jianfei Sun [3,*] and Zuojian Zhou [1,*]

1 Engineering Research Center of TCM Intelligence Health Service, School of Artificial Intelligence and Information Technology, Nanjing University of Chinese Medicine, Nanjing 210023, China; lican@njucm.edu.cn (C.L.); yihua.song@njucm.edu.cn (Y.S.); zhangxia@njucm.edu.cn (X.Z.)
2 Department of Clinical Medical Engineering, The First Affiliated Hospital of Nanjing Medical University, Nanjing 210029, China; hw66886688@126.com
3 State Key Laboratory of Bioelectronics and Jiangsu Key Laboratory of Biomaterials and Devices, School of Biological Sciences & Medical Engineering, Southeast University, Nanjing 210009, China
* Correspondence: sunzaghi@seu.edu.cn (J.S.); zhouzj@njucm.edu.cn (Z.Z.); Tel.: +86-182-5195-5667 (Z.Z.)
† These authors contributed equally to this work.

Abstract: Cell co-culture technology aims to study the communication mechanism between cells and to better reveal the interactions and regulatory mechanisms involved in processes such as cell growth, differentiation, apoptosis, and other cellular activities. This is achieved by simulating the complex organismic environment. Such studies are of great significance for understanding the physiological and pathological processes of multicellular organisms. As an emerging cell cultivation technology, 3D cell co-culture technology, based on microfluidic chips, can efficiently, rapidly, and accurately achieve cell co-culture. This is accomplished by leveraging the unique microchannel structures and flow characteristics of microfluidic chips. The technology can simulate the native microenvironment of cell growth, providing a new technical platform for studying intercellular communication. It has been widely used in the research of oncology, immunology, neuroscience, and other fields. In this review, we summarize and provide insights into the design of cell co-culture systems on microfluidic chips, the detection methods employed in co-culture systems, and the applications of these models.

Keywords: microchip; co-culture; microfluidic technology; cell cultivation; intercellular communication

Citation: Li, C.; He, W.; Song, Y.; Zhang, X.; Sun, J.; Zhou, Z. Advances of 3D Cell Co-Culture Technology Based on Microfluidic Chips. *Biosensors* **2024**, *14*, 336. https://doi.org/10.3390/bios14070336

Received: 16 April 2024
Revised: 30 June 2024
Accepted: 8 July 2024
Published: 10 July 2024

1. Introduction

Intercellular communication is a key characteristic of multicellular organisms [1]. Crucially, understanding the communication mechanisms between cells is essential for unveiling the physiological and pathological processes of multicellular organisms, which encompasses cell development and growth, immune interactions, cancer metastasis, cell differentiation, as well as tissue and organ formation [2,3]. Hence, it is necessary to establish a similar in vivo cultivation environment and system to research cell–cell interactions. Cell co-culture technology, based on cell cultivation, has emerged [4]. By co-culturing different types of cells in the same environment, researchers can simulate the in vivo environment to the greatest extent, enabling cells to communicate and support each other for growth and proliferation. Furthermore, this approach allows for the exploration of the mechanisms underlying cell growth, differentiation, apoptosis, and other processes. Additionally, researchers can investigate the mechanism of drug action and targets by detecting the relationships between different cytokines [5]. The cell co-culture system can be applied to many research areas [6], such as cell differentiation, the function and vitality of cells, cell proliferation and migration, metabolic mechanisms, the development of various on-chip organs for drug testing, disease modeling, and personalized medicine.

There are mainly two types of co-culture systems established based on cell co-culture technology. One is the direct co-culture system, which involves physical contact among multiple cells in co-culture; the other is the indirect co-culture system, where co-cultured cells share the same cultivation system without direct contact. The goal of cell co-culture technology is to control parameters like cultivation conditions and cell proportions through the cell co-culture system to promote cell interaction and growth. By constructing a cell co-culture environment, researchers can simulate the complex physiological environment in vivo. It helps researchers to induce cell differentiation, regulate cell proliferation, promote early embryonic development, and provide crucial support and assistance for research in biomedical fields [7]. In native tissues and organs, the microenvironments of cells and their interactions with the surrounding environment are crucial for maintaining the normal structure and function of tissues and organs. In the two-dimensional (2D) cell cultivation method [8,9], cells grow on a flat surface, which cannot simulate the growth state of cells in a 3D environment in vivo, leading to significant differences in cell behavior. This limitation hinders the accurate simulation of the true morphology of cells in in vivo tissues and prevents the achievement of authentic intercellular communication or interaction between cells and the extracellular matrix. Consequently, 2D culture exhibits limited accuracy in drug screening and tissue engineering applications [10]. Moreover, the current understanding of intercellular communication mechanisms is still limited, and intercellular communication in multicellular systems is extremely complex. Traditional 2D cultivation methods fall short of accurately simulating physiological microenvironments and effectively transmitting intercellular communication signals [11,12]. To address these limitations and better simulate the complex intercellular communication found in the real physiological microenvironments, researchers have proposed a three-dimensional (3D) in vitro cultivation method. This method enables cells to grow and differentiate in a 3D space. Compared with the 2D culture method, the 3D culture method [13] promotes cell aggregation and tissue formation through long-term cell cultivation. It also regulates cell morphology, behavior, and function in a 3D physiological environment by manipulating gene and protein expression, proliferation, differentiation, and migration.

Microfluidics technology is a micrometer-scale method designed for processing and manipulating fluids in small channels. This technology operates on scales consistent with mammalian cells, and its unique fluid dynamic manipulation system allows it to mimic a physiological environment more similar to in vivo conditions. Microfluidic technology can co-cultivate multiple cells, generate and control signal gradients, and perform dynamic perfusion cultivation through spatially controllable methods. This enables the precise manipulation of individual cells, the simulation of physiologically relevant microenvironments, and high-throughput analysis under different conditions [14–16]. Moreover, it is expected that this technology will contribute to better research on cell growth, cell differentiation, and the effects of drugs on cells, providing a more precise and reliable experimental platform for drug screening and tissue engineering applications. In a 3D microfluidic system, diverse types of cells are cultivated in separate interconnected chambers. Microfluidic devices allow cells to obtain nutrients and oxygen through fluid circulation, exposing them to spatial cues or signal gradients required for cell differentiation, growth, vitality, and proliferation [17,18]. Additionally, microfluidic platforms enable the analysis of dynamic cell–cell interactions under reproducible in vitro cultivation conditions [19]. In recent years, microfluidic systems have been developed and widely used in many areas, such as cancer research [20,21], drug screening [22,23], vascular modeling [24,25], and neuroscience [26], to better evaluate drug efficacy and the feasibility of tissue engineering applications. The cell co-culture technology based on microfluidic chips has become a research hotspot and has been widely applied in fields such as tumor metastasis and analysis [8,27,28], anti-cancer drug screening [29–33], drug absorption, and drug metabolism [29,34,35]. It is expected to be developed into a 3D cultivation model for in vitro physiological research. This article reviews and summarizes the design of cell co-culture system on-chip, the detection of an on-chip cell co-culture system, as well as current applications extended through microfluidic features. And finally, the potential value and future development trends of 3D microfluidic cell culture technology are discussed.

2. Design of Cell Co-Culture System On-Chip

A microfluidic system is an experimental platform that integrates functions such as actuation, manipulation, monitoring, reaction, detection, and analysis [14]. Microfluidic chips serve as the core components of the microfluidic system [36]. The effective structure contains fluid at the micrometer scale in at least one dimension. Due to its micrometer-scale structure, fluids exhibit specific properties distinct from macroscopic scales [37–39]. Additionally, they are compatible with 3D cell cultivation, which opens up new avenues for establishing disease models [17]. The cell co-culture model can be used to observe interactions between cells or cells and their surrounding microenvironment [40–42]. Different types of cells can be placed on the same interface [43], and different types of cells can also be placed on separate interfaces to study the effects of certain chemical factors that regulate cell behavior [44]. These two models are named the direct contact co-culture model and indirect contact co-culture model [42,45,46], respectively (see Figure 1). In the direct co-culture model (Figure 1a), communication between cells takes place via direct cell contact, as well as autocrine and paracrine signaling pathways (illustrated by black arrows). On the other hand, in the indirect co-culture system (Figure 1b), there is no physical interaction between cells, and communication occurs solely through autocrine and paracrine mechanisms.

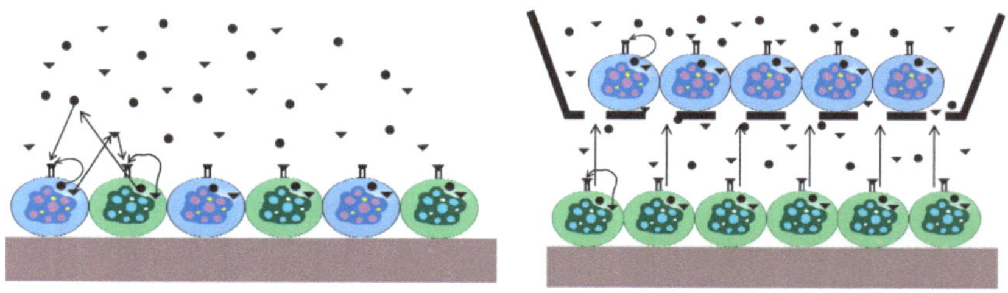

(a) (b)

Figure 1. Schematic representation of the interactions in the co-culture systems. (**a**) Direct co-culture system; (**b**) indirect co-culture system.

2.1. Direct Co-Culture of Cells

The direct contact co-culture of cells on microfluidic chips [47] typically involves the use of droplet technology, trap capture technology, and bioprinting technology to control different cells in the same confined space. Microdroplet technology is a micro technique that utilizes the interaction between flow shear force and surface tension to divide a continuous fluid into discrete microliters/nanoliters or smaller-volume droplets in micrometer-scale channels [48]. During the generation of droplets, two immiscible liquids act as continuous and discrete phases, respectively, and enter different microchannels under the drive of an injection pump with fixed volume flow rate. When the two fluids meet at the intersection, the discrete phase fluid continues to extend to form a "plug like" or "jet like" liquid column. Under the shear and compression of the continuous phase fluid, it fractures due to the instability of the free interface, and the "plug like" or "jet like" liquid column is sandwiched and dispersed in the continuous phase in the form of small-volume units to form droplets. Microdroplets have the characteristics of small volume, large surface area, fast speed, large flow rate, uniform size, closed system, and internal stability. A large number of uniform droplets with picoliter volume can separate the solution to various levels, and individual cells and molecules in each droplet can be visualized, barcoded, and analyzed. By adjusting the geometric structure, two-phase viscosity, flow rate, wettability, and interfacial tension of microfluidic channels, precise control of droplet size, morphology, uniformity, and other parameters can be achieved. Cell capture with microfluidic devices can be divided into contact capture and non-contact capture [49]. Contact capture refers to the direct contact between cells and the trap area during the capture process. This method does not require external energy, but relies on fluid dynamics. On the other hand, non-contact capture refers to the process in which cells do not come into contact with the trap area. However, this method requires external energy, such as electric and magnetic fields. Bioprinting technology can precisely control the arrangement of cells in three-dimensional space and has become the core technology for the fabrication of artificial tissues. Bioprinting works by depositing cells captured in bioink fluids through the use of an automated printhead [50]. In the past few years, microfluidics has been used to enhance droplet-based bioprinting, achieving higher printing accuracy and precision in cell deposition, enhancing the complexity of printed tissues, and promoting new biological applications. The setup with direct contact co-culture of cells allows for direct contact, making it suitable for studying interactions among all cells. The common methods for cell contact include using a shared chamber, where cell suspensions are mixed for culture, or employing a cell-stratified contact culture. In the cell suspension mixed culture approach, more than one cell suspension is directly mixed, and the combined solution is then inoculated into the chip cell culture chamber for co-culture. In the contact culture method, one type of cell is typically aggregated first to form a cell layer. Subsequently, other types of cells are inoculated to adhere to the surface of the initial layer, enabling direct intercellular cross-layer interaction and communication.

Dura et al. ([47], see Figure 2a) developed a microfluidic device featuring a trap cup array with a scanning electron micrograph image of the trap array for strategically capturing and pairing cells. Each cell trap consists of a single-cell capture cup (the back-side trap) and a double cell capture cup (the front-side trap). The support pillars on each side of the capture cups allow fluid to flow through the cups to guide cells into the traps. This system facilitates the efficient and definitive matching of lymphocytes within a specified duration of interaction, enabling precise evaluation of initial activation occurrences for every pair in controlled microenvironments. Additionally, the platform allows for the simultaneous capture of dynamic processes and static parameters from both partners, facilitating the profiling of lymphocyte interactions over hundreds of pairs in a single experiment with pairwise-correlated multi-parametric analysis. Chen et al. [51] presented a novel microfluidic device with two layers and multiple channels, specifically engineered for the co-culture of cells in direct contact with vessels. The microfluidic device used in the study has one upper microchannel and multiple lower microchannels, separated by a porous membrane of polyethylene terephthalate (PET) with a diameter of 8 μm (Figure 2b). With this apparatus, a co-culture model of the outer blood–retina barrier (oBRB) was developed to replicate the in vivo interaction among retinal pigment epithelial cells, Bruch membrane, and fenestrated choroids. The possibility of evaluating the integrity of the epithelial barrier on a microchip was proven by incorporating platinum electrodes to measure transepithelial electrical resistance (TEER). The design permits the co-culture of cells in direct contact, either between cells or between cells and vessels. Additionally, it can be customized to enable real-time assessment of the state of epithelial monolayers. Dudman et al. [52] employed microvalve technology for bioprinting (as shown in Figure 2c) to generate co-cultures of laminar mesenchymal stromal cells (MSCs) and chondrocytes. The aim was to examine if the addition of MSCs in autologous chondrocyte implantation (ACI) procedures could potentially promote increased synthesis of extracellular matrix (ECM) by chondrocytes. Bioprinting utilizing microvalves employs small-scale solenoid valves (microvalves) to consistently and repeatedly place cells suspended in media. In this research, a laminar co-culture was established by printing MSCs and chondrocytes sequentially into an insert-based transwell system. Revisions were made to the ratios of cell types in order to investigate the capacity of MSCs in promoting ECM production. Histological analysis and indirect immunofluorescence staining demonstrated the formation of dense tissue structures within the chondrocyte and MSC–chondrocyte cell co-cultures, along with the establishment of a proliferative region at the bottom of the tissue. This research presents an innovative approach that facilitates the efficient manufacturing of therapeutically relevant micro-tissue models. These models can be employed in in vitro research for ACI procedure optimization.

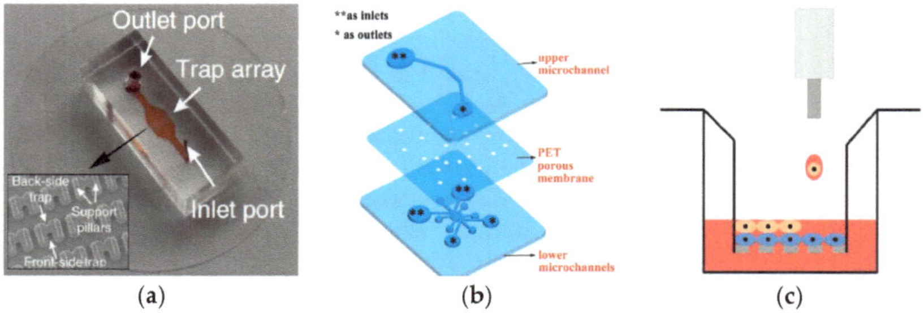

Figure 2. Schematic illustration of the direct co-culture of cell chips. (**a**) Image of the microfluidic cell pairing device with a scanning electron micrograph image of the trap array. Reproduced with permission from [47]. (**b**) Configuration of the multi-channel microfluidic device: exploded view; medium reservoirs are marked with asterisks (** as inlets, * as outlets when loading cells or media). Reproduced with permission from [51]. (**c**) Illustration of the co-culture cell printing platform. Reproduced with permission from [52].

The most obvious advantage of the direct contact co-culture method is its ability to demonstrate interactions among cells. For instance, when neural stem cells (NSCs) are co-cultured with microglia, factors secreted by microglia enhance the dopaminergic differentiation of human NSCs [53]. Similarly, co-cultured astrocytes promote neuronal differentiation of NSCs [54]. Moreover, other regulatory factors such as immune cytokines can be added to the co-culture system for cell–cell interaction research. For example, adding interleukin-33 (IL-33) to a mixed culture system containing primary mouse cortical neurons and glial cells revealed that IL-33 induces glial cells to release inflammatory mediators, thereby reducing neuronal mortality in the co-culture system [55].

Additionally, the co-culture system involving feeder cells is also considered a form of direct contact co-culture. In this system, other cells are plated on a monolayer of specific cells, such as granulosa cells, fibroblasts, or fallopian tube epithelial cells [56,57]. These feeder cells are treated with mitotic inhibitors, commonly known as mitomycin, to suppress cell division while preserving their ability to secrete growth factors. The survival and proliferation of other cells depend on the growth factors secreted by the feeder cells. In cell culture processes, the feeder cell layer acts as a promoter of growth and proliferation and an inhibitor of differentiation. This role is particularly important in the cultivation of embryonic stem cells (ESCs) [58,59]. Furthermore, this method often facilitates the formation of cell junctions among multiple cell types. For instance, by extending the co-culture of two cell types, a multicellular system comprising neurons, astrocytes, and microglia has been established. This system more realistically simulates the neuroinflammatory response in the body, providing a better understanding of the impact of cell crosstalk on neuroinflammation [5].

2.2. *Indirect Co-Culture of Cells*

The interaction or regulation among cells can occur not only through direct cell contact but also via chemical signals released in the microenvironment. In the latter case, where direct contact is to be avoided, an indirect co-culture system is required. Indirect contact co-culture of cells involves cultivating multiple cell types through chemical interactions within the culture medium, without direct contact. This method utilizes microvalves, hydrogel, semipermeable membranes, and narrow channels [60–63] to culture different types of cells in distinct areas of the chip. This indirect co-culture method eliminates the influence of direct cell contact and is suitable for paracrine and endocrine signal transduction research. On microfluidic chips, indirect co-culture of cells can be achieved through both co-chamber [64] and independent chamber [38,39,62,63] setups, mainly realized through microvalve isolation [45,65], channel isolation [38,39,66], and membrane isolation [67]. Microvalve isolation is primarily achieved through a pneumatic drive system that controls the microvalve to form a raised compartment, thereby connecting and isolating chambers [67]. Shi et al. [45] developed a vertically layered setup and a four-chamber setup for co-culturing central nervous system (CNS) neurons and glia (see Figure 3a). The cell compartments in the apparatus were separated by valve barriers that could be activated by pressure, facilitating regulated interaction between the two types of cells. This distinctive design enabled the close co-culturing of glial cells and neurons, the selective transfection of specific neuronal groups, and the real-time observation of neuronal interactions, including the growth of synapses. De et al. (see Figure 3b) reported a microfluidic device with multiple compartments [68], enabling the cultivation of three distinct cell populations in separate fluidic circuits. The chip consists of three perfusable compartments (500 μm wide, either 100 or 250 μm high, and 6 mm long) with distinct inlets and outlets (diameter of 2 mm), interconnected through a series of narrow and parallel microgrooves (either 2.5, 5, or 10 μm wide, 2.5 μm high, and 250 μm long) that can allow the separation between soma and neurites and promote unidirectional neurite elongation from one cell compartment to the adjacent one. The chip setups for cell culture contain the tube system (TS), the steel connector system (SCS) and the reservoir system (RS). The device permits cell migration across the compartments and their differentiation. The researchers showed that optimizing the device's geometric characteristics and cell culture parameters can enhance the attachment and growth of neuron-like human cells (SH-SY5Y cells), regulate the migration of neurons and Schwann cells between compartments, and facilitate prolonged studies on cell cultures. These discoveries present opportunities for plenty of in vitro co-culture research in neuroscience.

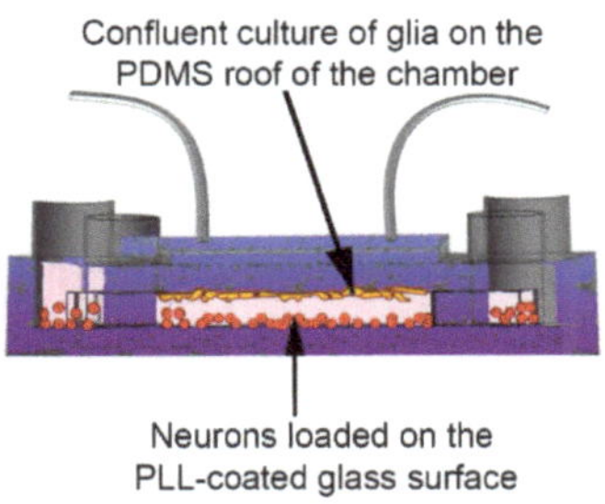

(a)

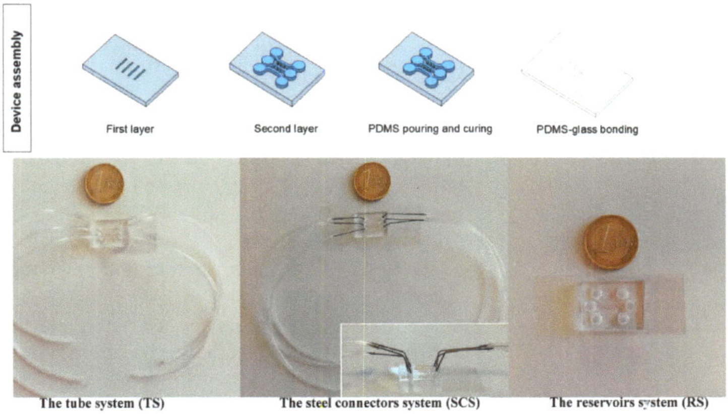

(b)

Figure 3. Schematic illustration of the indirect co-culture of cells chips. (**a**) The vertically layered microfluidic platforms with glia (orange spheres) and neuron (red spheres) indirect co-culture. Reproduced with permission from [45]. (**b**) A microfluidic device with multiple compartments for the cultivation of multi-cell populations. Reproduced with permission from [68].

Channel isolation in microfluidic cell co-culture systems can be categorized into independent chamber isolation [38,62,65,68] and shared chamber isolation [39,60]. Currently, the preferred method for studying intercellular communication in microfluidic cell co-culture systems relies on independent chamber isolation. Multiple chambers are isolated using techniques such as microvalves, microarray columns [60], or fluid dynamics [61]. The combination of microfluidic technology and fluid dynamics achieves co-culture of cells in a shared chamber by controlling the liquid flow rate and cell contact with the medium. Various cell co-culture models based on channel isolation have been developed, and research on intercellular communication on microfluidic chips has transitioned from cell population studies to single-cell analyses [69]. Utilizing precise laminar flow control, microfluidic chips with independent chambers and fluid mechanics can automatically and stably move along both sides of the main channel, being introduced into adjacent separation areas to form a non-contact co-culture model. This design is convenient for real-time observations of cell behavior [61,70]. Microfluidic chips combined with shared chambers and fluid dynamics typically consist of two parallel side channels and an intermediate channel [67,71]. The interface of the intermediate channel connects the left and right channels, and the edges of the different compartments are neat and free of cell debris or substances that adversely affect cell migration. Apart from cell population research, shared chamber heart-shaped flute designs [69–74] can be used for single-cell capture and analysis. The unique heart-shaped depression design allows effective pairing of different types of cells at the single-cell level, minimizing spatial constraints on cells. By adjusting the number and position of heart-shaped flutes, various functions can be achieved, such as cell pairing ratio control, cell pairing spacing, and the formation of various single-cell arrays.

Using porous membrane isolation, the two channels are connected by substrate membranes of different materials, such as polycarbonate membrane, polyethylene (PE) membrane, etc. In this setup, one type of cell is cultured at the bottom of the chip, while the other type of cell is cultured at the top of the membrane [75]. Alternatively, the microfluidic characteristics can be leveraged to inoculate cells on both sides of the membrane [76,77]. The layered structure on the chip supports long-term co-culture of cells. However, the single-channel layered co-culture model based on membrane isolation faces challenges in maintaining long-term stable stratification among cells. Over time, it may evolve into a random co-culture monolayer. Cells often tend to aggregate at the entrance and exit of microchannels, limiting its application in more physiologically relevant research. On the other hand, the dual-channel co-culture model based on membrane isolation addresses these issues by increasing the channel height, resulting in a more uniform cell layer and stable cell morphology.

In general, the direct co-culture method of cells is simple, usually requiring the placement of two or more types of cells in the same culture dish in specific proportions for direct interaction between cells. Due to the direct interaction between cells, this method can better retain the connection information between cells and make the cultured cells more similar to their in vivo state. It is suitable for studying the interaction between adjacent tissue cells, cellular interactions, and the induction of cell differentiation. However, due to the direct interaction between cells, separating the two types of cells becomes more difficult, making observation and subsequent detection inconvenient. Additionally, during direct co-culture, different cells may influence each other, affecting the analysis of experimental results. In indirect co-culture, cells do not directly interact but communicate through cell-secreted factors, facilitating subsequent cell separation and detection. Indirect co-culture can be used for studies on cytokine-induced cell differentiation and proliferation, simulating the liquid circulation environment in the body, and studying intracellular (autocrine) and intercellular (paracrine) interactions. However, due to the lack of direct contact, direct interactions between cells may not be observed. Special culture plates and chambers are required, which can be costly. Using microfluidic cell chips requires relatively complex operations, high technical expertise, and specialized equipment.

3. Detection of Cell Co-Culture System by Microfluidics

After constructing an on-chip cell co-culture system, it is crucial to evaluate the system or barrier to determine its suitability for studying intercellular communication mechanisms. This evaluation involves assessing both system or barrier permeability and intercellular interactions, as illustrated in Table 1.

Table 1. Detection of co-culture system.

Detection Classification	Detection Target	Detection Method or Marking	Ref.
System or barrier permeability assessment	The permeability of molecules	Glucose Rhodamine	[78] [79]
	Cell viability	Colorimetry Staining of live/dead cells Lactic dehydrogenase (LDH) Activity assay	[80,81] [82] [83,84]
	Electrophysiological activity	Transendothelial electrical resistance (TEER)	[85–87]
	Immunofluorescence of cellular marker substances	Actin Green fluorescence protein (GFP)	[88] [45,89,90]
	Formation of spherical bodies	Electron microscope	[91–93]
Intercellular interaction	Cell migration Cell differentiation Cellular fibrosis Cytotoxicity testing	Mass spectrometry analysis, qPCR, Immunofluorescence	[80,81,88,89,94] [95–98] [70,99–101] [102,103]

In existing research, the assessment of system or barrier permeability primarily involves detecting the permeability of molecules, evaluating cell viability, measuring electrophysiological activity, identifying cell markers, and observing cell morphogenesis using electron microscopy. Detection of permeability molecules, such as glucose [78] and rhodamine [79], enables the assessment of the hydrogel penetrability within the system. Cell viability is determined through methods like trypan blue colorimetry/thiazolyl blue colorimetry [80,81], live or dead cell staining [82], and lactic dehydrogenase (LDH) activity assay [83,84], providing insights into whether the system offers a conducive environment for cell co-culture. Electrophysiological activity detection, often applied to neural cells and neurons, utilizes techniques like patch clamp technology [61] and transmembrane resistance (TEER) [85–87] to measure and quantitatively evaluate barrier permeability. Cell markers [45,88–90] and cell morphology [91–93] are utilized for detecting the formation of a co-cultured 3D culture pattern. After verifying the feasibility of the cell co-culture system, the simulation of the native microenvironment enables further research on intercellular interactions. This includes studying cell migration [80,81,89,90,94], cell differentiation [95–98], cell fibrosis [70,99–101], and cell toxicity testing [102,103]. On-chip cell migration refers to movement influenced by signals from other cells, involving processes like angiogenesis and cancer metastasis. Migration distance is mainly observed through microscopy. Cell differentiation and fibrosis are marked by specific cells using immunofluorescence staining methods [45], and confocal microscopy is employed to observe and analyze the selective expression of cells in time and space, including changes in cell morphology and group dynamics. Toxicity testing involves detecting changes in signaling factors related to interactions, allowing the study of the impact of targeted drugs on the system. Jeong [104] analyzed the migration ability of fibroblasts towards the 3D tumor chamber by measuring the migration distance of the nucleus of fibroblasts in the culture medium channel. In this co-culture system, the sensitivity differences of paclitaxel drug therapy were studied by comparing the changes in the proportion of live and dead cells.

After building the on-chip cell co-culture system, it is crucial to evaluate the system's applicability or barrier function to ensure the accuracy and reliability of experimental results. During the evaluation process, key points such as cell growth and differentiation, intercellular interactions, simulation of the internal environment, barrier function, stability, and repeatability need to be focused on.

4. Application

In on-chip cell co-culture models, the most common studied system is the vascular system, followed by blood–brain barrier chips, gas–blood barrier chips, and other organoid chips. Let us explore the applications of microfluidic cell co-culture technology in angiogenesis chips, blood–brain barrier chips, gas–blood barrier chips, and organoid chips.

4.1. Angiogenesis

Angiogenesis chips are typically designed with microgaps, often combined with a 3D gel to serve as a scaffold for angiogenesis. This chip structure facilitates cell connectivity, promotes the formation of vascular networks, and creates a reliable three-dimensional microenvironment. Such designs offer valuable insights for clinical medicine and tissue engineering. In cardiovascular model establishment, simulating and studying fluid stress often involves inserting a membrane to replicate the three-dimensional cardiovascular environment. Sometimes, the inserted membrane becomes a key component for studying the biological mechanisms of blood vessels and heart valves. Co-culture systems on these chips have also been used to study the impact of certain types of cancer cells on angiogenesis [105–107]. Such studies have revealed a close relationship between vascular networks and the occurrence and development of tumors [108].

Liu et al. [79] conducted a study using a microfluidic chip to simulate the three-dimensional tumor microvascular structure and investigated the impact of antioxidants on malignant glioma cells in vitro. They utilized hydrogel to construct a 3D chamber for co-culturing endothelial cells and glioma cells, creating a simulated environment for tumor microvasculature (see Figure 4a). A macroporous gelatin transglutaminase hydrogel with favorable biomechanical properties for cell culture and nutrient renewal was employed in the study. In another investigation, Kim et al. [109] (see Figure 4b) observed that the angiogenesis of human umbilical vein endothelial cells (HUVECs) relies on co-culturing with human lung fibroblasts (LFs), as the formation of an interconnected vascular network was not observed in the HUVEC system not co-cultured with LFs. Ibrahim et al. [105] investigated the influence of stromal cell effects on the attachment and proliferation of tumor cells, along with the reciprocal consequences of tumor cells on vascular and mesothelial permeability. They employed an in vitro model of the vascularized human peritoneal omentum and ovarian tumor microenvironment (TME) to investigate metastases at both early and advanced stages. The results indicated that the growth of tumors resulted in a reduction in microvascular permeability through physical mechanisms, while simultaneously inducing an elevation in microvascular permeability via cytokine signaling. This emphasizes the sophistication and potential conflicting roles of tumor cells in the development of ascites. The developed system functions as a sturdy platform for investigating the interactions between tumor cells and stromal cells during the spread of ovarian cancer within the peritoneal cavity, presenting a novel in vitro vascularized model of the human peritoneum and ovarian cancer TME, which is shown in Figure 4c.

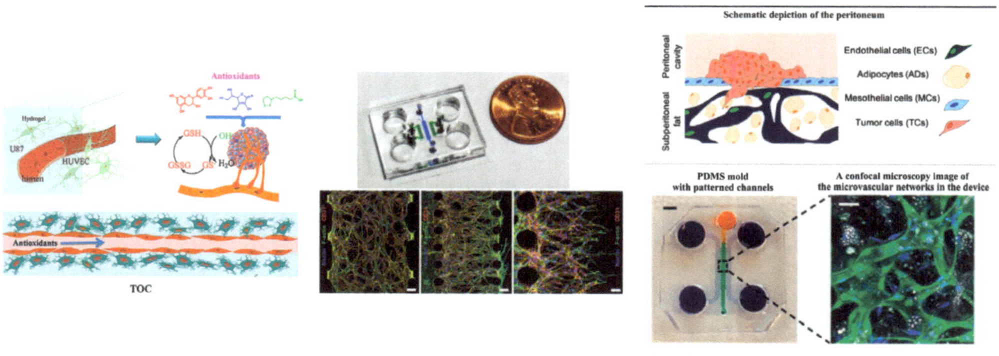

(**a**) (**b**) (**c**)

Figure 4. Angiogenesis chips. (**a**) The three-dimensional tumor microvascular structure simulated on a microfluidic chip for study of antioxidant effects on malignant glioma cells in vitro. Reproduced with permission from [79]. (**b**) Microfluidic chip design and confocal micrographs showing the overall architectures of vascular networks established by vasculogenic and angiogenic processes at day 4 (scale bars, 100 mm), as well as angiogenic sprouts grown for 2 days (scale bar, 50 mm). Reproduced with permission from [109]. (**c**) In vitro microvascular network model of the peritoneum produced with polydimethylsiloxane (PDMS) employing the technique of soft lithography. Reproduced with permission from [105].

4.2. Blood–Brain Barrier

The blood–brain barrier is a dense barrier structure composed of vascular endothelial cells, astrocytes, pericytes, and basement membrane in the brain, playing a crucial role in maintaining the stability of the central nervous system environment [110,111]. The most common co-culture model for the nervous system and blood–brain barrier involves astrocytes and endothelial cells, with chip designs often incorporating microchannel connections to simulate axonal guidance function [37,112,113]. Microfluidic structures for the blood–brain barrier on the chip come in two main types: planar and vertical. The planar model typically utilizes microcolumns or microchannel arrays as the boundary between the blood and lateral brain cavities (such as Figure 5a). These microstructures have gaps small enough to capture cells on either side, enabling co-culturing of both cell types on their respective sides, similar to vertically placed porous membranes.

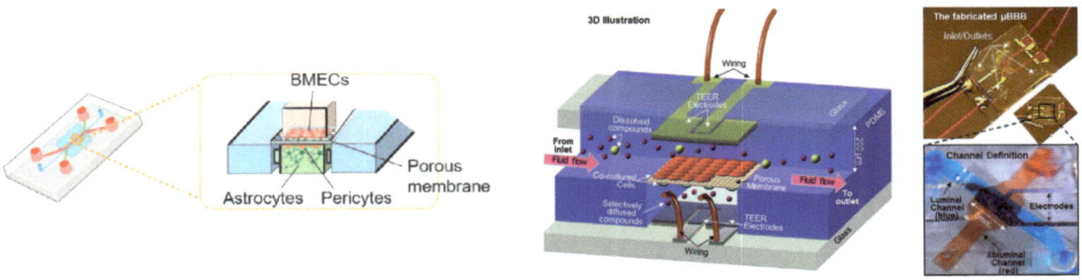

(**a**) (**b**)

Figure 5. Blood–brain barrier (BBB)-on-a-chip. (**a**) Microfluidic BBB-on-a-chip with the cross-sectional view. Reproduced with permission from [110]. (**b**) Structure and design of the multi-layered microfluidic device (μBBB). Reproduced with permission from [114].

The vertical model utilizes a porous membrane as the boundary between blood vessels and brain tissue ([114], see Figure 5b). In Figure 5b, the top channel contains brain microvascular endothelial cells (BMECs) and the bottom channel contains astrocytes and a media pool. A porous membrane is sandwiched in between these two channels. Positioned between these two microfluidic structures, one side accommodates blood vessels, while the other supports brain tissue. Chip designs for the blood–brain barrier often incorporate microchannel connections due to their ability to simulate axonal guidance functions. To establish an optimal in vitro model of the blood–brain barrier, it is crucial to replicate key physical characteristics of the cerebral capillary microenvironment including fluid flow, extracellular matrix, and the cylindrical geometric structure of normal brain microvessels. Astrocytes play a vital role in promoting the formation of the blood–brain barrier, assisting in the maturation of neurovascular endothelial cells, and facilitating a tighter connection for barrier integrity. Jeong et al. [83] introduced a multi-chamber microfluidic blood–brain barrier chip designed to recapitulate the key functions of the blood–brain barrier at the astrocyte capillary interface. The chip optimizes physiological shear stress and extracellular matrix conditions to enhance the formation of tight cell junctions. Notably, this design allows for the simultaneous conduct of up to 16 different tests on a single chip.

4.3. Blood–Gas Barrier

The blood–gas barrier is a complex tissue structure crucial for facilitating normal gas exchange in the body. It comprises the liquid layer on the surface of alveoli, type I alveolar cells, basement membrane, a thin layer of connective tissue, capillary basement membrane, and endothelium of pulmonary capillaries [115]. The formation of the blood–gas barrier is intricately linked to the construction of the alveolar–capillary interface. To recreate the connection between alveoli and blood vessels, alveolar cells and vascular endothelial cells are co-cultured, and the essential function of alveolar capillaries is simulated through gas–liquid exposure [116]. The membrane-based separation channel structure allows independent manipulation of fluid flow, as well as the transfer of cells and nutrients. Additionally, lung function involves cell stretching during respiratory movement, necessitating the application of mechanical forces on the chip to mimic the dynamic mechanical deformation of the alveolar–capillary interface caused by respiratory movement. Huh et al. [117] constructed a chip model of alveolar pulmonary capillary units (see Figure 6a), successfully replicating the dynamic mechanical deformation of the alveolar–capillary interface caused by respiratory movement. Human alveolar epithelial cells and human pulmonary microvascular endothelial cells are cultured on both sides of the membrane. As cells form layers, air is introduced into the epithelial compartment, creating a gas–liquid interface that more accurately simulates the inner layer of alveolar air space. To model the human small airway-on-a-chip, Benam et al. ([118]) used soft lithography to create a microfluidic device made of PDMS containing an upper channel with a height and width (both 1 mm) similar to the radius of a human bronchiole separated from a parallel lower microvascular channel (0.2 mm high, 1 mm wide) by a thin, porous (0.4 μm), polyester membrane coated on both sides with type I collagen (see Figure 6b). The primary human airway epithelial cells (hAECs) isolated from healthy donors or COPD patients were cultured on top of the membrane until confluent with medium flowing (60 μL/h) in both channels. To trigger lung airway epithelial differentiation, the apical medium was removed after five days and air was introduced to create an air–liquid interface (ALI), while retinoic acid (3 μg/ mL) was added to the medium flowing in the lower channel to prevent the development of a squamous phenotype. Three to five weeks later, primary human lung microvascular endothelial cells were seeded on the opposite side of the porous membrane and cultured at the same flow rate until confluent to create a tissue–tissue interface. Immunofluorescence confocal microscopic analysis revealed that these culture conditions resulted in the formation of a pseudo-stratified, mucociliary, airway epithelium and a planar microvascular endothelium on opposite sides of the same ECM-coated membrane.

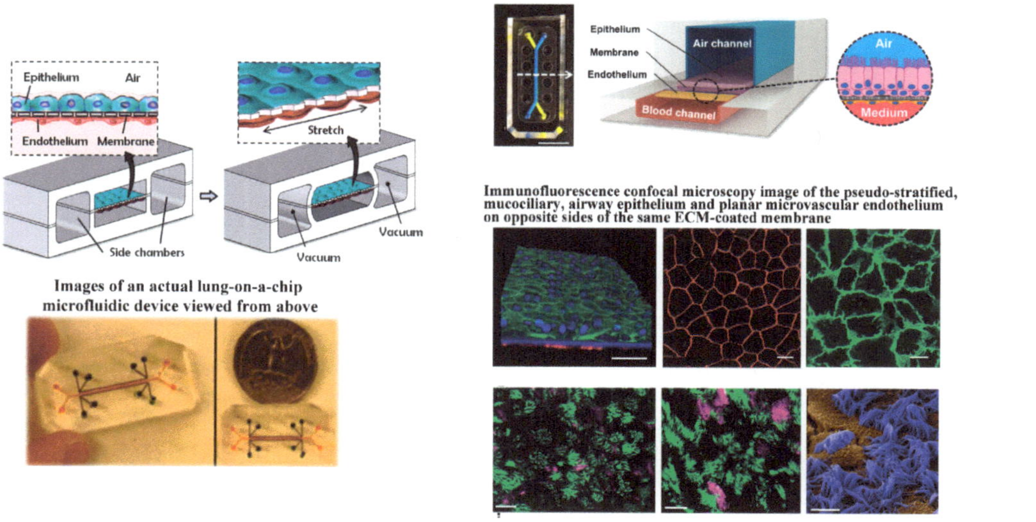

(a) (b)

Figure 6. The design of blood–gas barrier by microdevice. (**a**) The microfabricated lung mimic device utilizes compartmentalized PDMS microchannels to form an alveolar–capillary barrier on a thin porous flexible PDMS membrane coated with ECM. Reproduced with permission from [117]. (b) The human small airway-on-a-chip. Reproduced with permission from [118].

4.4. Other Organoid Chips

The organoid chip is an innovative biomedical technology designed to replicate the physiological processes of various organs in the human body by cultivating human tissues or organ cells on the chip. This approach offers a more realistic, rapid, and accurate model for disease treatment and the development of new drugs [23,30,118–120]. Initially proposed by Sin et al. [121], the organoid chip aims to construct and simulate the microenvironment of different tissues and organs in the human body, with continuous advancements and developments. In addition to the organoid chips mentioned earlier, recent progress has expanded the repertoire to include kidney chips [122], liver chips [123], intestinal chips [124,125], bone chips [126], tumor chips [127], and multi-organ chip [128], broadening the applications of these human organoid chips.

The liver sinusoid wave chip is a common model for the liver, primarily utilizing the co-culture of liver cells and endothelial cells to simulate the structure of liver sinusoids [129]. The microfluidic structure of the liver model on the chip mainly involves membrane-isolated indirect co-culture. This model can be extended to various liver biology research and liver-related disease studies, such as drug-induced liver toxicology, cancer research, and the pathological effects of various hepatophilic infectious factors. Research has indicated that the flow of mediators in liver cell secretions and the dynamic interaction with collagen play a crucial role in maintaining primary liver cell function [129–131]. Therefore, liver cells were cultured on a microfluidic platform under the condition of flow perfusion of the culture medium and covered with collagen to investigate the interaction between medium flow, collagen production, and liver cell function. Ya et al. ([130], see Figure 7a) developed a lifelike bionic liver lobule chip (LLC), on which perfusable hepatic sinusoid networks were achieved using a microflow-guided angiogenesis methodology. To accurately replicate the structure of liver sinusoids, Du et al. ([132], see Figure 7b) developed an in vitro three-dimensional liver chip comprising four primary liver cell types under shear flow, aiming to mimic the liver microenvironment with precise cell composition and quantified physical interactions. Busche et al. ([133], see Figure 7c) established a novel, parallelized, and scalable microfluidic in vitro liver model demonstrating hepatocyte function. This model showcased fully automated cell culture preparation in the HepaChip microplate (HepaChip-MP) using a pipetting robot. The HepaChip-MP consists of 24 independent culture chambers. An automated dielectrophoresis process selectively assembles viable cells into elongated microtissues. Freshly isolated primary human hepatocytes (PHHs) and primary human liver endothelial cells (HuLECs) were successfully assembled as co-cultures, mimicking the liver sinusoid. The establishment of microtissues using the HepaChip-MP necessitates only small amounts of primary human cells. The system is expected to be integrated into routine procedures in cell culture labs, enabling comprehensive investigations on liver biology and its potential applications in preclinical drug development. Zheng et al. [134] developed a 3D dynamic multi-cellular liver-on-a-chip device (3D-DMLoC) to replicate the microenvironment of liver tissue in vivo (see Figure 7d). The device incorporated functionalities like simulated hepatic sinusoid, perisinusoidal space, and continuous liquid perfusion, resulting in the formation of 3D cell spheroids. The HepaRG cells and HUVECs were co-cultured for 7 days within this chip. The observed liver toxicity was correlated with acute hepatocyte injury, which was indicated by the ratios of secreted AST/ALT contents. The liver-on-a-chip device demonstrated successful development and validation, providing a more accurate reproduction of the in vivo physiological microenvironment of the liver. This platform holds promise for the easy, efficient, and accurate screening of potential hepatotoxic chemicals in the future.

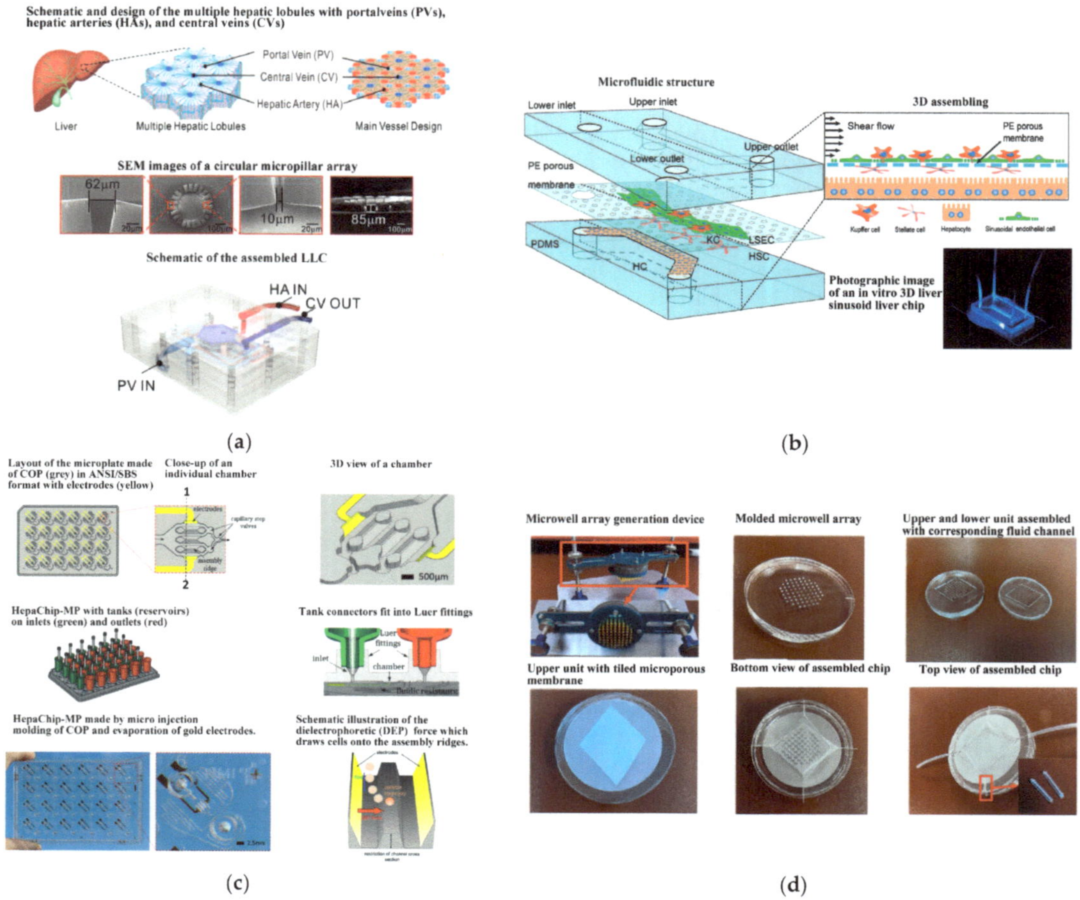

Figure 7. Schematic components of the liver lobule chip (LLC). (**a**) Schematic, design, and characterization of the multiple hepatic lobules with portalveins (PVs), hepatic arteries (HAs), and central veins (CVs). Reproduced with permission from [130], (**b**) Schematic of the in vitro 3D liver sinusoid liver chip. Reproduced with permission from [132]. (**c**) The HepaChip-MP design. Reproduced with permission from [133]. (**d**) The 3D-DMLoC system. Reproduced with permission from [134].

Since 2019, Jalili-Firoozinezhad et al. [124] and Puschhof et al. [125] have explored the development of intestinal chips, emphasizing the interaction between the gut and microorganisms. The former employed stretchable materials to replicate the rhythmic peristalsis and contraction of the intestine, as depicted in Figure 8a. The latter focused on the interaction between intestinal epithelium and microorganisms to investigate the influence of gut microbiota on health and disease. Shah et al. [135] presented a modular, microfluidics-based model (HuMiX, human–microbial crosstalk), which allows co-culture of human and microbial cells under conditions representative of the gastrointestinal human–microbe interface. The model integrates oxygen sensors (optodes) for the real-time monitoring of the dissolved oxygen concentrations within the device. In addition, they also fabricated a specially designed version of HuMiX, which allows the insertion of a commercial chopstick-style electrode to monitor TEER for the characterization of cell growth and differentiation within the device (see Figure 8b). In 2021, Ao et al. [131] engineered a brain chip (see Figure 9a) by co-culturing glial cells and immune cells, aiming to simulate the tissue structure and function and offering novel research perspectives for brain diseases. Pediaditakis et al. ([136], see Figure 9b) leveraged the Organs-on-Chips technology to develop a human brain chip representative of the substantia nigra area of the brain containing dopaminergic neurons, astrocytes, microglia, pericytes, and microvascular brain endothelial cells, cultured under fluid flow. The αSyn fibril-induced model was capable of reproducing several key aspects of Parkinson's disease, including accumulation of phosphorylated αSyn (pSer129-αSyn), mitochondrial impairment, neuroinflammation, and compromised barrier function. This model may enable research into the dynamics of cell–cell interactions in human synucleinopathies and serve as a testing platform for target identification and validation of novel therapeutics. Chou et al. [137] showed a vascularized human bone-marrow-on-a-chip (see Figure 10a) that supports the differentiation and maturation of multiple blood cell lineages over 4 weeks while improving CD34+ cell maintenance, and it recapitulates aspects of marrow injury, including myeloerythroid toxicity after clinically relevant exposures to chemotherapeutic drugs and ionizing radiation as well as marrow recovery after drug-induced myelosuppression. The chip comprises a fluidic channel filled with a fibrin gel in which CD34+ cells and bone-marrow-derived stromal cells are co-cultured, a parallel channel lined by human vascular endothelium and perfused with culture medium, and a porous membrane separating the two channels. As an in vitro model of hematopoietic dysfunction, the bone-marrow-on-a-chip may serve as a human-specific alternative to animal testing for the study of bone marrow pathophysiology. Subsequently, in 2022, Glaser et al. constructed a vascularized bone marrow chip [126], illustrated in Figure 10b. They established an in vitro simulated vascularized bone marrow microenvironment using microfluidics and stem cell technology. This model features a dynamic and permeable vascular network, faithfully reproducing bone marrow function and providing a novel platform for understanding human bone marrow function and the mechanisms of action of related drugs.

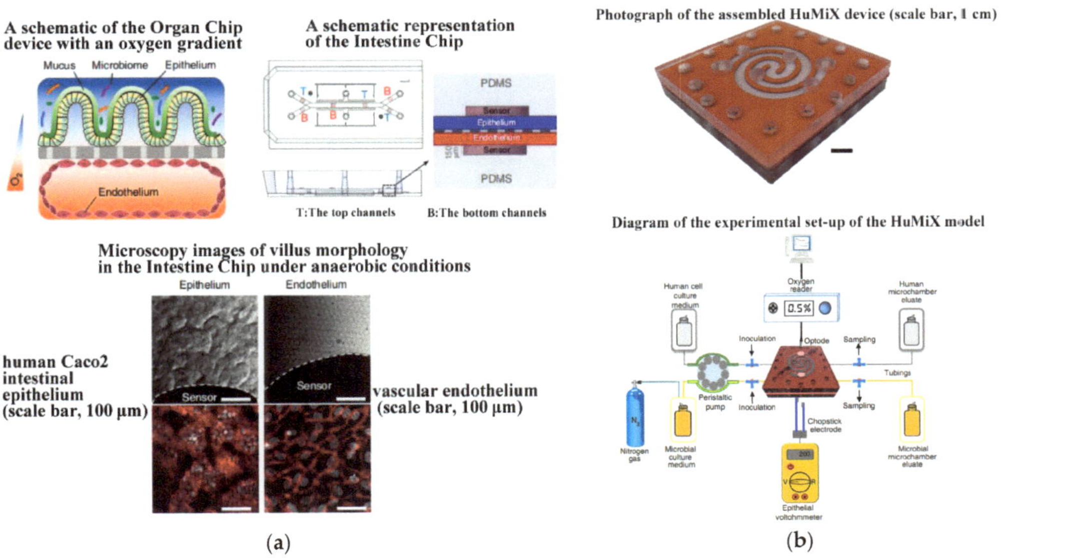

Figure 8. Gut chips. (**a**) Oxygen-sensitive human intestine chip microfluidic culture device. Reproduced with permission from [124]. (**b**) The in vitro model (HuMiX) of the gastrointestinal human–microbe interface. Reproduced with permission from [135].

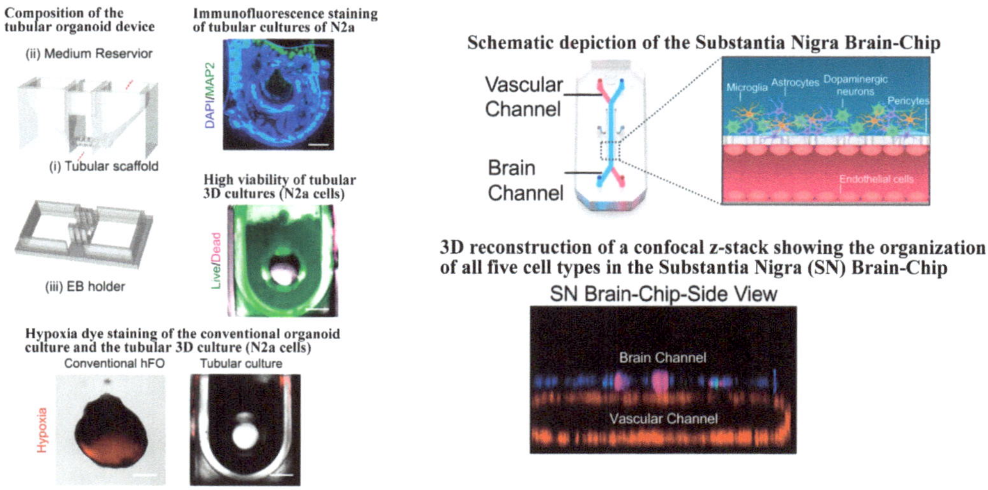

(**a**) (**b**)

Figure 9. Brain chips based on microfluidics. (**a**) Tubular organoid device design and validation. Reproduced with permission from [131]. (**b**) The human substantia nigra brain chip. Reproduced with permission from [136].

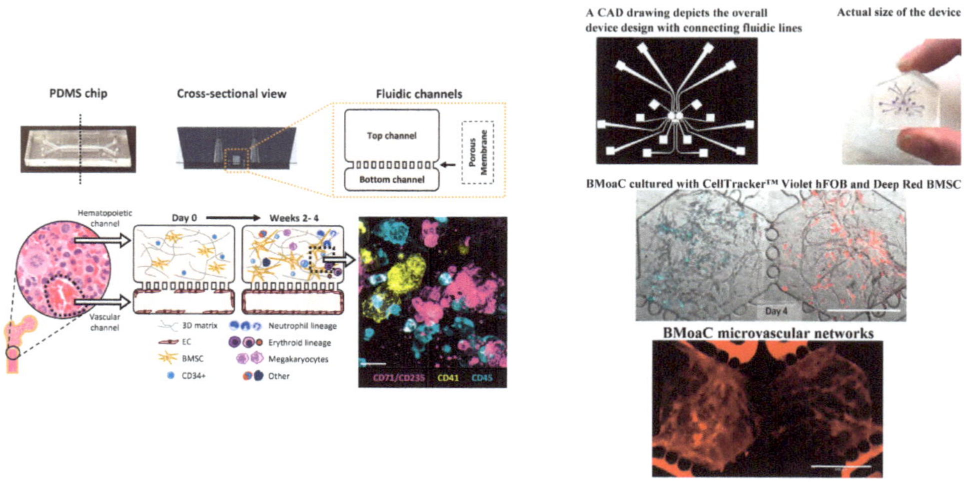

(**a**) (**b**)

Figure 10. Bone chips based on microfluidics. (**a**) Primary human bone marrow chip supports in vitro hematopoiesis over 4 weeks in culture and improves CD34+ progenitor survival and colony forming capacity. Reproduced with permission from [137]. (**b**) Vascularized bone marrow chip. Reproduced with permission from [126].

A single-organ chip can replicate the physiological and pathological processes of a specific organ in the human body. However, a true biological system emerges from the interaction of multiple organs. Therefore, the development of multiple-organ chips is considered a crucial direction for future advancements. In recent years, scientists have created various organ chips combined to form human chips (Human-on-a-Chip) [132,133,136] ([138], see Figure 11a). Maschmeyer et al. [139] first introduced a microphysiological four-organ chip system that enables a reproducible 28-day co-culture of four tissues (intestine, liver, skin, and kidney tissue). Barrier integrity, continuous molecular transport against gradients, and metabolic activity could be demonstrated for the four-organ chip (4OC) co-cultures, thus making it a perfect platform for further in vitro ADME (absorption, distribution, metabolism, and excretion) and repeated dose toxicity testing. It enhances the simulation of interactions and complex physiological and pathological processes between multiple organs in the human body, thereby improving the efficiency and accuracy of drug screening and evaluation. Table 2 summarizes the typical applications of microfluidic chips and their co-cultured cell/organ types in this review.

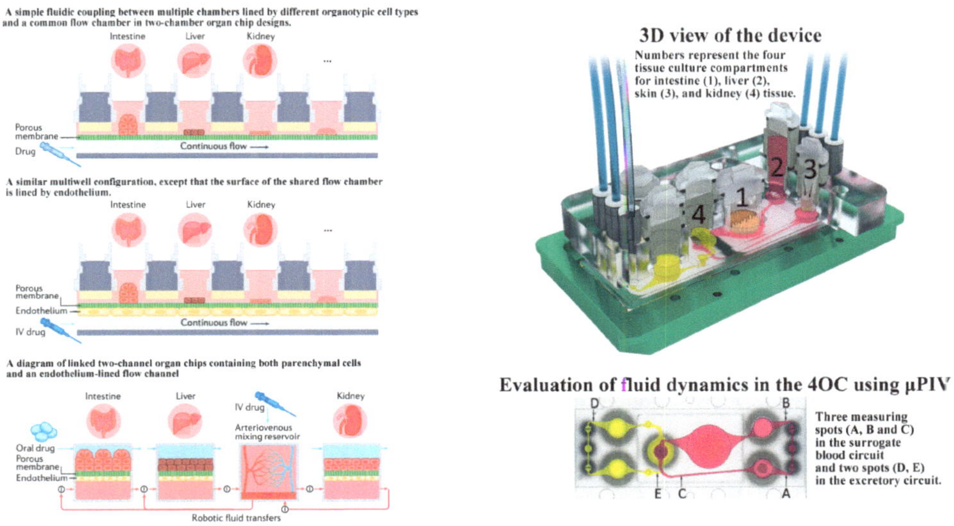

(a) (b)

Figure 11. Human chips. (**a**) Schematics showing different multi-organ human body-on-chip formats. Reproduced with permission from [138]. (**b**) The microfluidic four-organ chip device. Reproduced with permission from [139].

Table 2. Application of microchips.

Application		Co-Culture Type	Ref.
Angiogenesis		Endothelial cells, Glioma cells	[79]
		HUVECs, LFs	[109]
		Endothelial cells, Adipocytes, Mesothelial cells, Tumor cells	[105]
Blood–brain Barrier		Astrocytes, Endothelial cells	[110,114]
Blood–gas Barrier		Human alveolar epithelial cells, Human pulmonary microvascular endothelial cells	[117]
		HAECs, primary human lung microvascular endothelial cells	[118]
Other Organoid Chips	liver chips	liver cells, endothelial cells	[130,132–134]
	intestinal chips	intestinal epithelium, microorganisms	[125,135]
	bone chips	CD34+cells, bone-marrow-derived stromal cells, human vascular endothelium	[137]
	brain chips	glial cells and immune cells	[131]
		dopaminergic neurons, astrocytes, microglia, pericytes, and microvascular brain endothelial cells	[136]
	multi-organ chip	intestine, liver, skin, and kidney tissue	[139]

Organ chips are extensively applied in drug screening and toxicity assessment [140,141]. In drug development, these chips provide a rapid and accurate method to evaluate drug toxicity and efficacy, offering a more reliable and efficient screening approach for new drug development. Concerning toxicity assessment, organ chips better emulate the human body's responses to various chemical substances, enhancing the precision and efficiency of drug toxicity evaluation. This contributes to a more scientific and reliable method for safety assessments of chemical substances. Moreover, artificial intelligence technology [142] is employed to design and optimize the structure and function of organ chips, improving their simulation effectiveness and accuracy. This technology allows precise regulation of cell activity in organ chips, enabling advancements in drug screening, toxicity testing, and research on disease pathogenesis [143]. As human chip technology continues to develop, organ chips will increasingly play a vital role in fields such as life sciences, medicine, and drug research.

5. Conclusions

Cell co-culture technology based on microfluidic chips can replicate the native microenvironment, capturing the complexity of metabolism and regulation. This technology serves as a novel platform for studying cell–cell communication and holds significant value in uncovering the physiological and pathological processes of multicellular organisms. Currently, the design of cell co-culture systems on microfluidic chips predominantly involves two approaches: direct contact co-culture and indirect contact co-culture. The direct co-culture method typically adopts a co-chamber design, while the indirect contact co-culture employs both co-chambers and independent chambers that utilize microvalve isolation, channel isolation, and membrane isolation. After constructing a cell co-culture system on the chip, the functionality of the co-culture system or barriers can be evaluated through various methods, including permeability molecule detection, cell viability assessment, electrophysiological activity detection, cell marker detection, and electron microscopy observation of cell morphology. These evaluations help determine the system's suitability for studying intercellular communication mechanisms. Once the feasibility of the co-culture system is confirmed, it becomes a valuable tool for simulating the native microenvironment and conducting subsequent research on cell communication mechanisms. This includes investigating processes such as cell migration, differentiation, fibrosis, and toxicity detection. With advancements in microfluidic cell co-culture technology, diverse chip models have been constructed. Notably, the vascular system has been the most modeled tissue, closely followed by the blood–brain barrier, blood–gas barrier, and other organoid models. The establishment of co-culture models on microfluidic chips creates structures similar to in vivo tissues or organs, addressing limitations of traditional two-dimensional cell culture. These models find applications in basic in vitro research and various fields such as targeted drug screening and toxicity detection.

As technology progresses, on-chip cell co culture technology is evolving from simple multicellular models towards organoids. This shift aims to fully simulate organ-level functions necessary for physiological homeostasis and complex disease processes. However, organoid models are the latest technology in human tissue experimental research. Compared to traditional models, they are still in the exploratory stage. The stability, repeatability, scalability, and precise control of microenvironmental conditions have become issues that need to be overcome in the development of organoid co-culture technology. To create relevant co-culture systems for cell interaction research, it is necessary to integrate organoid models with standardized microdevices. Looking ahead, the development of the organoid chip model paves the way for constructing a human system on-chip through fluid connections. This advanced model has the potential to simulate interactions and physiological reactions of multiple organs at the system level, proving effective in diverse fields including medicine, life sciences, and environmental sciences.

Author Contributions: Conceptualization, C.L. and W.H.; methodology, C.L., W.H. and Z.Z.; software, Y.S. and X.Z.; validation, Y.S.; formal analysis, X.Z.; investigation, C.L., W.H. and J.S.; data curation, C.L., W.H., J.S. and Z.Z.; writing—original draft preparation, C.L. and W.H.; writing—review and editing, Y.S. and X.Z.; visualization, C.L.; supervision, J.S. and Z.Z.; project administration, J.S. and Z.Z.; funding acquisition, C.L. All authors have read and agreed to the published version of the manuscript.

Funding: This research was funded by the National Natural Science Foundation of China, grant number 82205312, and the supporting project of Nanjing University of Chinese Medicine for the National Natural Science Foundation of China, grant number XPT82205312.

Institutional Review Board Statement: Not applicable.

Informed Consent Statement: Not applicable.

Data Availability Statement: Not applicable.

Acknowledgments: We acknowledge the financial support extended by Nanjing University of Chinese Medicine, China.

Conflicts of Interest: The authors declare no conflicts of interest.

References

1. Du, H.; Bartleson, J.M.; Butenko, S.; Alonso, V.; Liu, W.F.; Winer, D.A.; Butte, M.J. Tuning immunity through tissue mechanotransduction. *Nat. Rev. Immunol.* **2023**, *23*, 174–188. [CrossRef] [PubMed]
2. Vu, T.Q.; de Castro, R.M.; Qin, L. Bridging the gap: Microfluidic devices for short and long distance cell-cell communication. *Lab Chip* **2017**, *17*, 1009–1023. [CrossRef] [PubMed]
3. Vera, D.; García-Díaz, M.; Torras, N.; Álvarez, M.; Villa, R.; Martinez, E. Engineering Tissue Barrier Models on Hydrogel Microfluidic Platforms. *ACS Appl. Mater. Interfaces* **2021**, *13*, 13920–13933. [CrossRef]
4. Yuan, J.; Li, X.; Yu, S. Cancer organoid co-culture model system: Novel approach to guide precision medicine. *Front. Immunol.* **2023**, *13*, 1061388. [CrossRef]
5. Goshi, N.; Morgan, R.K.; Lein, P.J.; Seker, E. A primary neural cell culture model to study neuron, astrocyte, and microglia interactions in neuroinflammation. *J. Neuroinflamm.* **2020**, *17*, 155. [CrossRef] [PubMed]
6. Arora, D.; Gupta, P.; Jaglan, S.; Roullier, C.; Grovel, O.; Bertrand, S. Expanding the chemical diversity through microorganisms co-culture: Current status and outlook. *Biotechnol. Adv.* **2020**, *40*, 107521. [CrossRef]
7. Ribeiro-Filho, A.C.; Levy, D.; Ruiz, J.L.M.; Mantovani, M.D.C.; Bydlowski, S.P. Traditional and Advanced Cell Cultures in Hematopoietic Stem Cell Studies. *Cells* **2019**, *8*, 1628. [CrossRef]
8. Yin, S.; Lu, R.; Liu, C.; Zhu, S.; Wan, H.; Lin, Y.; Wang, Q.; Qu, X.; Li, J. Composite Microfluidic Petri Dish-Chip (MPD-Chip) without Protein Coating for 2D Cell Culture. *Langmuir* **2023**, *39*, 15643–15652. [CrossRef]
9. Ravi, M.; Paramesh, V.; Kaviya, S.R.; Anuradha, E.; Solomon, F.D. 3D cell culture systems: Advantages and applications. *J. Cell. Physiol.* **2015**, *230*, 16–26. [CrossRef]
10. Edmondson, R.; Broglie, J.J.; Adcock, A.F.; Yang, L. Three-dimensional cell culture systems and their applications in drug discovery and cell-based biosensors. *Assay Drug Dev. Technol.* **2014**, *12*, 207–218. [CrossRef]
11. Guo, F.; French, J.B.; Li, P.; Zhao, H.; Chan, C.Y.; Fick, J.R.; Benkovic, S.J.; Huang, T.J. Probing cell-cell communication with microfluidic devices. *Lab Chip* **2013**, *13*, 3152–3162. [CrossRef] [PubMed]
12. Nahavandi, S.; Tang, S.Y.; Baratchi, S.; Soffe, R.; Nahavandi, S.; Kalantar-zadeh, K.; Mitchell, A.; Khoshmanesh, K. Microfluidic platforms for the investigation of intercellular signaling mechanisms. *Small* **2014**, *10*, 4810–4826. [CrossRef] [PubMed]
13. Kirsch, M.; Rach, J.; Handke, W.; Seltsam, A.; Pepelanova, I.; Strauß, S.; Vogt, P.; Scheper, T.; Lavrentieva, A. Comparative Analysis of Mesenchymal Stem Cell Cultivation in Fetal Calf Serum, Human Serum, and Platelet Lysate in 2D and 3D Systems. *Front. Bioeng. Biotechnol.* **2021**, *8*, 598389. [CrossRef]
14. Sackmann, E.K.; Fulton, A.L.; Beebe, D.J. The present and future role of microfluidics in biomedical research. *Nature* **2014**, *507*, 181–189. [CrossRef] [PubMed]
15. Filippi, M.; Buchner, T.; Yasa, O.; Weirich, S.; Katzschmann, R.K. Microfluidic Tissue Engineering and Bio-Actuation. *Adv. Mater.* **2022**, *34*, 2108427. [CrossRef] [PubMed]
16. Sart, S.; Ronteix, G.; Jain, S.; Amselem, G.; Baroud, C.N. Cell Culture in Microfluidic Droplets. *Chem. Rev.* **2022**, *122*, 7061–7096. [CrossRef] [PubMed]
17. de Jongh, R.; Spijkers, X.M.; Pasteuning-Vuhman, S.; Vulto, P.; Pasterkamp, R.J. Neuromuscular junction-on-a-chip: ALS disease modeling and read-out development in microfluidic devices. *J. Neurochem.* **2021**, *157*, 393–412. [CrossRef] [PubMed]
18. Natarajan, A.; Sethumadhavan, A.; Krishnan, U.M. Toward Building the Neuromuscular Junction: In Vitro Models To Study Synaptogenesis and Neurodegeneration. *ACS Omega* **2019**, *4*, 12969–12977. [CrossRef] [PubMed]
19. Habibey, R.; Rojo Arias, J.E.; Striebel, J.; Busskamp, V. Microfluidics for Neuronal Cell and Circuit Engineering. *Chem. Rev.* **2022**, *122*, 14842–14880. [CrossRef]

20. Mehta, P.; Rahman, Z.; Ten Dijke, P.; Boukany, P.E. Microfluidics meets 3D cancer cell migration. *Trends Cancer* **2022**, *8*, 683–697. [CrossRef]
21. Li, C.; He, W.; Wang, N.; Xi, Z.; Deng, R.; Liu, X.; Kang, R.; Xie, L.; Liu, X. Application of microfluidics in detection of circulating tumor cells. *Front. Bioeng. Biotechnol.* **2022**, *10*, 907232. [CrossRef] [PubMed]
22. Limongi, T.; Guzzi, F.; Parrotta, E.; Candeloro, P.; Scalise, S.; Lucchino, V.; Gentile, F.; Tirinato, L.; Coluccio, M.L.; Torre, B.; et al. Microfluidics for 3D Cell and Tissue Cultures: Microfabricative and Ethical Aspects Updates. *Cells* **2022**, *11*, 1699. [CrossRef] [PubMed]
23. Saorin, G.; Caligiuri, I.; Rizzolio, F. Microfluidic organoids-on-a-chip: The future of human models. Semin. *Cell Dev. Biol.* **2023**, *144*, 41–54. [CrossRef] [PubMed]
24. Knoška, J.; Adriano, L.; Awel, S.; Beyerlein, K.R.; Yefanov, O.; Oberthuer, D.; Murillo, G.E.P.; Roth, N.; Sarrou, I.; Villanueva-Perez, P.; et al. Ultracompact 3D microfluidics for time-resolved structural biology. *Nat. Commun.* **2020**, *11*, 657. [CrossRef] [PubMed]
25. Ma, J.; Yan, S.; Miao, C.; Li, L.; Shi, W.; Liu, X.; Luo, Y.; Liu, T.; Lin, B.; Wu, W.; et al. Paper Microfluidics for Cell Analysis. *Adv. Healthc. Mater.* **2019**, *8*, 1801084. [CrossRef] [PubMed]
26. Liu, R.; Meng, X.; Yu, X.; Wang, G.; Dong, Z.; Zhou, Z.; Qi, M.; Yu, X.; Ji, T.; Wang, F. From 2D to 3D Co-Culture Systems: A Review of Co-Culture Models to Study the Neural Cells Interaction. *Int. J. Mol. Sci.* **2022**, *23*, 13116. [CrossRef] [PubMed]
27. Palikuqi, B.; Nguyen, D.T.; Li, G.; Schreiner, R.; Pellegata, A.F.; Liu, Y.; Redmond, D.; Geng, F.; Lin, Y.; Gómez-Salinero, J.M.; et al. Adaptable haemodynamic endothelial cells for organogenesis and tumorigenesis. *Nature* **2020**, *585*, 426–432. [CrossRef] [PubMed]
28. Khoo, B.L.; Grenci, G.; Lim, Y.B.; Lee, S.C.; Han, J.; Lim, C.T. Expansion of patient-derived circulating tumor cells from liquid biopsies using a CTC microfluidic culture device. *Nat. Protoc.* **2018**, *13*, 34–58. [CrossRef] [PubMed]
29. Park, S.E.; Georgescu, A.; Huh, D. Organoids-on-a-chip. *Science* **2019**, *364*, 960–965. [CrossRef]
30. Hofer, M.; Lutolf, M.P. Engineering organoids. *Nat. Rev. Mater.* **2021**, *6*, 402–420. [CrossRef]
31. Nashimoto, Y.; Okada, R.; Hanada, S.; Arima, Y.; Nishiyama, K.; Miura, T.; Yokokawa, R. Vascularized cancer on a chip: The effect of perfusion on growth and drug delivery of tumor spheroid. *Biomaterials* **2020**, *229*, 119547. [CrossRef] [PubMed]
32. Kim, M.; Mun, H.; Sung, C.O.; Cho, E.J.; Jeon, H.J.; Chun, S.M.; Jung, D.J.; Shin, T.H.; Jeong, G.S.; Kim, D.K.; et al. Patient-derived lung cancer organoids as in vitro cancer models for therapeutic screening. *Nat. Commun.* **2019**, *10*, 3991. [CrossRef] [PubMed]
33. Schuster, B.; Junkin, M.; Kashaf, S.S.; Romero-Calvo, I.; Kirby, K.; Matthews, J.; Weber, C.R.; Rzhetsky, A.; White, K.P.; Tay, S. Automated microfluidic platform for dynamic and combinatorial drug screening of tumor organoids. *Nat. Commun.* **2020**, *11*, 5271–5283. [CrossRef] [PubMed]
34. Youhanna, S.; Lauschke, V.M. The Past, Present and Future of Intestinal In Vitro Cell Systems for Drug Absorption Studies. *J. Pharm. Sci.* **2021**, *110*, 50–65. [CrossRef] [PubMed]
35. Moarefian, M.; Davalos, R.V.; Tafti, D.K.; Achenie, L.E.; Jones, C.N. Modeling iontophoretic drug delivery in a microfluidic device. *Lab Chip* **2020**, *20*, 3310–3321. [CrossRef] [PubMed]
36. Mark, D.; Haeberle, S.; Roth, G.; von Stetten, F.; Zengerle, R. Microfluidic lab-on-a-chip platforms: Requirements, characteristics and applications. *Chem. Soc. Rev.* **2010**, *39*, 1153–1182. [CrossRef] [PubMed]
37. Amirifar, L.; Shamloo, A.; Nasiri, R.; de Barros, N.R.; Wang, Z.Z.; Unluturk, B.D.; Libanori, A.; Ievglevskyi, O.; Diltemiz, S.E.; Sances, S.; et al. Brain-on-a-chip: Recent advances in design and techniques for microfluidic models of the brain in health and disease. *Biomaterials* **2022**, *285*, 121531. [CrossRef] [PubMed]
38. Adriani, G.; Ma, D.; Pavesi, A.; Kamm, R.D.; Goh, E.L. A 3D neurovascular microfluidic model consisting of neurons, astrocytes and cerebral endothelial cells as a blood-brain barrier. *Lab Chip* **2017**, *17*, 448–459. [CrossRef] [PubMed]
39. Bang, S.; Lee, S.; Choi, N.; Kim, H.N. Emerging Brain-Pathophysiology-Mimetic Platforms for Studying Neurodegenerative Diseases: Brain Organoids and Brains-on-a-Chip. *Adv. Healthc. Mater.* **2021**, *10*, 2002119–2002145. [CrossRef]
40. Noel, G.; Baetz, N.W.; Staab, J.F.; Donowitz, M.; Kovbasnjuk, O.; Pasetti, M.F.; Zachos, N.C. A primary human macrophage-enteroid co-culture model to investigate mucosal gut physiology and host-pathogen interactions. *Sci. Rep.* **2017**, *7*, 45270. [CrossRef]
41. Rahman, S.; Ghiboub, M.; Donkers, J.M.; van de Steeg, E.; van Tol, E.A.F.; Hakvoort, T.B.M.; de Jonge, W.J. The Progress of Intestinal Epithelial Models from Cell Lines to Gut-On-Chip. *Int. J. Mol. Sci.* **2021**, *22*, 13472. [CrossRef] [PubMed]
42. Borciani, G.; Montalbano, G.; Baldini, N.; Cerqueni, G.; Vitale-Brovarone, C.; Ciapetti, G. Co-culture systems of osteoblasts and osteoclasts: Simulating in vitro bone remodeling in regenerative approaches. *Acta Biomater.* **2020**, *108*, 22–45. [CrossRef] [PubMed]
43. Hsieh, F.H.; Sharma, P.; Gibbons, A.; Goggans, T.; Erzurum, S.C.; Haque, S.J. Human airway epithelial cell determinants of survival and functional phenotype for primary human mast cells. *Proc. Natl. Acad. Sci. USA* **2005**, *102*, 14380–14385. [CrossRef] [PubMed]
44. Hou, A.; Hou, K.; Huang, Q.; Lei, Y.; Chen, W. Targeting Myeloid-Derived Suppressor Cell, a Promising Strategy to Overcome Resistance to Immune Checkpoint Inhibitors. *Front. Immunol.* **2020**, *11*, 783. [CrossRef]
45. Shi, M.; Majumdar, D.; Gao, Y.; Brewer, B.M.; Goodwin, C.R.; McLean, J.A.; Li, D.; Webb, D.J. Glia co-culture with neurons in microfluidic platforms promotes the formation and stabilization of synaptic contacts. *Lab Chip* **2013**, *13*, 3008–3021. [CrossRef] [PubMed]

46. Bayik, D.; Lathia, J.D. Cancer stem cell-immune cell crosstalk in tumour progression. *Nat. Rev. Cancer* **2021**, *21*, 526–536. [CrossRef] [PubMed]
47. Dura, B.; Dougan, S.K.; Barisa, M.; Hoehl, M.M.; Lo, C.T.; Ploegh, H.L.; Voldman, J. Profiling lymphocyte interactions at the single-cell level by microfluidic cell pairing. *Nat. Commun.* **2015**, *6*, 5940. [CrossRef]
48. Liu, J.; Zhang, B.; Wang, L.; Peng, J.; Wu, K.; Liu, T. The development of droplet-based microfluidic virus detection technology for human infectious diseases. *Anal. Methods* **2024**, *16*, 971–978. [CrossRef]
49. Deng, Y.; Guo, Y.; Xu, B. Recent development of microfluidic technology for cell trapping in single cell analysis A review. *Processes* **2020**, *8*, 1253. [CrossRef]
50. Zhang, P.; Liu, C.; Modavi, C.; Abate, A.; Chen, H. Printhead on a chip: Empowering droplet-based bioprinting with microfluidics. *Trends Biotechnol.* **2023**, *42*, 353–388. [CrossRef]
51. Chen, L.J.; Raut, B.; Nagai, N.; Abe, T.; Kaji, H. Prototyping a Versatile Two-Layer Multi-Channel Microfluidic Device for Direct-Contact Cell-Vessel Co-Culture. *Micromachines* **2020**, *11*, 79. [CrossRef]
52. Dudman, J.; Ferreira, A.M.; Gentile, P.; Wang, X.; Dalgarno, K. Microvalve Bioprinting of MSC-Chondrocyte Co-Cultures. *Cells* **2021**, *10*, 3329. [CrossRef] [PubMed]
53. Schmidt, S.I.; Bogetofte, H.; Ritter, L.; Agergaard, J.B.; Hammerich, D.; Kabiljagic, A.A.; Wlodarczyk, A.; Lopez, S.G.; Sørensen, M.D.; Jørgensen, M.L.; et al. Microglia-Secreted Factors Enhance Dopaminergic Differentiation of Tissue- and iPSC-Derived Human Neural Stem Cells. *Stem Cell Rep.* **2021**, *16*, 281–294. [CrossRef]
54. Liu, Y.; Wang, L.; Long, Z.; Zeng, L.; Wu, Y. Protoplasmic astrocytes enhance the ability of neural stem cells to differentiate into neurons in vitro. *PLoS ONE* **2012**, *7*, e38243. [CrossRef] [PubMed]
55. Kempuraj, D.; Khan, M.M.; Thangavel, R.; Xiong, Z.; Yang, E.; Zaheer, A. Glia maturation factor induces interleukin-33 release from astrocytes: Implications for neurodegenerative diseases. *J. Neuroimmune Pharmacol.* **2013**, *8*, 643–650. [CrossRef] [PubMed]
56. Saadeldin, I.M.; Abdelfattah-Hassan, A.; Swelum, A.A. Feeder Cell Type Affects the Growth of In Vitro Cultured Bovine Trophoblast Cells. *Biomed. Res. Int.* **2017**, 1061589. [CrossRef]
57. Le-Bel, G.; Cortez Ghio, S.; Guérin, L.P.; Bisson, F.; Germain, L.; Guérin, S.L. Irradiated Human Fibroblasts as a Substitute Feeder Layer to Irradiated Mouse 3T3 for the Culture of Human Corneal Epithelial Cells: Impact on the Stability of the Transcription Factors Sp1 and NFI. *Int. J. Mol. Sci.* **2019**, *20*, 6296. [CrossRef] [PubMed]
58. Trettner, S.; Findeisen, A.; Taube, S.; Horn, P.A.; Sasaki, E.; zur Nieden, N.I. Osteogenic induction from marmoset embryonic stem cells cultured in feeder-dependent and feeder-independent conditions. *Osteoporos. Int.* **2014**, *25*, 1255–1266. [CrossRef] [PubMed]
59. López-Fagundo, C.; Livi, L.L.; Ramchal, T.; Darling, E.M.; Hoffman-Kim, D. A biomimetic synthetic feeder layer supports the proliferation and self-renewal of mouse embryonic stem cells. *Acta Biomater.* **2016**, *39*, 55–64. [CrossRef]
60. Li, R.; Zhang, X.; Lv, X.; Geng, L.; Li, Y.; Qin, K.; Deng, Y. Microvalve controlled multi-functional microfluidic chip for divisional cell co-culture. *Anal. Biochem.* **2017**, *539*, 48–53. [CrossRef]
61. Yang, K.; Park, H.J.; Han, S.; Lee, J.; Ko, E.; Kim, J.; Lee, J.S.; Yu, J.H.; Song, K.Y.; Cheong, E.; et al. Recapitulation of in vivo-like paracrine signals of human mesenchymal stem cells for functional neuronal differentiation of human neural stem cells in a 3D microfluidic system. *Biomaterials* **2015**, *63*, 177–188. [CrossRef] [PubMed]
62. Li, P.; Cui, F.; Chen, H.; Yang, Y.; Li, G.; Mao, H.; Lyu, X. A Microfluidic Cell Co-Culture Chip for the Monitoring of Interactions between Macrophages and Fibroblasts. *Biosensors* **2022**, *13*, 70. [CrossRef] [PubMed]
63. Sun, Y.; Cai, B.; Wei, X.; Wang, Z.; Rao, L.; Meng, Q.F.; Liao, Q.; Liu, W.; Guo, S.; Zhao, X. A valve-based microfluidic device for on-chip single cell treatments. *Electrophoresis* **2019**, *40*, 961–968. [CrossRef] [PubMed]
64. Park, J.; Wetzel, I.; Marriott, I.; Dréau, D.; D'Avanzo, C.; Kim, D.Y.; Tanzi, R.E.; Cho, H. A 3D human triculture system modeling neurodegeneration and neuroinflammation in Alzheimer's disease. *Nat. Neurosci.* **2018**, *21*, 941–951. [CrossRef] [PubMed]
65. Sakthivel, K.; O'Brien, A.; Kim, K.; Hoorfar, M. Microfluidic analysis of heterotypic cellular interactions: A review of techniques and applications. *Trends Anal. Chem.* **2019**, *117*, 166–185. [CrossRef]
66. Zahavi, E.E.; Ionescu, A.; Gluska, S.; Gradus, T.; Ben-Yaakov, K.; Perlson, E. A compartmentalized microfluidic neuromuscular co-culture system reveals spatial aspects of GDNF functions. *J. Cell Sci.* **2015**, *128*, 1241–1252. [PubMed]
67. Chung, H.H.; Mireles, M.; Kwarta, B.J.; Gaborski, T.R. Use of porous membranes in tissue barrier and co-culture models. *Lab Chip* **2018**, *18*, 1671–1689. [CrossRef]
68. De Vitis, E.; La Pesa, V.; Gervaso, F.; Romano, A.; Quattrini, A.; Gigli, G.; Moroni, L.; Polini, A. A microfabricated multi-compartment device for neuron and Schwann cell differentiation. *Sci. Rep.* **2021**, *11*, 7019–7030. [CrossRef] [PubMed]
69. Son, M.; Wang, A.G.; Kenna, E.; Tay, S. High-throughput co-culture system for analysis of spatiotemporal cell-cell signaling. *Biosens. Bioelectron.* **2023**, *225*, 115089–115098. [CrossRef] [PubMed]
70. He, J.; Chen, W.; Deng, S.; Xie, L.; Feng, J.; Geng, J.; Jiang, D.; Dai, H.; Wang, C. Modeling alveolar injury using microfluidic co-cultures for monitoring bleomycin-induced epithelial/fibroblastic cross-talk disorder. *RSC Adv.* **2017**, *7*, 42738–42749. [CrossRef]
71. Saygili, E.; Yildiz-Ozturk, E.; Green, M.J.; Ghaemmaghami, A.M.; Yesil-Celiktas, O. Human lung-on-chips: Advanced systems for respiratory virus models and assessment of immune response. *Biomicrofluidics* **2021**, *15*, 021501. [CrossRef]
72. Zhu, J.; Wang, Y.; Chen, P.; Su, H.; Du, W.; Liu, B.F. Highly efficient microfluidic device for cell trapping and pairing towards cell-cell communication analysis. *Sens. Actuators B Chem.* **2019**, *283*, 685–692. [CrossRef]
73. Liu, Y.; Fan, Z.; Qiao, L.; Liu, B. Advances in microfluidic strategies for single-cell research. *Trends Analyt. Chem.* **2022**, *157*, 116822. [CrossRef]

74. Anggraini, D.; Ota, N.; Shen, Y.; Tang, T.; Tanaka, Y.; Hosokawa, Y.; Li, M.; Yalikun, Y. Recent advances in microfluidic devices for single-cell cultivation: Methods and applications. *Lab Chip* **2022**, *22*, 1438–1468. [CrossRef]
75. Jie, M.; Li, H.F.; Lin, L.; Zhang, J.; Lin, J.M. Integrated microfluidic system for cell co-culture and simulation of drug metabolism. *RSC Adv.* **2016**, *6*, 54564–54572. [CrossRef]
76. Vivas, A.; IJspeert, C.; Pan, J.Y.; Vermeul, K.; van den Berg, A.; Passier, R.; Keller, S.S.; van der Meer, A.D. Generation and Culture of Cardiac Microtissues in a Microfluidic Chip with a Reversible Open Top Enables Electrical Pacing, Dynamic Drug Dosing and Endothelial Cell Co-Culture. *Adv. Mater. Technol.* **2022**, *7*, 2101355. [CrossRef]
77. Yoo, J.; Jung, Y.; Char, K.; Jang, Y. Advances in cell coculture membranes recapitulating in vivo microenvironments. *Trends Biotechnol.* **2023**, *41*, 214–227. [CrossRef]
78. Liu, Y.; Kongsuphol, P.; Chiam, S.Y.; Zhang, Q.X.; Gourikutty, S.B.N.; Saha, S.; Biswas, S.K.; Ramadan, Q. Adipose-on-a-chip: A dynamic microphysiological in vitro model of the human adipose for immune-metabolic analysis in type II diabetes. *Lab Chip* **2019**, *19*, 241–253. [CrossRef]
79. Liu, H.; Jie, M.; He, Z.; Li, H.F.; Lin, J.M. Study of antioxidant effects on malignant glioma cells by constructing a tumor-microvascular structure on microchip. *Anal. Chim. Acta* **2017**, *978*, 1–9. [CrossRef]
80. Sato, K.; Kikuchi, S.; Yoshida, E.; Ishii, R.; Sasaki, N.; Tsunoda, K.; Sato, K. Patterned Co-culture of Live Cells on a Microchip by Photocrosslinking with Benzophenone. *Anal. Sci.* **2016**, *32*, 113–116. [CrossRef]
81. Sun, W.; Chen, Y.; Wang, Y.; Luo, P.; Zhang, M.; Zhang, H.; Hu, P. Interaction study of cancer cells and fibroblasts on a spatially confined oxygen gradient microfluidic chip to investigate the tumor microenvironment. *Analyst* **2018**, *143*, 5431–5437. [CrossRef]
82. Chen, H.J.; Miller, P.; Shuler, M.L. A pumpless body-on-a-chip model using a primary culture of human intestinal cells and a 3D culture of liver cells. *Lab Chip* **2018**, *18*, 2036–2046. [CrossRef]
83. Hattersley, S.M.; Sylvester, D.C.; Dyer, C.E.; Stafford, N.D.; Haswell, S.J.; Greenman, J. A microfluidic system for testing the responses of head and neck squamous cell carcinoma tissue biopsies to treatment with chemotherapy drugs. *Ann. Biomed. Eng.* **2012**, *40*, 1277–1288. [CrossRef]
84. Zhang, Y.; Chen, S.; Fan, F.; Xu, N.; Meng, X.L.; Zhang, Y.; Lin, J.M. Neurotoxicity mechanism of aconitine in HT22 cells studied by microfluidic chip-mass spectrometry. *J. Pharm. Anal.* **2023**, *13*, 88–98. [CrossRef]
85. Jeong, S.; Kim, S.; Buonocore, J.; Park, J.; Welsh, C.J.; Li, J.; Han, A. A Three-Dimensional Arrayed Microfluidic Blood-Brain Barrier Model With Integrated Electrical Sensor Array. *IEEE Trans. Biomed. Eng.* **2018**, *65*, 431–439. [CrossRef]
86. Yeste, J.; García-Ramírez, M.; Illa, X.; Guimerà, A.; Hernández, C.; Simó, R.; Villa, R. A compartmentalized microfluidic chip with crisscross microgrooves and electrophysiological electrodes for modeling the blood-retinal barrier. *Lab Chip* **2017**, *18*, 95–105. [CrossRef]
87. Esch, M.B.; Ueno, H.; Applegate, D.R.; Shuler, M.L. Modular, pumpless body-on-a-chip platform for the co-culture of GI tract epithelium and 3D primary liver tissue. *Lab Chip* **2016**, *16*, 2719–2729. [CrossRef]
88. Li, X.; Fan, B.; Cao, S.; Chen, D.; Zhao, X.; Men, D.; Yue, W.; Wang, J.; Chen, J. A microfluidic flow cytometer enabling absolute quantification of single-cell intracellular proteins. *Lab Chip* **2017**, *17*, 3129–3137. [CrossRef]
89. Lee, J.M.; Seo, H.I.; Bae, J.H.; Chung, B.G. Hydrogel microfluidic co-culture device for photothermal therapy and cancer migration. *Electrophoresis* **2017**, *38*, 1318–1324. [CrossRef]
90. Wang, X.; Li, X.; Ding, J.; Long, X.; Zhang, H.; Zhang, X.; Jiang, X.; Xu, T. 3D bioprinted glioma microenvironment for glioma vascularization. *J. Biomed. Mater. Res. A* **2021**, *109*, 915–925. [CrossRef]
91. Ko, J.; Ahn, J.; Kim, S.; Lee, Y.; Lee, J.; Park, D.; Jeon, N.L. Tumor spheroid-on-a-chip: A standardized microfluidic culture platform for investigating tumor angiogenesis. *Lab Chip* **2019**, *19*, 2822–2833. [CrossRef]
92. Moshksayan, K.; Kashaninejad, N.; Warkiani, M.E.; Lock, J.G.; Moghadas, H.; Firoozabadi, B.; Saidi, M.S.; Nguyen, N.-T. Spheroids-on-a-chip: Recent advances and design considerations in microfluidic platforms for spheroid formation and culture. *Sens. Actuators B Chem.* **2018**, *263*, 151–176. [CrossRef]
93. Ma, Y.; Pan, J.Z.; Zhao, S.P.; Lou, Q.; Zhu, Y.; Fang, Q. Microdroplet chain array for cell migration assays. *Lab Chip* **2016**, *16*, 4658–4665. [CrossRef]
94. Lin, L.; He, Z.; Jie, M.; Lin, J.M.; Zhang, J. 3D microfluidic tumor models for biomimetic engineering of glioma niche and detection of cell morphology, migration and phenotype change. *Talanta* **2021**, *234*, 122702. [CrossRef]
95. Song, H.; Wang, Y.; Rosano, J.M.; Prabhakarpandian, B.; Garson, C.; Pant, K.; Lai, E. A microfluidic impedance flow cytometer for identification of differentiation state of stem cells. *Lab Chip* **2013**, *13*, 2300–2310. [CrossRef]
96. Tan, H.Y.; Trier, S.; Rahbek, U.L.; Dufva, M.; Kutter, J.P.; Andresen, T.L. A multi-chamber microfluidic intestinal barrier model using Caco-2 cells for drug transport studies. *PLoS ONE* **2018**, *13*, e0197101. [CrossRef]
97. Liu, Y.; Chen, X.; Chen, J.; Luo, Y.; Chen, Z.; Lin, D.; Zhang, J.; Liu, D. Gel-Free Single-Cell Culture Arrays on a Microfluidic Chip for Highly Efficient Expansion and Recovery of Colon Cancer Stem Cells. *ACS Biomater. Sci. Eng.* **2022**, *8*, 3623–3632. [CrossRef]
98. Tehranirokh, M.; Kouzani, A.Z.; Francis, P.S.; Kanwar, J.R. Microfluidic devices for cell cultivation and proliferation. *Biomicrofluidics* **2013**, *7*, 051502. [CrossRef]
99. Sacchi, M.; Bansal, R.; Rouwkema, J. Bioengineered 3D Models to Recapitulate Tissue Fibrosis. *Trends Biotechnol.* **2020**, *38*, 623–636. [CrossRef]
100. van Grunsven, L.A. 3D in vitro models of liver fibrosis. *Adv. Drug Deliv. Rev.* **2017**, *121*, 133–146. [CrossRef]

101. Jang, K.J.; Otieno, M.A.; Ronxhi, J.; Lim, H.K.; Ewart, L.; Kodella, K.R.; Petropolis, D.B.; Kulkarni, G.; Rubins, J.E.; Conegliano, D.; et al. Reproducing human and cross-species drug toxicities using a Liver-Chip. *Sci. Transl. Med.* **2019**, *11*, eaax5516. [CrossRef]
102. Materne, E.M.; Ramme, A.P.; Terrasso, A.P.; Serra, M.; Alves, P.M.; Brito, C.; Sakharov, D.A.; Tonevitsky, A.G.; Lauster, R.; Marx, U. A multi-organ chip co-culture of neurospheres and liver equivalents for long-term substance testing. *J. Biotechnol.* **2015**, *205*, 36–46. [CrossRef]
103. Caruso, G.; Musso, N.; Grasso, M.; Costantino, A.; Lazzarino, G.; Tascedda, F.; Gulisano, M.; Lunte, S.M.; Caraci, F. Microfluidics as a Novel Tool for Biological and Toxicological Assays in Drug Discovery Processes: Focus on Microchip Electrophoresis. *Micromachines* **2020**, *11*, 593. [CrossRef]
104. Jeong, S.Y.; Lee, J.H.; Shin, Y.; Chung, S.; Kuh, H.J. Co-Culture of Tumor Spheroids and Fibroblasts in a Collagen Matrix-Incorporated Microfluidic Chip Mimics Reciprocal Activation in Solid Tumor Microenvironment. *PLoS ONE* **2016**, *11*, e0159013. [CrossRef]
105. Ibrahim, L.I.; Hajal, C.; Offeddu, G.S.; Gillrie, M.R.; Kamm, R.D. Omentum-on-a-chip: A multicellular, vascularized microfluidic model of the human peritoneum for the study of ovarian cancer metastases. *Biomaterials* **2022**, *288*, 121728. [CrossRef]
106. Oh, S.; Ryu, H.; Tahk, D.; Ko, J.; Chung, Y.; Lee, H.K.; Jeon, N.L. "Open-top" microfluidic device for in vitro three-dimensional capillary beds. *Lab Chip* **2017**, *17*, 3405–3414. [CrossRef]
107. Wang, S.; Mao, S.; Li, M.; Li, H.F.; Lin, J.M. Near-physiological microenvironment simulation on chip to evaluate drug resistance of different loci in tumour mass. *Talanta* **2019**, *191*, 67–73. [CrossRef]
108. Guimarães, C.F.; Cruz-Moreira, D.; Caballero, D.; Pirraco, R.P.; Gasperini, L.; Kundu, S.C.; Reis, R.L. Shining a Light on Cancer-Photonics in Microfluidic Tumor Modeling and Biosensing. *Adv. Healthc. Mater.* **2023**, *12*, e2201442. [CrossRef]
109. Kim, S.; Lee, H.; Chung, M.; Jeon, N.L. Engineering of functional, perfusable 3D microvascular networks on a chip. *Lab Chip* **2013**, *13*, 1489–1500. [CrossRef]
110. Liang, Y.; Yoon, J.Y. In situ sensors for blood-brain barrier (BBB) on a chip. *Sens. Actuators Rep.* **2021**, *3*, 100031. [CrossRef]
111. Benz, F.; Liebner, S. Structure and Function of the Blood–Brain Barrier (BBB). *Handb. Exp. Pharmacol.* **2020**, *273*, 3–31.
112. Park, J.; Koito, H.; Li, J.; Han, A. Microfluidic compartmentalized co-culture platform for CNS axon myelination research. *Biomed. Microdevices* **2009**, *11*, 1145–1153. [CrossRef] [PubMed]
113. Dinh, N.D.; Chiang, Y.Y.; Hardelauf, H.; Baumann, J.; Jackson, E.; Waide, S.; Sisnaiske, J.; Frimat, J.-P.; van Thriel, C.; Janasek, D.; et al. Microfluidic construction of minimalistic neuronal co-cultures. *Lab Chip* **2013**, *13*, 1402–1412. [CrossRef] [PubMed]
114. Booth, R.; Kim, H. Characterization of a microfluidic in vitro model of the blood-brain barrier (μBBB). *Lab Chip* **2012**, *12*, 1784–1792. [CrossRef] [PubMed]
115. Leiby, K.L.; Raredon, M.S.B.; Niklason, L.E. Bioengineering the Blood-gas Barrier. *Compr. Physiol.* **2020**, *10*, 415–452 [PubMed]
116. Sellgren, K.L.; Butala, E.J.; Gilmour, B.P.; Randell, S.H.; Grego, S. A biomimetic multicellular model of the airways using primary human cells. *Lab Chip* **2014**, *14*, 3349–3358. [CrossRef] [PubMed]
117. Huh, D.; Matthews, B.D.; Mammoto, A.; Montoya-Zavala, M.; Hsin, H.Y.; Ingber, D.E. Reconstituting organ-level lung functions on a chip. *Science* **2010**, *328*, 1662–1668. [CrossRef] [PubMed]
118. Benam, K.H.; Villenave, R.; Lucchesi, C.; Varone, A.; Hubeau, C.; Lee, H.H.; E Alves, S.; Salmon, M.; Ferrante, T.C.; Weaver, J.C.; et al. Small airway-on-a-chip enables analysis of human lung inflammation and drug responses in vitro. *Nat. Methods* **2016**, *13*, 151–157. [CrossRef] [PubMed]
119. Wiedenmann, S.; Breunig, M.; Merkle, J.; von Toerne, C.; Georgiev, T.; Moussus, M.; Schulte, L.; Seufferlein, T.; Sterr, M.; Lickert, H.; et al. Single-cell-resolved differentiation of human induced pluripotent stem cells into pancreatic duct-like organoids on a microwell chip. *Nat. Biomed. Eng.* **2021**, *5*, 897–913. [CrossRef]
120. Kim, S.K.; Kim, Y.H.; Park, S.; Cho, S.W. Organoid engineering with microfluidics and biomaterials for liver, lung disease, and cancer modeling. *Acta Biomater.* **2021**, *132*, 37–51. [CrossRef]
121. Sin, A.; Chin, K.C.; Jamil, M.F.; Kostov, Y.; Rao, G.; Shuler, M.L. The design and fabrication of three-chamber microscale cell culture analog devices with integrated dissolved oxygen sensors. *Biotechnol. Prog.* **2004**, *20*, 338–345. [CrossRef] [PubMed]
122. Weinberg, E.; Kaazempur-Mofrad, M.; Borenstein, J. Concept and computational design for a bioartificial nephron-on-a-chip. *Int. J. Artif. Organs* **2008**, *31*, 508–514. [CrossRef] [PubMed]
123. Lee, S.A.; Kang, E.; Ju, J.; Kim, D.S.; Lee, S.H. Spheroid-based three-dimensional liver-on-a-chip to investigate hepatocyte-hepatic stellate cell interactions and flow effects. *Lab Chip* **2013**, *13*, 3529–3537. [CrossRef] [PubMed]
124. Jalili-Firoozinezhad, S.; Gazzaniga, F.S.; Calamari, E.L.; Camacho, D.M.; Fadel, C.W.; Bein, A.; Swenor, B.; Nestor, B.; Cronce, M.J.; Tovaglieri, A.; et al. A complex human gut microbiome cultured in an anaerobic intestine-on-a-chip. *Nat. Biomed. Eng.* **2019**, *3*, 520–531. [CrossRef]
125. Puschhof, J.; Pleguezuelos-Manzano, C.; Clevers, H. Organoids and organs-on-chips: Insights into human gut-microbe interactions. *Cell Host Microbe* **2021**, *29*, 867–878. [CrossRef] [PubMed]
126. Glaser, D.E.; Curtis, M.B.; Sariano, P.A.; Rollins, Z.A.; Shergill, B.S.; Anand, A.; Deely, A.M.; Shirure, V.S.; Anderson, L.; Lowen, J.M.; et al. Organ-on-a-chip model of vascularized human bone marrow niches. *Biomaterials* **2022**, *280*, 121245–121289. [CrossRef] [PubMed]
127. Jung, D.J.; Shin, T.H.; Kim, M.; Sung, C.O.; Jang, S.J.; Jeong, G.S. A one-stop microfluidic-based lung cancer organoid culture platform for testing drug sensitivity. *Lab Chip* **2019**, *19*, 2854–2865. [CrossRef] [PubMed]

128. Novak, R.; Ingram, M.; Marquez, S.; Das, D.; Delahanty, A.; Herland, A.; Maoz, B.M.; Jeanty, S.S.F.; Somayaji, M.R.; Burt, M.; et al. Robotic fluidic coupling and interrogation of multiple vascularized organ chips. *Nat. Biomed. Eng.* **2020**, *4*, 407–420. [CrossRef]
129. Kang, Y.B.; Sodunke, T.R.; Lamontagne, J.; Cirillo, J.; Rajiv, C.; Bouchard, M.J.; Noh, M. Liver sinusoid on a chip: Long-term layered co-culture of primary rat hepatocytes and endothelial cells in microfluidic platforms. *Biotechnol. Bioeng.* **2015**, *112*, 2571–2582. [CrossRef]
130. Ya, S.; Ding, W.; Li, S.; Du, K.; Zhang, Y.; Li, C.; Liu, J.; Li, F.; Li, P.; Luo, T.; et al. On-Chip Construction of Liver Lobules with Self-Assembled Perfusable Hepatic Sinusoid Networks. *ACS Appl. Mater. Interfaces* **2021**, *13*, 32640–32652. [CrossRef]
131. Ao, Z.; Cai, H.; Wu, Z.; Song, S.; Karahan, H.; Kim, B.; Lu, H.-C.; Kim, J.; Mackie, K.; Guo, F. Tubular human brain organoids to model microglia-mediated neuroinflammation. *Lab Chip* **2021**, *21*, 2751–2762. [CrossRef] [PubMed]
132. Du, Y.; Li, N.; Yang, H.; Luo, C.; Gong, Y.; Tong, C.; Gao, Y.; Lü, S.; Long, M. Mimicking liver sinusoidal structures and functions using a 3D-configured microfluidic chip. *Lab Chip* **2017**, *17*, 782–794. [CrossRef]
133. Busche, M.; Tomilova, O.; Schütte, J.; Werner, S.; Beer, M.; Groll, N.; Hagmeyer, B.; Pawlak, M.; Jones, P.D.; Schmees, C.; et al. HepaChip-MP–a twenty-four chamber microplate for a continuously perfused liver coculture model. *Lab Chip* **2020**, *20*, 2911–2926. [CrossRef]
134. Zheng, Y.B.; Ma, L.D.; Wu, J.L.; Wang, Y.M.; Meng, X.S.; Hu, P.; Liang, Q.L.; Xie, Y.Y.; Luo, G. Design and fabrication of an integrated 3D dynamic multicellular liver-on-a-chip and its application in hepatotoxicity screening. *Talanta* **2022**, *241*, 123262. [CrossRef]
135. Shah, P.; Fritz, J.V.; Glaab, E.; Desai, M.S.; Greenhalgh, K.; Frachet, A.; Niegowska, M.; Estes, M.; Jäger, C.; Seguin-Devaux, C.; et al. A microfluidics-based in vitro model of the gastrointestinal human–microbe interface. *Nat. Commun.* **2016**, *7*, 11535. [CrossRef] [PubMed]
136. Pediaditakis, I.; Kodella, K.R.; Manatakis, D.V.; Le, C.Y.; Hinojosa, C.D.; Tien-Street, W.; Manolakos, E.S.; Vekrellis, K.; Hamilton, G.A.; Ewart, L.; et al. Modeling alpha-synuclein pathology in a human brain-chip to assess blood-brain barrier disruption. *Nat. Commun.* **2021**, *12*, 5907–6023. [CrossRef] [PubMed]
137. Chou, D.B.; Frismantas, V.; Milton, Y.; David, R.; Pop-Damkov, P.; Ferguson, D.; MacDonald, A.; Bölükbaşı, V.; Joyce, C.E.; Teixeira, L.S.M.; et al. On-chip recapitulation of clinical bone marrow toxicities and patient-specific pathophysiology. *Nat. Biomed. Eng.* **2020**, *4*, 394–406. [CrossRef]
138. Ingber, D.E. Human organs-on-chips for disease modelling, drug development and personalized medicine. *Nat. Rev. Genet.* **2022**, *23*, 467–491. [CrossRef]
139. Maschmeyer, I.; Lorenz, A.K.; Schimek, K.; Hasenberg, T.; Ramme, A.P.; Hübner, J.; Lindner, M.; Drewell, C.; Bauer, S.; Thomas, A.; et al. A four-organ-chip for interconnected long-term co-culture of human intestine, liver, skin and kidney equivalents. *Lab Chip* **2015**, *15*, 2688–2699. [CrossRef]
140. Smith, V.M.; Nguyen, H.; Rumsey, J.W.; Long, C.J.; Shuler, M.L.; Hickman, J.J. A Functional Human-on-a-Chip autoimmune disease model of myasthenia gravis for development of therapeutics. *Front. Cell Dev. Biol.* **2021**, *9*, 745897. [CrossRef]
141. Wang, Y.I.; Carmona, C.; Hickman, J.J.; Shuler, M.L. Multiorgan microphysiological systems for drug development: Strategies, advances, and challenges. *Adv. Healthc. Mater.* **2018**, *7*, 1701000. [CrossRef] [PubMed]
142. Deng, S.; Li, C.; Cao, J.; Cui, Z.; Du, J.; Fu, Z.; Yang, H.; Chen, P. Organ-on-a-chip meets artificial intelligence in drug evaluation. *Theranostics* **2023**, *13*, 4526. [CrossRef] [PubMed]
143. Scheinpflug, J.; Pfeiffenberger, M.; Damerau, A.; Schwarz, F.; Textor, M.; Lang, A.; Schulze, F. Journey into bone models: A review. *Genes* **2018**, *9*, 247. [CrossRef] [PubMed]

biosensors

MDPI

Review

Advancing Tissue Culture with Light-Driven 3D-Printed Microfluidic Devices

Xiangke Li, Meng Wang, Thomas P. Davis, Liwen Zhang * and Ruirui Qiao *

Australian Institute for Bioengineering and Nanotechnology, The University of Queensland, Brisbane, QLD 4072, Australia
* Correspondence: liwen.zhang@uq.edu.au (L.Z.); r.qiao@uq.edu.au (R.Q.)

Abstract: Three-dimensional (3D) printing presents a compelling alternative for fabricating microfluidic devices, circumventing certain limitations associated with traditional soft lithography methods. Microfluidics play a crucial role in the biomedical sciences, particularly in the creation of tissue spheroids and pharmaceutical research. Among the various 3D printing techniques, light-driven methods such as stereolithography (SLA), digital light processing (DLP), and photopolymer inkjet printing have gained prominence in microfluidics due to their rapid prototyping capabilities, high-resolution printing, and low processing temperatures. This review offers a comprehensive overview of light-driven 3D printing techniques used in the fabrication of advanced microfluidic devices. It explores biomedical applications for 3D-printed microfluidics and provides insights into their potential impact and functionality within the biomedical field. We further summarize three light-driven 3D printing strategies for producing biomedical microfluidic systems: direct construction of microfluidic devices for cell culture, PDMS-based microfluidic devices for tissue engineering, and a modular SLA-printed microfluidic chip to co-culture and monitor cells.

Keywords: microfluidics; 3D printing; tissue culture

Citation: Li, X.; Wang, M.; Davis, T.P.; Zhang, L.; Qiao, R. Advancing Tissue Culture with Light-Driven 3D-Printed Microfluidic Devices. *Biosensors* **2024**, *14*, 301. https://doi.org/10.3390/bios14060301

Received: 29 April 2024
Revised: 4 June 2024
Accepted: 6 June 2024
Published: 8 June 2024

1. Introduction

Microfluidic systems, which enable the manipulation of minute fluid volumes (10^{-9} to 10^{-18} L) within channels sized from tens to hundreds of microns, have garnered considerable attention in scientific research and application [1]. Microfluidics offers several distinct advantages. Microfluidics can reduce sample and reagent quantities for cost savings and provide exceptional resolution and sensitivity for precise control of microemulsion generation [2]. The diminutive dimensions of microfluidic channels, characterized by high surface-area-to-volume ratios, promote enhanced thermal homogeneity at the reaction site, facilitating swift heat transfer [3]. These benefits allow microfluidics to be applied in biomedical applications such as tissue engineering, DNA purification, PCR activity, and medical diagnostics including pregnancy monitoring and glucose measurement [4].

Microfluidic devices have significant advantages over traditional batch methods in tissue culture. Traditional methods usually suffer from low throughput, high reagent consumption, and limited microenvironment control. By contrast, microfluidic devices can reduce reagent usage, enhance control of the cellular microenvironment, and perform high-throughput experiments [5]. First, microfluidic devices allow precise control of various environmental parameters, such as nutrient delivery, oxygen content, pH, and shear stress [6]. These advantages allow researchers to design customized microenvironments that closely resemble specific tissues or organs in the body. With this precise control, researchers can better simulate physiological conditions and, thus, obtain more biologically meaningful experimental results. Second, microfluidic devices offer unique advantages for high-throughput screening, as they can construct multiple chambers or channels. These capabilities allow for parallel experiments and high-throughput screening of multiple conditions or samples [7], significantly increasing experimental efficiency and accelerating the process of research and drug development. In addition, microfluidic devices are conveniently miniaturized and automated. Due to their compact structure, experiments can be greatly scaled down, reducing the amount of reagents and samples used [8]. At the same time, these devices can be connected to an automation system to improve fluid handling, sample and data collection operations, and experiment repeatability. Microfluidic devices can also be integrated with sensors, imaging systems, and other analysis tools to achieve real-time monitoring and analysis of cell reactions and behaviors [9]. This integration allows researchers to collect dynamic and time-resolved data that can reveal important insights into complex biological phenomena, including precise spatial and temporal control of the delivery of signaling molecules, drugs, and other stimuli to cells and tissues as well as gradient effects, cell-to-cell interactions, and dynamic biological responses [10]. By creating physiologically relevant models that closely mimic in vivo conditions, these benefits allow for the study of cellular behavior, drug responses, and disease mechanisms in a controlled environment [11]. As such, microfluidic devices play a crucial role in advancing biomedical research and fostering innovation in tissue culture methodologies. However, persistent challenges impede microfluidics integration across diverse fields, such as complex sample preparation, external high-tech equipment, and high-resolution requirements [12,13]. These complexities collectively pose significant hurdles to attaining large-scale commercial production.

Three-dimensional printing, also known as additive manufacturing, refers to the process of creating three-dimensional objects by adding materials layer by layer based on a digital model. Recent advancements in three-dimensional (3D) printing have profoundly impacted microfabrication, tissue engineering, drug delivery, and medical device development [14,15]. Notably, 3D printing techniques have been applied to manufacturing complex microfluidic devices. They offer a cost-effective alternative to the time-consuming molding process and cleanroom facility requirements.

However, traditional 3D printing techniques such as Fused Deposition Modeling (FDM), which extrude thermoplastic filaments layer by layer to create objects, often involve elevated processing temperatures, presenting challenges in using sensitive materials and biocompatible substances. Since FDM printing has a relatively lower resolution than other 3D printing methods, it inevitably reduces precision in microfluidic structure fabrication. Additionally, printed objects by Selective Laser Sintering and Selective Laser Melting (SLM) exhibit surface roughness, which impacts microfluidic devices' fluidic behavior [16]. Addressing these challenges is essential for broadening the potential uses and capabilities of 3D printing technologies in the field of microfluidics.

Light-driven 3D printing methods, which employ photocuring to solidify liquid resins, have emerged as a promising solution for fabricating microfluidic systems [17]. These techniques, including Stereolithography (SLA), Digital Light Processing (DLP), and Inkjet printing, offer high resolutions, mild operational conditions, rapid production rates, relatively smooth surfaces, and a wide selection of resins. These benefits make light-driven 3D printing particularly well-suited to fabricating microfluidic devices, which are expected to have high-resolution channels, optical transparency, smooth internal channel surfaces, and robust mechanical stability to tolerate high pressure [18,19]. The emergence of new technology has significantly enhanced 3D-printed microfluidic devices' research and applications in the field of biomedicine, opening new pathways for innovation and practicality.

In this review, we introduce the current progress of light-based 3D-printed microfluidics and discuss their implications for pivotal areas such as cell culture, tissue engineering, as well as the co-culturing and monitoring of cells.

2. Introduction of Microfluidics

Microfluidic chips consist of a network of tiny channels etched or molded into materials such as siloxane and PDMS. They are also used for tasks such as mixing fluids, pumping, sorting particles, and regulating biochemical conditions [17]. These chips offer excellent gas permeability, surface modification capabilities, and biocompatibility, making them invaluable for assessing cell viability in vitro [20]. The transparency of microfluidic chips also allows for easier observation of fluid dynamics using optical microscopes [21]. Currently, soft lithography is the primary technique for fabricating microfluidic chips using elastomer masks, stamps, and molds for its rapid prototyping advantages [22,23]. Typically, photolithography is used to manufacture master molds for microfluidics with a pre-polymer such as PDMS cured on top of the mold. Once cured, a PDMS-negative stamp of the mold is created and permanently bonded to glass (Figure 1) [24].

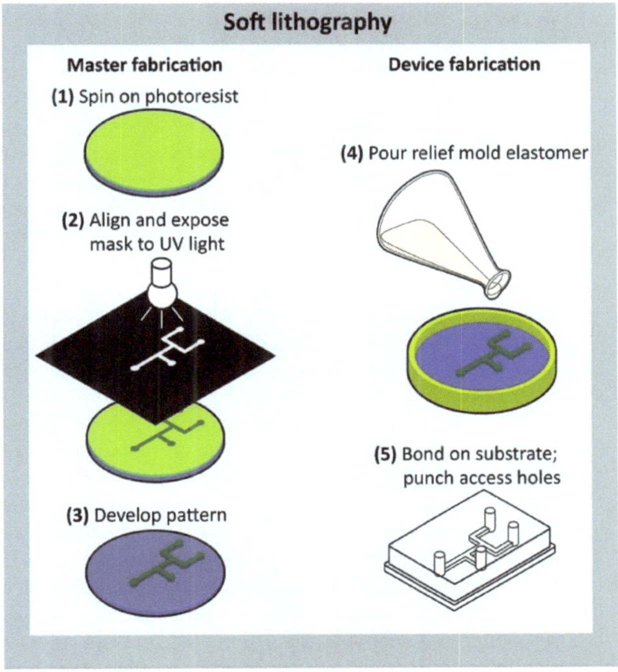

Figure 1. Rapid prototyping using soft lithography. Soft lithography is a multistep process in which a master mold is created, followed by curing a prepolymer substrate, peeling it off, bonding it to a substrate, and punching access holes [20]. Copyright 2017 Elsevier.

These microfluidic chips can achieve high-resolution features at the micrometer and even nanometer scales. PDMS, a widely used polymer in microfluidics, is favored for its biocompatibility, ready availability, transparency, hydrophobic properties, gas permeability, and elastomeric nature [21]. As such, microfluidic chips manufactured through soft lithography are widely employed in the biological and medical fields, including genetic engineering, proteomics, medical diagnostics, cell culture, drug research and development, and biosensors for biochemical and pathogen detection [22]. However, soft lithography faces limitations related to material durability and chemical resistance. For instance, soft lithography-fabricated PDMS devices typically lack robustness, which can lead to flow profile issues due to leakage and uneven pressure distribution [25]. Additionally, PDMS has drawbacks such as lower mechanical robustness and chemical resistance than some 3D-printed resins [26,27]. In terms of mechanical characteristics, the elastic modulus of PDMS (measurement of stiffness) is between 1.32 and 2.97 MPa, and the tensile strength (stretch value before breaking) is between 3.51 and 5.13 MPa. These numbers can change depending on how much curing agent is used and the temperature at the time it was manufactured [28]. The absorption of small molecules by PDMS can also influence microfluidic experiments, especially in the areas of drug discovery, proteomic analysis, and cell culture, where the compounds being studied are present in very low concentrations [29]. Therefore, exploring other fabrication techniques becomes essential to overcoming these challenges and expanding microfluidic technology applications. Notably, 3D printing allows for rapid prototyping, scalability, and the production of complex geometries that are difficult to achieve with PDMS [30].

3. 3D-Printed Microfluidic Devices

Currently, 3D printing has become an advanced approach to creating microfluidic devices. Three-dimensional printing methods, such as FDM, PolyJet (PJ), SLA, etc., have been successfully employed to fabricate fluidic channels. Compared to soft lithography techniques, which have limited design complexity and customization, 3D printing offers faster prototyping and design flexibility for intricate and customized microfluidic devices.

SLA is one of the most widely used light-based 3D printing methods primarily due to its accessibility, swift printing capabilities, and production of smooth, accurate structures [31]. As shown in Figure 2a, this process involves the successive curing of liquid polymers layer by layer, employing laser UV light and a controlled built platform. The built platform plays a crucial role in supporting and positioning each layer of the object during printing, ensuring accuracy and structural integrity. At the same time, UV light allows precise liquid resin polymerization, solidifying it with each layer. This process results in a finely detailed and cohesive three-dimensional structure [32]. The UV laser performs two key functions in SLA printing: designing patterns and curing liquid resin. Generally, a UV laser source and a scanning mirror are used to design a raster pattern for printed models. Alternatively, the 2D pattern can be exposed to photo-curable resin using a UV source and a digital micromirror device to control printed shapes [33–35]. Using free surface or restricted surface methods for printing 3D materials is also feasible. Both methods involve the photopolymerization of liquid resins under UV light irradiation. An essential component in various SLA processes is an absorber, which reduces the light's penetration into printed layers and prevents the polymerization of unpatented void features [13]. Hence, SLA's precision and high-resolution output make it cost-effective for fabricating complex structures, especially microfluidic devices for tissue-related research [36–39]. L. Ding et al. used an SLA 3D printer to rapidly print modular microfluidic systems for detaching and separating mesenchymal stem cells (MSCs) from microcarriers (MCs) [40]. Direct SLA printing was used to create each module, resulting in inexpensive and easy-to-manufacture high-precision 3D objects [40].

DLP is another light-driven 3D printing technique using a digital light projector as a light source to cure photopolymers (Figure 2b) [41]. The UV light from the projector cures the photopolymer resin layer all at once, setting DLP apart from SLA where laser points trace each layer [42]. This simultaneous layer curing significantly accelerates printing speed while maintaining intricate designs and high accuracy. L. Wang et al. used DLP technology to manufacture polymer-based microfluidic chips, allowing for the rapid realization of a wide range of functional microstructures outside of a cleanroom and minimal masking requirements [43].

In photopolymer inkjet printing, inkjet print heads play a crucial role in expelling liquid photopolymers from built platforms. This material undergoes immediate curing and solidification upon exposure to UV lamps, facilitating layer-by-layer construction (Figure 2c) [44]. Photopolymer inkjet printing offers fewer constraints than alternative methods, making it highly suitable for creating patterned environments for drug-testing organoids and tissue models [45]. Lidia Donvito et al. developed an inkjet 3D-printed droplet microfluidic device from acrylonitrile and wax [46]. Inkjet printing eliminates misalignment issues that plague conventional manufacturing processes, making it possible to manufacture microfluidic chips in a single procedure. Moreover, it permits quick prototyping at minimal cost.

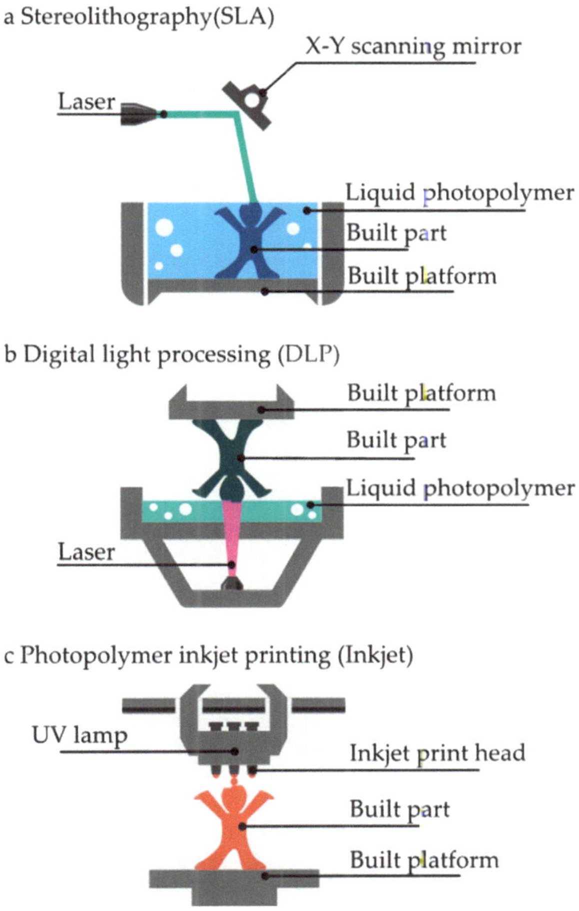

Figure 2. Schematic showing three types of light-induced 3D printing technologies.

In summary, Table 1 provides a comparative overview that considers the operational principles, material usage, advantages, and limitations of three 3D printing techniques for microfluidic device fabrication.

Table 1. Summary of three main light-driven 3D printing techniques for fabricating microfluidic devices.

3D Printing Technique	Energy Source	Materials	Advantages	Disadvantages
SLA	UV	Photocurable resin/polymer	Easy-to-make large pieces, allow uncured material, and high precision	Post-curing and support removal required
DLP	UV	Photocurable resin/polymer	Laying precision, high resolution, and reusing uncured photopolymer	Consumables insecurity and unsuitable for large structures
Inkjet	UV	Photocurable resin/polymer	Multi-material, fast-build printing	Laborious to remove channel support materials

4. Light-Driven 3D-Printed Microfluidics for Tissue Culture

In recent years, there has been a notable transformation in the methodologies employed for fabricating microfluidic devices. This change is attributed to the emergence of light-driven 3D printing, a new and promising method for overcoming some of the drawbacks of traditional soft lithography techniques [47]. Light-driven 3D printing offers an appealing option for creating microfluidic devices, mainly due to its cost-effectiveness, precision, high resolution, and ability to reproduce complex designs consistently [42,48]. Within the field of 3D-printed microfluidic devices, direct printing, mold-based, modular, and hybrid manufacturing procedures are the four primary production types. Microfluidic devices with integrated inlets, outlets, and microchannels are fabricated via 3D printing techniques. Novel techniques have also emerged, such as replica molding using 3D-printed molds. This approach involves the creation of microfluidic structures using 3D-printed molds, followed by bottom-side channel sealing with materials such as PDMS or glass slides. Another innovative concept is similar to assembling Lego® blocks and involves piecing together 3D-printed microfluidic modules. Another notable advancement is a hybrid solution, where the lower layer of a 3D-printed micro-channel is connected to a transparent top layer, offering a unique combination of functionalities. The advantages of additive technology make 3D-printed microfluidics particularly useful in tissue engineering [47], including the construction of culture platforms. High-precision light-driven 3D printing allows microfluidics to become more readily available [49].

In tissue culture, 3D-printed microfluidic devices exhibit multiple functions and significantly enhance 3D cell culture techniques for various cell types, including mammalian cell lines, stem cells, and primary cells [50]. First, mammalian cell lines, such as HeLa cells and Chinese hamster ovary (CHO) cells, are widely used in biomedical research for applications as diverse as drug testing, gene expression studies, and cancer research. However, in traditional 2D cultures, these cells grow in monolayers, limiting their interaction with other cells and extracellular matrix components. By contrast, 3D-printed microfluidic devices provide a more complex environment that promotes better cell communication and function, thereby enhancing the physiological relevance of the studies conducted. Second, stem cells, including embryonic stem cells (ESCs) and induced pluripotent stem cells (iPSCs), greatly benefit from 3D culture systems. These cells require a highly controlled environment to maintain their pluripotency and differentiate them into specific cell types. Three-dimensionally printed microfluidic devices facilitate these complex cultures by providing controlled microenvironments and nutrient flow, which support the formation of multicellular spheroids and organoids and provide essential cues for differentiation and growth [51]. This level of control supports stem cell proliferation, differentiation, and the formation of complex tissue structures. It is especially critical for tissue engineering and regenerative medicine applications. In addition, primary cells that are directly isolated from tissue usually retain more of the physiological properties of their original tissue than immortalized cell lines. For instance, hepatocytes and cardiomyocytes benefit from the 3D architecture provided by these devices, which mimics in vivo conditions better than traditional 2D cultures. Specifically, hepatocytes are the primary functional cells of the liver and play a key role in drug metabolism and detoxification research. However, in 2D culture, liver cells often lose their phenotype and function very quickly. Three-dimensionally printed microfluidic devices can reconstruct the liver microenvironment, support liver cell function, and enable long-term studies of liver physiology and disease modeling [52].

Three-dimensional cell cultures, such as spheroids and organoids, offer a more physiologically relevant environment, promoting better tissue-like structures and functions [53]. Spheroids are simple cellular aggregates formed by a single cell type or a combination of multicellular cells typically grown as free-floating aggregates [54]. On the other hand, organoids are more sophisticated models created from iPSCs or by the self-organization of tissue-derived cells such as cancer cells or stem cells [54]. Organoids are highly complicated because the variety of cell types comprise their makeup better reflects the structure and function of the organ. Organoids are more complicated than spheroids and may be cultured for extended periods. Furthermore, 3D-printed microfluidic devices can be applied to organ-on-a-chip (OoCs) technologies, which simulate the functions of entire organs or organ systems on a micro-scale [51]. This model provides the scientific community with a cutting-edge tool by combining engineering and biological expertise. Therefore, 3D-printed microfluidic devices can replicate physiological conditions with high precision, offering a powerful tool for drug testing and disease modeling [55].

4.1. Light 3D-Printed Microfluidics for Spheroid Perfusion Culture

Spheroid perfusion culture is a cell cultivation method that involves growing three-dimensional cell clusters, or spheroids, in a controlled environment. In this technique, a continuous flow of culture media circulates through the spheroids, supplying nutrients and oxygen while removing waste products. This approach enhances cell viability and functionality, mimicking the natural tissue conditions more closely than traditional static cultures. Generally, microfluidic spheroid cultures use channels in microstructures to hold 3D cell clusters, enabling controlled nutrient perfusion for advanced cellular studies. This setup allows for the subsequent development of spheroids or organoids under medium perfusion conditions [56–59]. Since spheroids or organoids must be immobilized and fixed in microfluidic devices for extended periods ranging from days to weeks, monitoring downstream biological assays via visualization methods is necessary. Addressing these challenges requires careful consideration of biocompatibility, suitability for bioimaging, and facilitation of cell retrieval in the design and production of 3D-printed devices for tissue culture.

In the context of spheroid cell culture and analysis, the unique operating principles and resolution restrictions of each 3D printing method are worth considering. To support multicellular spheroid cultures, Ong et al. reported the first example of a 3D-printed microfluidic perfusion cell culture device to directly immobilize 3D multicellular spheroids while keeping them alive and functioning [60]. They assessed two popular 3D printing technologies (SLA and PolyJet printing) to fabricate microfluidic devices containing cell-immobilized microstructures inside a microfluidic network. Their results revealed that SLA outperformed PolyJet printing, primarily in terms of resolution and post-processing ease. Notably, the 3D-printed microfluidic perfusion culture device comprised two separate components, a top layer, and a bottom mounting section (Figure 3a). Unlike a unibody device, this modular 3D-printed microfluidic device could be easily disassembled, allowing users to retrieve tissue samples without the need for specialized tools. In contrast to prototyping the device in a single build, this segmented design simplified the removal of support materials from the microfluidic network, making it possible to reuse the 3D-printed device. Notably, conventional 3D printing materials such as acrylonitrile butadiene styrene and polycarbonate exhibit limited transparency despite their prevalence [61–63]. This challenge was addressed by integrating a viewing window into the mounting base, which enables the visual examination of biological samples using sophisticated microscopy techniques, such as light and fluorescence microscopy. Additionally, delicately engineered PDMS membranes were introduced to serve as efficient gaskets during the integration of the top and mounting base components, ensuring the robust sealing of microfluidic channels. As depicted in Figure 3b, the schematic of the cell culture chamber's central section features an array of microstructures arranged in a circular pattern. These microstructures physically entrap cells introduced via the seed intake channel. The cell entrapment method ensures that the spaces between these microstructures are smaller than the size of a single cell or cell aggregate, as documented in previous studies [56,64,65]. A different channel network was employed to facilitate the perfusion of the culture medium after the cells were seeded into the device. Notably, the cells entrapped within the device were cultured using a pump-free perfusion technique [66]. Furthermore, the entire 3D-printed microfluidic perfusion culture apparatus can be placed within a sterile secondary container and positioned within a controlled environment featuring a 37°C temperature and a CO_2 incubator. This design enables the apparatus to function as a standalone device, eliminating the need for additional pumps and tubing. Figure 3c,d demonstrate the cell-immobilized microstructure array's ability to effectively capture and retain both parental and metastatic oral squamous cell carcinoma (OSCC) tumor spheroids within the cell culture compartment for up to 72 h while maintaining high levels of viability and functionality. This microfluidic device, fabricated using SLA printing, facilitates the integration and substantial streamlining of the microfluidic culture system's setup and operation. It broadens the system's potential utility for investigating medication effectiveness, metabolic processes, and toxicity assessments.

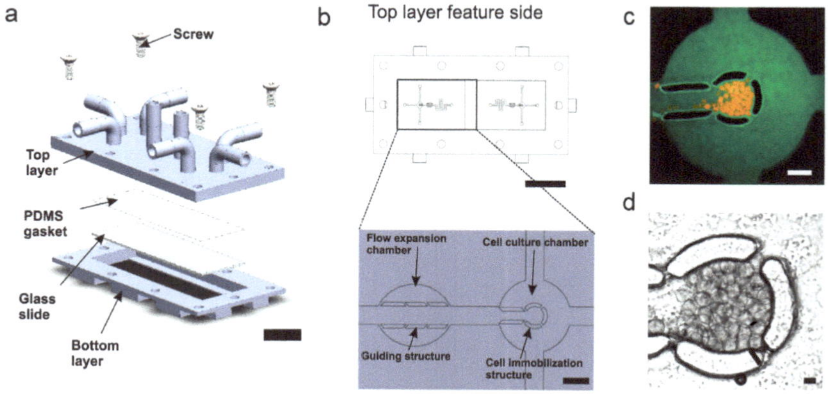

Figure 3. A 3D-printed microfluidic device for a spheroid culture system. (**a**) An exploded diagram of the device. (**b**) A 3D-printed top-layer device with microstructures for directing flow expansion and immobilizing cells (enlarged view). Imaging using fluorescence (**c**) and light transmission (**d**) showed spheroids of metastatic HN137 OSCC, which were trapped inside a cell culture chamber. To better view the cell culture chamber, FITC-tagged BSA was added to the culture media in (**c**). Scale bars: (**a**,**b**) 1 cm; (**b**) magnified view, 1 mm; (**c**) 500 µm; (**d**) 100 µm [60]. Copyright 2017 IOP Publishing.

4.2. Light 3D-Printed Replica Molding Process for Constructing Tissue-Engineered Skeletal Muscles

Functional skeletal muscle tissue substitutes show potential for treating a range of muscular diseases and injuries. Currently, photolithography, MultiJet Printing (MJP), and SLA are used to construct tissue-engineered skeletal muscles. Bian et al. manufactured polydimethylsiloxane (PDMS) molds via photolithography to create sizeable neonatal rat skeletal muscle tissue networks that exhibited consistent and controllable structural properties [67]. This technology requires costly cleanroom facilities, complex and error-prone procedures [68], and time-consuming post-processing [69]. Contrary to the photolithography method, Afshar et al. used a MJP plastic 3D printer to create a reusable mold, which was then used to produce a 96-well platform for large-scale production of 3D human skeletal muscle microtissues. Despite its effectiveness, MJP printing still has limitations related to the resolution of microfabrication techniques [70]. Its other limitations include high costs for high-end printers and slow, complex processes requiring specialized training, which restrict the broader adoption of these traditional fabrication methods in tissue engineering [71]. To address these challenges, employing readily available and affordable 3D printing materials, along with innovative replica molding techniques, are effective strategies for producing microdevices and scaling up at a low cost. Replica molding is a type of soft lithography for creating microfluidic chips by duplicating a mold's morphology and structure [72]. Within a day, hundreds of 3D culture devices of varying sizes and geometries can be manufactured in-house using replica molding techniques. PBAT (Polybutylene Adipate Terephthalate) is fully biodegradable polyester. BASF, a leading manufacturer, produces PBAT under the Ecoflex® (BASF, Ludwigshafen, Germany) brand name, which includes Ecoflex F Blend C1200 [73]. Ecoflex®, an economical, flexible, and highly resilient silicone material often employed in soft robotics, serves as a key component in this replica molding approach.

The material is notable for its impressive stretchability, which can be stretched to over nine times its original length before breaking [74].

The SLA printing technique can be used to fabricate replica moldings with smaller dimensions, excellent repeatability, and high throughput. Notably, SLA-printed objects exhibit both millimetric and micrometric features, a precision level difficult to achieve with conventional photolithography, particularly at the sub-millimeter scale [75,76]. To reproduce such structures in PDMS for high throughput applications of 3D culture devices, A. Iuliano et al. used the elastic polymer Ecoflex as a moldable replica substrate to construct tissue-engineered skeletal muscles (TESMS) in vitro [77]. The PBAT replica molding process was instrumental in generating negative molds following the initial 3D printing phase. Subsequently, the final positive PDMS structures could be effortlessly detached when the mold was stretched (Figure 4a). The creation of 3D culture chambers, featuring a volumetric capacity of 15 μL and T-shaped pillars measuring 500 μm in diameter and 2.5 mm in height, exemplifies a notable advancement in microfluidic device design. Most importantly, there is no discernible disparity between the original 3D-printed structures and PDMS replicas. Afterward, TESMs were cultivated in 48-well plates. Following 7 days of differentiation, engineered tissues in the 3D culture systems exhibited a consistent organization, characterized by the presence of long, aligned, and multinucleated myofibers positively stained for the sarcomeric protein titin. Confocal imaging revealed a typical striated titin pattern in the myofibers. The PBAT replica strategy has potential use in a wide range of applications due to its affordability and simplicity. It can be used as a compliant substrate for specific cell culture requirements and in developing supportive devices for other load-bearing tissues, such as the heart and tendons.

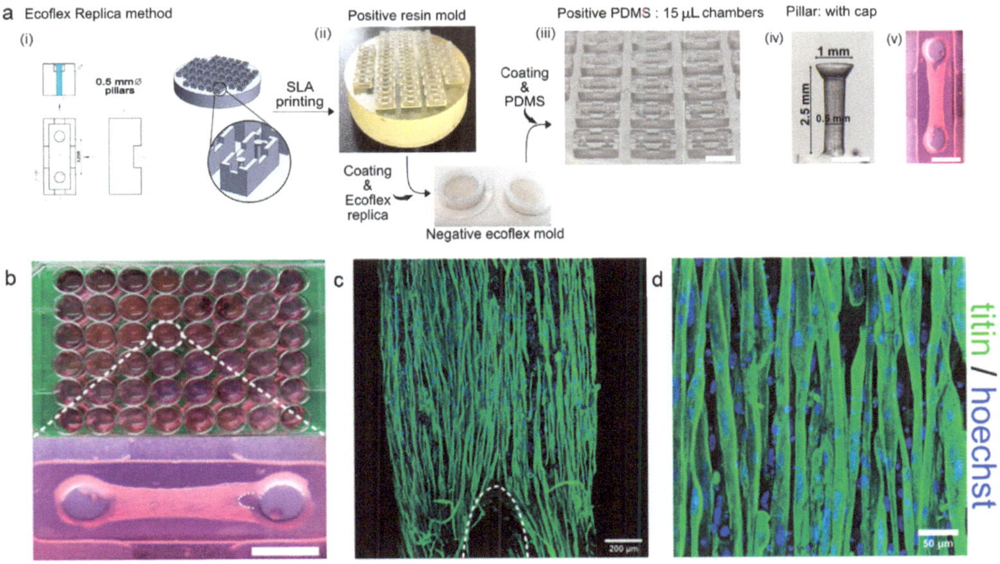

Figure 4. The Ecoflex® Replica technique. Single-chamber technical drawing and 3D CAD model of the positive master mold (**i**); SLA-printed resin positive master mold (**ii**), Ecoflex® negative mold replicas (**iii**), positive PDMS replica detail (**iii**), and T-shaped pillar with conical cap (**iv**). (**b**) TESMs grown in 15 μL Ecoflex® Replica chambers were all contained on a 48-well plate. (**c**,**d**) Even in the smallest tissue, myogenic progenitors differentiated into long, multinucleated myofibers with an ordered titin pattern. The loop-like shape at the TESM's far end, indicated by a dashed curved line, is the result of stress on the tissue. Scale bars: (**a**) 5 mm (**iii**); 1 mm (**iv**,**v**); (**b**) 1 mm; (**c**) 200 μm; (**d**) 50 μm [77]. Copyright 2020 Wiley.

4.3. Light 3D-Printed Insert-Chip Microfluidics for Co-Culturing Cells

In recent years, in vitro modeling systems have been developed for mimicking cellular interactions. These modeling systems simulate the tissue microenvironment, illuminate human physiology, and exploit the underlying processes of disease [78,79]. To mimic in vivo microenvironments and barrier tissues, Transwell (TW) cell culture inserts are a conventional approach for investigating cell barriers that involves Transwell platforms, where cells are seeded on opposing sides of a porous polymer membrane [80–83]. Nevertheless, TW cell culture is static, making it challenging to develop models of vascular and epithelial tissues [84–86]. By contrast, microfluidic devices, known as "Organs-on-a-Chip" (OoCs), offer a distinctive approach by allowing the co-culturing of cells while maintaining controlled fluid flow—a departure from the static nature of TWs [87]. OoCs, essentially microfluidic chips containing biomimetic models of physiological organs, regulate fluid flow and offer valuable insights into interactions between different organs [88]. However, traditional OoCs are complex and integrated systems that involve time-consuming fabrication processes and demand specialized knowledge [9].

Currently, a transformative approach to overcoming limitations associated with TW inserts and conventional OoCs is using 3D-printed modular microfluidic systems [88]. Rauti et al. used modular structure-based SLA 3D printing techniques to customize microfluidic chips for co-culturing and monitoring various cell types under flow conditions. This work introduces a novel insert-chip, a microfluidic device with the functionalities of the OoCs platform, that facilitates cell co-culturing, exposure to flow, and observation of interactions [88]. These SLA-printed chips can be seamlessly integrated into standard cell culture platforms, including conventional well plate platforms such as microelectrode array platforms. Not only can these chips benefit from different technologies, but they can also cut down both development time and cost [89]. Compared to the laborious manufacturing steps of "conventional" OoCs, SLA-printed devices have their designs updated quickly to better suit their intended use and drastically decrease the manufacturing period from several days to a few hours.

The prevalent choice for fabricating microfluidic devices, such as OoCs, uses PDMS because of its acknowledged attributes of biocompatibility, transparency, and advantageous gas permeability [90]. However, a notable limitation of PDMS involves the absorption of hydrophobic substances [29]. Different from PDMS-manufactured chips, these 3D-printed chips are fabricated using non-absorbing materials. A transparent dental resin insert-chip was printed in the shape of a cylinder (Figure 5a) in which cells could grow over the porous membrane and the chip itself could store data. Each insert-chip has a cell culture chamber with a configurable exterior diameter of up to 25 mm, an internal diameter of 17 mm, and a medium capacity of up to 2 mL. The inlet and outlet channels of the chip's upper portion enable connection to a regulated fluid flow system. These channels have external and internal channel dimensions of 2.5 mm and 1.5 mm, respectively. To overcome limitations related to high-magnification imaging on standard dual-channel OoCs, the researchers devised a membrane that can be extracted from the chip using tweezers (Figure 5a). Subsequently, the membrane can be placed on a glass coverslip for standard immunocytochemistry processes and microscopic examination. Figure 5b shows a confocal reconstruction with magnified images of SY-SH5Y, U87, and HUVEC from bottom to top.

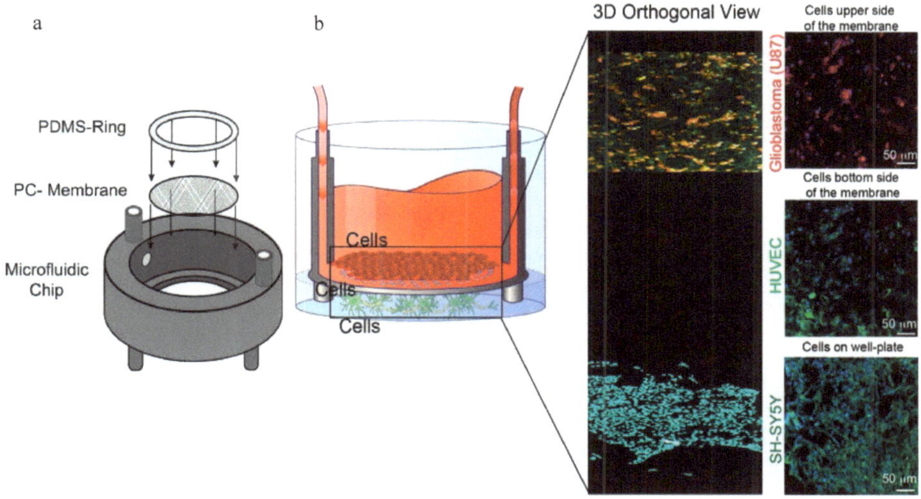

Figure 5. An insert-chip kind of layout. (**a**) An exploded look at the insert-chip reveals its three main parts: the 3D-printed base, the porous PC membrane, and the PDMS ring. (**b**) Tri-culture setup; the insert-chip can cultivate three distinct kinds of cells (on top of the membrane, at the bottom of the membrane, and at the bottom of the well). Three-dimensional confocal view of the system, with magnified images of SY-SH5Y (stained for actin in green and DAPI in blue) growing on the well-plate, U87 (stained for GFAP in red and DAPI in blue) on the top of the membrane, and HUVEC (stained for CD31 in green and DAPI in blue) on the bottom [88]. Copyright 2021 AIP Publishing.

5. Summary

In this review, the definitions, theories, and advantages of three principal light-driven 3D printing technologies are provided: SLA, DLP, and photopolymer inkjet printing. We also discuss three specific strategies for using light-driven 3D printing to fabricate microfluidic devices, such as (1) utilizing SLA to directly print microfluidic device channels; (2) creating microfluidic devices using molds produced via SLA printing; (3) employing a modular approach to assemble microfluidic chips. Furthermore, the biomedical applications of these printed microfluidic devices are explored, including spheroid perfusion culture, the replica molding process for constructing tissue-engineered skeletal muscles, and insert-chip microfluidics designed to enhance cell co-cultures.

In the coming years, we anticipate that light-driven 3D printing technologies will be the primary choice for manufacturing microfluidic devices. However, materials and technical development must first achieve higher resolution, optimal optical properties, and enhanced biocompatibility. Although light-driven 3D printing offers benefits in manufacturing microfluidic devices, several limitations must be resolved. UV absorbers and photoinitiators in SLA and DLP resins may cause cytotoxicity, which has risks for cell-based applications. The type of resin selected significantly impacts cytotoxicity and gas permeability. Recent advancements in developing PDMS-based resins have led to improved low-viscosity resins that retain PDMS properties, such as optical clarity, gas permeability, and biocompatibility, enabling automated fabrication of microfluidic devices via 3D printing [91]. In addition, surface treatment methods can significantly enhance the optical quality of directly printed microfluidic devices [92]. Several methods have been used to enhance surface smoothness, including (1) mechanical surface treatments such as sand polishing [93]; (2) chemical polishing treatments [94]; (3) polymer coatings such as spray-coated clear acrylic [95].

While current DLP printers have high resolution, they fail to achieve the ultra-fine resolution required for fabricating intricate microchannels essential to microfluidic devices. Additionally, a significant technical challenge is controlling light penetration and exposure during the 3D printing process, which is vital for ensuring the quality and reliability of microfluidic channels. UV overexposure can cause channel blockages while inadequate curing time causes the leakage of photoinitiators and unreacted monomers. Unlike traditional SLA, TPP uses femtosecond laser pulses to initiate polymerization at the focal point, allowing for voxel-by-voxel construction of 3D structures at a sub-micron scale [96]. This process enables the creation of complex geometries with smooth surfaces and high aspect ratios below 100 nm precision, which are essential for precise microfluidic applications [97]. Therefore, incorporating TPP into microfluidics fabrication can significantly enhance the precision and functionality of these devices. As technology continues to advance, we can expect even more sophisticated and application-specific microfluidic devices to emerge, further pushing the boundaries of what is possible in tissue culture and biomedical research. Although light-driven 3D printing has not become mainstream in producing microfluidic chips, it is creating new opportunities for businesses and research institutions to significantly impact global healthcare.

Author Contributions: X.L. and M.W., Conceptualization, investigation, writing the original draft. T.P.D., L.Z. and R.Q., Conceptualization, supervision, writing, review, and editing the original draft. All authors have read and agreed to the published version of the manuscript.

Funding: This research was funded by the National Health and Medical Research Council (APP1196850, 1197373 and 2019056), Queensland-Chinese Academy of Sciences (Q-CAS) Collaborative Science Fund (QCAS2022016), UQ Amplify Women's Academic Research Equity (UQAWARE), UQ Foundation of Excellence Research Award (UQFERA), and Advance Queensland Women's Research Assistance Program (AQWRAP). The authors also wish to acknowledge the Queensland node of the NCRIS-enabled Australian National Fabrication Facility (ANFF) and their facilities.

Conflicts of Interest: The authors declare no conflict of interest.

References

1. Whitesides, G.M. The origins and the future of microfluidics. *Nature* **2006**, *442*, 368–373. [CrossRef]
2. Chiu, Y.-L.; Chan, H.F.; Phua, K.K.L.; Zhang, Y.; Juul, S.; Knudsen, B.R.; Ho, Y.-P.; Leong, K.W. Synthesis of Fluorosurfactants for Emulsion-Based Biological Applications. *ACS Nano* **2014**, *8*, 3913–3920. [CrossRef]
3. Dittrich, P.S.; Manz, A. Lab-on-a-chip: Microfluidics in drug discovery. *Nat. Rev. Drug Discov.* **2006**, *5*, 210–218. [CrossRef]
4. Pattanayak, P.; Singh, S.K.; Gulati, M.; Vishwas, S.; Kapoor, B.; Chellappan, D.K.; Anand, K.; Gupta, G.; Jha, N.K.; Gupta, P.K.; et al. Microfluidic chips: Recent advances, critical strategies in design, applications and future perspectives. *Microfluid. Nanofluid.* **2021**, *25*, 1–28. [CrossRef]
5. Velve-Casquillas, G.; Le Berre, M.; Piel, M.; Tran, P.T. Microfluidic tools for cell biological research. *Nano Today* **2010**, *5*, 28–47. [CrossRef]
6. Tehranirokh, M.; Kouzani, A.Z.; Francis, P.S.; Kanwar, J.R. Microfluidic devices for cell cultivation and proliferation. *Biomicrofluidics* **2013**, *7*, 51502. [CrossRef]
7. De Stefano, P.; Bianchi, E.; Dubini, G. The impact of microfluidics in high-throughput drug-screening applications. *Biomicrofluidics* **2022**, *16*, 031501. [CrossRef]
8. Lin, C.-C.; Wang, J.-H.; Wu, H.-W.; Lee, G.-B. Microfluidic Immunoassays. *J. Assoc. Lab. Autom.* **2010**, *15*, 253–274. [CrossRef]
9. Coluccio, M.L.; Perozziello, G.; Malara, N.; Parrotta, E.; Zhang, P.; Gentile, F.; Limongi, T.; Raj, P.M.; Cuda, G.; Candeloro, P.; et al. Microfluidic platforms for cell cultures and investigations. *Microelectron. Eng.* **2019**, *208*, 14–28. [CrossRef]
10. Cardoso, B.D.; Castanheira, E.M.S.; Lanceros-Mendez, S.; Cardoso, V.F. Recent Advances on Cell Culture Platforms for In Vitro Drug Screening and Cell Therapies: From Conventional to Microfluidic Strategies. *Adv. Healthc. Mater.* **2023**, *12*, e2202936. [CrossRef]
11. Clancy, A.; Chen, D.; Bruns, J.; Nadella, J.; Stealey, S.; Zhang, Y.; Timperman, A.; Zustiak, S.P. Hydrogel-based microfluidic device with multiplexed 3D in vitro cell culture. *Sci. Rep.* **2022**, *12*, 17781. [CrossRef]
12. Battat, S.; Weitz, D.A.; Whitesides, G.M. An outlook on microfluidics: The promise and the challenge. *Lab Chip* **2022**, *22*, 530–536. [CrossRef]
13. Nielsen, J.B.; Hanson, R.L.; Almughamsi, H.M.; Pang, C.; Fish, T.R.; Woolley, A.T. Microfluidics: Innovations in Materials and Their Fabrication and Functionalization. *Anal. Chem.* **2020**, *92*, 150–168. [CrossRef]
14. Pavan Kalyan, B.; Kumar, L. 3D Printing: Applications in Tissue Engineering, Medical Devices, and Drug Delivery. *AAPS PharmSciTech* **2022**, *23*, 92. [CrossRef]
15. Tasoglu, S.; Folch, A. Editorial for the Special Issue on 3D Printed Microfluidic Devices. *Micromachines* **2018**, *9*, 609. [CrossRef]
16. Arefin, A.M.E.; Khatri, N.R.; Kulkarni, N.; Egan, P.F. Polymer 3D Printing Review: Materials, Process, and Design Strategies for Medical Applications. *Polymers* **2021**, *13*, 1499. [CrossRef]

17. Villegas, M.; Cetinic, Z.; Shakeri, A.; Didar, T.F. Fabricating smooth PDMS microfluidic channels from low-resolution 3D printed molds using an omniphobic lubricant-infused coating. *Anal. Chim. Acta* **2018**, *1000*, 248–255. [CrossRef]
18. Yadavali, S.; Jeong, H.-H.; Lee, D.; Issadore, D. Silicon and glass very large scale microfluidic droplet integration for terascale generation of polymer microparticles. *Nat. Commun.* **2018**, *9*, 1222. [CrossRef]
19. Luo, Z.; Zhang, H.; Chen, R.; Li, H.; Cheng, F.; Zhang, L.; Liu, J.; Kong, T.; Zhang, Y.; Wang, H. Digital light processing 3D printing for microfluidic chips with enhanced resolution via dosing-and zoning-controlled vat photopolymerization. *Microsyst. Nanoeng.* **2023**, *9*, 103. [CrossRef]
20. Walsh, D.I.; Kong, D.S.; Murthy, S.K.; Carr, P.A. Enabling Microfluidics: From Clean Rooms to Makerspaces. *Trends Biotechnol.* **2017**, *35*, 383–392. [CrossRef]
21. Whitesides, G.M.; Ostuni, E.; Takayama, S.; Jiang, X.; Ingber, D.E. Soft Lithography in Biology and Biochemistry. *Annu. Rev. Biomed. Eng.* **2001**, *3*, 335–373. [CrossRef]
22. Yeo, L.Y.; Chang, H.-C.; Chan, P.P.Y.; Friend, J.R. Microfluidic Devices for Bioapplications. *Small* **2011**, *7*, 12–48. [CrossRef]
23. Xia, Y.; Whitesides, G.M. Soft Lithography. *Angew. Chem. Int. Ed.* **1998**, *37*, 550–575. [CrossRef]
24. Melin, J.; Quake, S.R. Microfluidic Large-Scale Integration: The Evolution of Design Rules for Biological Automation. *Annu. Rev. Biophys. Biomol. Struct.* **2007**, *36*, 213–231. [CrossRef]
25. Kim, K.; Park, S.W.; Yang, S.S. The optimization of PDMS-PMMA bonding process using silane primer. *BioChip J.* **2010**, *4*, 148–154. [CrossRef]
26. Ko, Y.H.; Lee, S.H.; Leem, J.W.; Yu, J.S. High transparency and triboelectric charge generation properties of nano-patterned PDMS. *RSC Adv.* **2014**, *4*, 10216–10220. [CrossRef]
27. Lee, J.; Kim, M. Polymeric Microfluidic Devices Fabricated Using Epoxy Resin for Chemically Demanding and Day-Long Experiments. *Biosensors* **2022**, *12*, 838. [CrossRef]
28. Ariati, R.; Sales, F.; Souza, A.; Lima, R.A.; Ribeiro, J. Polydimethylsiloxane Composites Characterization and Its Applications: A Review. *Polymers* **2021**, *13*, 4258. [CrossRef]
29. Toepke, M.W.; Beebe, D.J. PDMS absorption of small molecules and consequences in microfluidic applications. *Lab Chip* **2006**, *6*, 1484–1486. [CrossRef]
30. Wang, X.; Jiang, M.; Zhou, Z.; Gou, J.; Hui, D. 3D printing of polymer matrix composites: A review and prospective. *Compos. Part B Eng.* **2017**, *110*, 442–458. [CrossRef]
31. O'Neill, P.F.; Ben Azouz, A.; Vázquez, M.; Liu, J.; Marczak, S.; Slouka, Z.; Chang, H.C.; Diamond, D.; Brabazon, D. Advances in three-dimensional rapid prototyping of microfluidic devices for biological applications. *Biomicrofluidics* **2014**, *8*, 052112. [CrossRef]
32. Amin, R.; Knowlton, S.; Hart, A.; Yenilmez, B.; Ghaderinezhad, F.; Katebifar, S.; Messina, M.; Khademhosseini, A.; Tasoglu, S. 3D-printed microfluidic devices. *Biofabrication* **2016**, *8*, 022001. [CrossRef]
33. Waldbaur, A.; Rapp, H.; Länge, K.; Rapp, B.E. Let there be chip—Towards rapid prototyping of microfluidic devices: One-step manufacturing processes. *Anal. Methods* **2011**, *3*, 2681. [CrossRef]
34. Melchels, F.P.W.; Feijen, J.; Grijpma, D.W. A review on stereolithography and its applications in biomedical engineering. *Biomaterials* **2010**, *31*, 6121–6130. [CrossRef]
35. Grogan, S.P.; Chung, P.H.; Soman, P.; Chen, P.; Lotz, M.K.; Chen, S.; D'Lima, D.D. Digital micromirror device projection printing system for meniscus tissue engineering. *Acta Biomater.* **2013**, *9*, 7218–7226. [CrossRef]
36. Utada, A.S.; Lorenceau, E.; Link, D.R.; Kaplan, P.D.; Stone, H.A.; Weitz, D.A. Monodisperse double emulsions generated from a microcapillary device. *Science* **2005**, *308*, 537–541. [CrossRef]
37. Lim, T.W.; Son, Y.; Jeong, Y.J.; Yang, D.-Y.; Kong, H.-J.; Lee, K.-S.; Kim, D.-P. Three-dimensionally crossing manifold micro-mixer for fast mixing in a short channel length. *Lab Chip* **2011**, *11*, 100–103. [CrossRef]
38. Nielson, R.; Kaehr, B.; Shear, J.B. Microreplication and Design of Biological Architectures Using Dynamic-Mask Multiphoton Lithography. *Small* **2009**, *5*, 120–125. [CrossRef]
39. Zhang, L.; Forgham, H.; Shen, A.; Wang, J.; Zhu, J.; Huang, X.; Tang, S.-Y.; Xu, C.; Davis, T.P.; Qiao, R. Nanomaterial integrated 3D printing for biomedical applications. *J. Mater. Chem. B* **2022**, *10*, 7473–7490. [CrossRef]
40. Ding, L.; Razavi Bazaz, S.; Asadniaye Fardjahromi, M.; Mckinnirey, F.; Saputro, B.; Banerjee, B.; Vesey, G.; Ebrahimi Warkiani, M. A modular 3D printed microfluidic system: A potential solution for continuous cell harvesting in large-scale bioprocessing. *Bioresour. Bioprocess.* **2022**, *9*, 64. [CrossRef]
41. Zhang, J.; Hu, Q.; Wang, S.; Tao, J.; Gou, M. Digital Light Processing Based Three-dimensional Printing for Medical Applications. *Int. J. Bioprint* **2020**, *6*, 242. [CrossRef]
42. Prabhakar, P.; Sen, R.K.; Dwivedi, N.; Khan, R.; Solanki, P.R.; Srivastava, A.K.; Dhand, C. 3D-Printed Microfluidics and Potential Biomedical Applications. *Front. Nanotechnol.* **2021**, *3*, 609355. [CrossRef]
43. Wang, L.; Kodzius, R.; Yi, X.; Li, S.; Hui, Y.S.; Wen, W. Prototyping chips in minutes: Direct Laser Plotting (DLP) of functional microfluidic structures. *Sens. Actuators B Chem.* **2012**, *168*, 214–222. [CrossRef]
44. Ho, C.M.B.; Ng, S.H.; Li, K.H.H.; Yoon, Y.-J. 3D printed microfluidics for biological applications. *Lab Chip* **2015**, *15*, 3627–3637. [CrossRef]
45. Tse, C.C.W.; Smith, P.J. Inkjet Printing for Biomedical Applications. In *Cell-Based Microarrays: Methods and Protocols*; Ertl, P., Rothbauer, M., Eds.; Springer: New York, NY, USA, 2018; pp. 107–117.

46. Donvito, L.; Galluccio, L.; Lombardo, A.; Morabito, G.; Nicolosi, A.; Reno, M. Experimental validation of a simple, low-cost, T-junction droplet generator fabricated through 3D printing. *J. Micromech. Microeng.* **2015**, *25*, 035013. [CrossRef]
47. Mehta, V.; Rath, S.N. 3D printed microfluidic devices: A review focused on four fundamental manufacturing approaches and implications on the field of healthcare. *Bio-Des. Manuf.* **2021**, *4*, 311–343. [CrossRef]
48. Lee, J.M.; Zhang, M.; Yeong, W.Y. Characterization and evaluation of 3D printed microfluidic chip for cell processing. *Microfluid. Nanofluid.* **2016**, *20*, 5. [CrossRef]
49. Lerman, M.J.; Lembong, J.; Gillen, G.; Fisher, J.P. 3D printing in cell culture systems and medical applications. *Appl. Phys. Rev.* **2018**, *5*, 041109. [CrossRef]
50. Fang, Y.; Eglen, R.M. Three-Dimensional Cell Cultures in Drug Discovery and Development. *SLAS Discov.* **2017**, *22*, 456–472. [CrossRef]
51. Li, X.J.; Valadez, A.V.; Zuo, P.; Nie, Z. Microfluidic 3D cell culture: Potential application for tissue-based bioassays. *Bioanalysis* **2012**, *4*, 1509–1525. [CrossRef]
52. Kapalczynska, M.; Kolenda, T.; Przybyla, W.; Zajaczkowska, M.; Teresiak, A.; Filas, V.; Ibbs, M.; Blizniak, R.; Luczewski, L.; Lamperska, K. 2D and 3D cell cultures—A comparison of different types of cancer cell cultures. *Arch. Med. Sci.* **2018**, *14*, 910–919.
53. Salinas-Vera, Y.M.; Valdes, J.; Perez-Navarro, Y.; Mandujano-Lazaro, G.; Marchat, L.A.; Ramos-Payan, R.; Nunez-Olvera, S.I.; Perez-Plascencia, C.; Lopez-Camarillo, C. Three-Dimensional 3D Culture Models in Gynecological and Breast Cancer Research. *Front. Oncol.* **2022**, *12*, 826113. [CrossRef]
54. Urzi, O.; Gasparro, R.; Costanzo, E.; De Luca, A.; Giavaresi, G.; Fontana, S.; Alessandro, R. Three-Dimensional Cell Cultures: The Bridge between In Vitro and In Vivo Models. *Int. J. Mol. Sci.* **2023**, *24*, 12046. [CrossRef]
55. Wu, X.; Shi, W.; Liu, X.; Gu, Z. Recent advances in 3D-printing-based organ-on-a-chip. *EngMedicine* **2024**, *1*, 100003. [CrossRef]
56. Lee, P.J.; Hung, P.J.; Lee, L.P. An artificial liver sinusoid with a microfluidic endothelial-like barrier for primary hepatocyte culture. *Biotechnol. Bioeng.* **2007**, *97*, 1340–1346. [CrossRef]
57. Bhushan, A.; Senutovitch, N.; Bale, S.S.; Mccarty, W.J.; Hegde, M.; Jindal, R.; Golberg, I.; Berk Usta, O.; Yarmush, M.L.; Vernetti, L.; et al. Towards a three-dimensional microfluidic liver platform for predicting drug efficacy and toxicity in humans. *Stem Cell Res. Ther.* **2013**, *4* (Suppl. S1), S16. [CrossRef]
58. Tan, G.-D.S.; Toh, G.W.; Birgersson, E.; Robens, J.; Van Noort, D.; Leo, H.L. A thin-walled polydimethylsiloxane bioreactor for high-density hepatocyte sandwich culture. *Biotechnol. Bioeng.* **2013**, *110*, 1663–1673. [CrossRef]
59. Wang, Y.; Toh, Y.C.; Li, Q.; Nugraha, B.; Zheng, B.; Lu, T.B.; Gao, Y.; Ng, M.M.; Yu, H. Mechanical compaction directly modulates the dynamics of bile canaliculi formation. *Integr. Biol.* **2013**, *5*, 390–401. [CrossRef]
60. Ong, L.J.Y.; Islam, A.; Dasgupta, R.; Iyer, N.G.; Leo, H.L.; Toh, Y.-C. A 3D printed microfluidic perfusion device for multicellular spheroid cultures. *Biofabrication* **2017**, *9*, 045005. [CrossRef]
61. Waheed, S.; Cabot, J.M.; Macdonald, N.P.; Lewis, T.; Guijt, R.M.; Paull, B.; Breadmore, M.C. 3D printed microfluidic devices: Enablers and barriers. *Lab Chip* **2016**, *16*, 1993–2013. [CrossRef]
62. Rogers, C.I.; Qaderi, K.; Woolley, A.T.; Nordin, G.P. 3D printed microfluidic devices with integrated valves. *Biomicrofluidics* **2015**, *9*, 016501. [CrossRef]
63. Chan, H.N.; Shu, Y.; Xiong, B.; Chen, Y.; Chen, Y.; Tian, Q.; Michael, S.A.; Shen, B.; Wu, H. Simple, Cost-Effective 3D Printed Microfluidic Components for Disposable, Point-of-Care Colorimetric Analysis. *ACS Sens.* **2016**, *1*, 227–234. [CrossRef]
64. Mathur, A.; Loskill, P.; Shao, K.; Huebsch, N.; Hong, S.; Marcus, S.G.; Marks, N.; Mandegar, M.; Conklin, B.R.; Lee, L.P.; et al. Human iPSC-based Cardiac Microphysiological System For Drug Screening Applications. *Sci. Rep.* **2015**, *5*, 8883. [CrossRef]
65. Toh, Y.-C.; Zhang, C.; Zhang, J.; Khong, Y.M.; Chang, S.; Samper, V.D.; Van Noort, D.; Hutmacher, D.W.; Yu, H. A novel 3D mammalian cell perfusion-culture system in microfluidic channels. *Lab Chip* **2007**, *7*, 302. [CrossRef]
66. Ong, L.J.Y.; Chong, L.H.; Jin, L.; Singh, P.K.; Lee, P.S.; Yu, H.; Ananthanarayanan, A.; Leo, H.L.; Toh, Y.-C. A pump-free microfluidic 3D perfusion platform for the efficient differentiation of human hepatocyte-like cells. *Biotechnol. Bioeng.* **2017**, *114*, 2360–2370. [CrossRef]
67. Bian, W.; Bursac, N. Engineered skeletal muscle tissue networks with controllable architecture. *Biomaterials* **2009**, *30*, 1401–1412. [CrossRef]
68. Kwon, Y.T.; Kim, Y.S.; Kwon, S.; Mahmood, M.; Lim, H.R.; Park, S.W.; Kang, S.O.; Choi, J.J.; Herbert, R.; Jang, Y.C.; et al. All-printed nanomembrane wireless bioelectronics using a biocompatible solderable graphene for multimodal human-machine interfaces. *Nat. Commun.* **2020**, *11*, 3450. [CrossRef]
69. Kantaros, A.; Ganetsos, T.; Petrescu, F.I.T.; Ungureanu, L.M.; Munteanu, I.S. Post-Production Finishing Processes Utilized in 3D Printing Technologies. *Processes* **2024**, *12*, 595. [CrossRef]
70. Afshar, M.E.; Abraha, H.Y.; Bakooshli, M.A.; Davoudi, S.; Thavandiran, N.; Tung, K.; Ahn, H.; Ginsberg, H.J.; Zandstra, P.W.; Gilbert, P.M. A 96-well culture platform enables longitudinal analyses of engineered human skeletal muscle microtissue strength. *Sci. Rep.* **2020**, *10*, 6918. [CrossRef]
71. Kalman, B.; Picart, C.; Boudou, T. Quick and easy microfabrication of T-shaped cantilevers to generate arrays of microtissues. *Biomed. Microdevices* **2016**, *18*, 43. [CrossRef]
72. Kajtez, J.; Buchmann, S.; Vasudevan, S.; Birtele, M.; Rocchetti, S.; Pless, C.J.; Heiskanen, A.; Barker, R.A.; Martínez-Serrano, A.; Parmar, M.; et al. 3D-Printed Soft Lithography for Complex Compartmentalized Microfluidic Neural Devices. *Adv. Sci.* **2020**, *7*, 2001150. [CrossRef]

73. Myalenko, D.; Fedotova, O. Physical, Mechanical, and Structural Properties of the Polylactide and Polybutylene Adipate Terephthalate (PBAT)-Based Biodegradable Polymer during Compost Storage. *Polymers* **2023**, *15*, 1619. [CrossRef]
74. Shepherd, R.F.; Ilievski, F.; Choi, W.; Morin, S.A.; Stokes, A.A.; Mazzeo, A.D.; Chen, X.; Wang, M.; Whitesides, G.M. Multigait soft robot. *Proc. Natl. Acad. Sci. USA* **2011**, *108*, 20400–20403. [CrossRef]
75. Osaki, T.; Uzel, S.G.M.; Kamm, R.D. On-chip 3D neuromuscular model for drug screening and precision medicine in neuromuscular disease. *Nat. Protoc.* **2020**, *15*, 421–449. [CrossRef]
76. Agrawal, G.; Aung, A.; Varghese, S. Skeletal muscle-on-a-chip: An in vitro model to evaluate tissue formation and injury. *Lab Chip* **2017**, *17*, 3447–3461. [CrossRef]
77. Iuliano, A.; Wal, E.; Ruijmbeek, C.; Groen, S.; Pijnappel, W.; de Greef, J.; Saggiomo, V. Coupling 3D Printing and Novel Replica Molding for In House Fabrication of Skeletal Muscle Tissue Engineering Devices. *Adv. Mater. Technol.* **2020**, *5*, 2000344. [CrossRef]
78. Nikolic, M.; Sustersic, T.; Filipovic, N. In vitro Models and On-Chip Systems: Biomaterial Interaction Studies With Tissues Generated Using Lung Epithelial and Liver Metabolic Cell Lines. *Front. Bioeng. Biotechnol.* **2018**, *6*, 120. [CrossRef]
79. Nikolakopoulou, P.; Rauti, R.; Voulgaris, D.; Shlomy, I.; Maoz, B.M.; Herland, A. Recent progress in translational engineered in vitro models of the central nervous system. *Brain* **2020**, *143*, 3181–3213. [CrossRef]
80. Stone, N.L.; England, T.J.; O'Sullivan, S.E. A Novel Transwell Blood Brain Barrier Model Using Primary Human Cells. *Front. Cell Neurosci.* **2019**, *13*, 230. [CrossRef]
81. Dogan, A.A.; Dufva, M. Customized 3D-printed stackable cell culture inserts tailored with bioactive membranes. *Sci. Rep.* **2022**, *12*, 3694. [CrossRef]
82. Mc Carthy, D.J.; Malhotra, M.; O'Mahony, A.M.; Cryan, J.F.; O'Driscoll, C.M. Nanoparticles and the Blood-Brain Barrier: Advancing from In-Vitro Models Towards Therapeutic Significance. *Pharm. Res.* **2015**, *32*, 1161–1185. [CrossRef]
83. Pohlit, H.; Bohlin, J.; Katiyar, N.; Hilborn, J.; Tenje, M. Technology platform for facile handling of 3D hydrogel cell culture scaffolds. *Sci. Rep.* **2023**, *13*, 12829. [CrossRef]
84. Booth, R.; Kim, H. Characterization of a microfluidic in vitro model of the blood-brain barrier (μBBB). *Lab Chip* **2012**, *12*, 1784. [CrossRef]
85. Tan, H.-Y.; Trier, S.; Rahbek, U.L.; Dufva, M.; Kutter, J.P.; Andresen, T.L. A multi-chamber microfluidic intestinal barrier model using Caco-2 cells for drug transport studies. *PLoS ONE* **2018**, *13*, e0197101. [CrossRef]
86. Frost, T.S.; Jiang, L.; Lynch, R.M.; Zohar, Y. Permeability of Epithelial/Endothelial Barriers in Transwells and Microfluidic Bilayer Devices. *Micromachines* **2019**, *10*, 533. [CrossRef]
87. Huh, D.; Matthews, B.D.; Mammoto, A.; Montoya-Zavala, M.; Hsin, H.Y.; Ingber, D.E. Reconstituting organ-level lung functions on a chip. *Science* **2010**, *328*, 1662–1668. [CrossRef]
88. Rauti, R.; Ess, A.; Le Roi, B.; Kreinin, Y.; Epshtein, M.; Korin, N.; Maoz, B.M. Transforming a well into a chip: A modular 3D-printed microfluidic chip. *APL Bioeng.* **2021**, *5*, 026103. [CrossRef]
89. Qiu, J.; Gao, Q.; Zhao, H.; Fu, J.; He, Y. Rapid Customization of 3D Integrated Microfluidic Chips via Modular Structure-Based Design. *ACS Biomater. Sci. Eng.* **2017**, *3*, 2606–2616. [CrossRef]
90. Leung, C.M.; de Haan, P.; Ronaldson-Bouchard, K.; Kim, G.-A.; Ko, J.; Rho, H.S.; Chen, Z.; Habibovic, P.; Jeon, N.L.; Takayama, S.; et al. A guide to the organ-on-a-chip. *Nat. Rev. Methods Primers* **2022**, *2*, 33. [CrossRef]
91. Fleck, E.; Keck, C.; Ryszka, K.; DeNatale, E.; Potkay, J. Low-Viscosity Polydimethylsiloxane Resin for Facile 3D Printing of Elastomeric Microfluidics. *Micromachines* **2023**, *14*, 773. [CrossRef]
92. Namgung, H.; Kaba, A.M.; Oh, H.; Jeon, H.; Yoon, J.; Lee, H.; Kim, D. Quantitative Determination of 3D-Printing and Surface-Treatment Conditions for Direct-Printed Microfluidic Devices. *BioChip J.* **2022**, *16*, 82–98. [CrossRef]
93. Beckwith, A.L.; Borenstein, J.T.; Velasquez-Garcia, L.F. Monolithic, 3D-Printed Microfluidic Platform for Recapitulation of Dynamic Tumor Microenvironments. *J. Microelectromech. Syst.* **2018**, *27*, 1009–1022. [CrossRef]
94. He, Y.; Xue, G.H.; Fu, J.Z. Fabrication of low cost soft tissue prostheses with the desktop 3D printer. *Sci. Rep.* **2014**, *4*, 6973. [CrossRef]
95. Tang, C.K.; Vaze, A.; Rusling, J.F. Automated 3D-printed unibody immunoarray for chemiluminescence detection of cancer biomarker proteins. *Lab Chip* **2017**, *17*, 484–489. [CrossRef]
96. Zhang, L.; Wang, C.; Zhang, C.; Xue, Y.; Ye, Z.; Xu, L.; Hu, Y.; Li, J.; Chu, J.; Wu, D. High-Throughput Two-Photon 3D Printing Enabled by Holographic Multi-Foci High-Speed Scanning. *Nano Lett.* **2024**, *24*, 2671–2679. [CrossRef]
97. Hohmann, J.K.; Renner, M.; Waller, E.H.; von Freymann, G. Three-Dimensional μ-Printing: An Enabling Technology. *Adv. Opt. Mater.* **2015**, *3*, 1488–1507. [CrossRef]

biosensors

MDPI

Article

Analytical Validation of a Spiral Microfluidic Chip with Hydrofoil-Shaped Pillars for the Enrichment of Circulating Tumor Cells

Begum Sen-Dogan [1,*], Mehmet Alper Demir [1], Buket Sahin [1], Ender Yildirim [1,2,3], Gizem Karayalcin [1], Sebnem Sahin [1], Ege Mutlu [1], Taylan Berkin Toral [1], Ebru Ozgur [1], Ozge Zorlu [1] and Haluk Kulah [1,3,4]

1 Mikro Biyosistemler A.S., 06530 Ankara, Turkey
2 Department of Mechanical Engineering, Middle East Technical University, 06800 Ankara, Turkey
3 METU MEMS Center, 06530 Ankara, Turkey
4 Department of Electrical and Electronics Engineering, Middle East Technical University, 06800 Ankara, Turkey
* Correspondence: begum.sendogan@mikrobiyo.com.tr

Abstract: The isolation of circulating tumor cells (CTCs) from peripheral blood with high efficiency remains a challenge hindering the utilization of CTC enrichment methods in clinical practice. Here, we propose a microfluidic channel design for the size-based hydrodynamic enrichment of CTCs from blood in an epitope-independent and high-throughput manner. The microfluidic channel comprises a spiral-shaped part followed by a widening part, incorporating successive streamlined pillars, that improves the enrichment efficiency. The design was tested against two benchmark designs, a spiral microfluidic channel and a spiral microfluidic channel followed by a widening channel without the hydrofoils, by processing 5 mL of healthy blood samples spiked with 100 MCF-7 cells. The results proved that the design with hydrofoil-shaped pillars perform significantly better in terms of recovery (recovery rate of 67.9% compared to 23.6% in spiral and 56.7% in spiral with widening section), at a cost of slightly lower white blood cell (WBC) depletion (depletion rate of 94.2% compared to 98.6% in spiral and 94.2% in spiral with widening section), at 1500 μL/min flow rate. For analytical validation, the design was further tested with A549, SKOV-3, and BT-474 cell lines, yielding recovery rates of 62.3 ± 8.4%, 71.0 ± 6.5%, and 82.9 ± 9.9%, respectively. The results are consistent with the size and deformability variation in the respective cell lines, where the increasing size and decreasing deformability affect the recovery rate in a positive manner. The analysis before and after the microfluidic chip process showed that the process does not affect cell viability.

Keywords: microfluidic channel; computational fluid dynamics; circulating tumor cell (CTC) separation; inertial hydrodynamics

Citation: Sen-Dogan, B.; Demir, M.A.; Sahin, B.; Yildirim, E.; Karayalcin, G.; Sahin, S.; Mutlu, E.; Toral, T.B.; Ozgur, E.; Zorlu, O.; et al. Analytical Validation of a Spiral Microfluidic Chip with Hydrofoil-Shaped Pillars for the Enrichment of Circulating Tumor Cells. *Biosensors* **2023**, *13*, 938. https://doi.org/10.3390/bios13100938

Received: 29 August 2023
Revised: 13 October 2023
Accepted: 17 October 2023
Published: 19 October 2023

1. Introduction

Cancer metastasis, which is the primary cause of mortality in cancer patients, is a multistage process including the detachment of tumor cells from the primary site, intravasation into the bloodstream, arresting at a secondary site, extravasation, and colonization to form metastatic tumors [1]. Detached cancer cells that enter the bloodstream are called circulating tumor cells (CTCs). Due to their key role in the metastatic process, CTCs are considered the most promising liquid biopsy biomarker for cancer diagnosis, prognosis, prediction, stratification, and pharmacodynamics [2–4]. CTCs are especially critical for obtaining information on tumor evolution during therapy monitoring, as frequent tissue biopsies are difficult or sometimes impossible to obtain from patients for evaluating the response to cancer therapy. Additionally, CTCs represent a dynamic and heterogeneous cell population arising from multiple metastatic lesions that may change significantly during the disease and course of the therapy. Therefore, each blood sample provides a snapshot of the disease status, which is not possible with standard tissue biopsy. With the current advances in molecular analysis techniques including single-cell analysis at the genomic, transcriptomic, and proteomic levels, the potential of CTC liquid biopsies for the improvement in cancer diagnosis and therapy materializes at a faster pace than before [5–7]. Currently, there are around 200 active clinical trials on CTCs with increasing interrogation of CTC molecular characterizations including mutations, epigenetic changes, multigene expression, and protein expression analyses, besides the usual CTC enumeration as the primary measure [8].

Despite its great potential in improving cancer care, the isolation of extremely rare CTCs from peripheral blood cells with high efficiency is still the main challenge that impedes their use in routine clinical practice. Due to their rarity, analysis of CTCs from blood samples without a prior enrichment step is too costly and time-consuming, if not impossible, as the blood volume required to analyze these is relatively high (≥7.5 mL, according to the FDA-approved CellSearch platform) and contains billions of cells which would require too many reagents, slides, and other materials for standard downstream analysis techniques. Over the past decade, a number of technologies have been developed to isolate or enrich CTCs based on their biological and/or physical properties that are distinct from normal hematological cells [9–12]. Among these, microfluidic approaches for the size-based enrichment of CTCs offer the advantage of epitope-independent, label-free isolation of CTCs from whole blood, providing a more heterogeneous CTC output.

Microfluidic technologies improve the separation efficiency of CTCs due to their innate properties like the small size that is comparable to the size of biological particles, which enables the separation of the CTCs solely based on hydrodynamic forces. Such examples include pinched-flow fractionation [13], deterministic lateral displacement [14,15], asymmetrical curved microchannels [16,17], or spiral microchannels [18], which have been adopted so far to isolate CTCs from blood. Different strategies have been adapted to these methods to further enhance isolation efficiency. For instance, hydrodynamic separation in spiral microchannels have been featured by utilizing a sheath flow [19], by slanting the channel top wall [20], or by reversing the secondary flow direction [21].

In this study, we present a novel spiral microfluidic channel design with a widening outlet section which also includes successive hydrofoil structures, enhancing the size-based separation efficiency of CTCs. The operation of the system does not rely on the existence of a sheath flow, simplifying the overall operation and increasing the throughput. The performance of the system was verified numerically and optimized experimentally. The analytical validation of the system was carried out by spiking experiments using blood samples from healthy donors. Results indicate a high CTC recovery rate independent of the type of cancer cells spiked with no damage to cells in terms of integrity and viability.

2. Materials and Methods

2.1. Microfluidic Design

Archimedean spirals are often used for the inertial focusing of particles in a stream by utilizing the drag forces induced by Dean vortices across the channel along with the lift forces. The balance between the lift forces (F_L in Equation (1)) and the Dean forces (F_D in Equation (2)) causes the particles to migrate to a focal position across the channel.

$$F_L = \left(4\rho U^2 a^2 C_L\right)/D_h \quad (1)$$

$$F_D = 3\pi\mu a U_D \quad (2)$$

In Equations (1) and (2), a is the particle diameter; U is the flow velocity; μ and ρ are the viscosity and density of the liquid, respectively; and D_h is the hydraulic diameter given for rectangular channels by $D_h = (2wh)/(w+h)$, where w and h are the width and the height of the channel, respectively.

It should be noted that the lift force on a particle is affected by the lift coefficient C_L, which is a function of the Reynolds number of the flow ($\mathrm{Re} = \rho U D_h/\mu$) and the location of the particle in the channel. F_D can be interpreted as the Stokes drag on the particle in the Dean vortex with average Dean velocity, U_D, which can be estimated using the relation $U_D = 1.8 \times 10^{-4}\mathrm{De}^{1.63}$ proposed by Ookawara et al. [22], where De is the Dean number given by

$$\mathrm{De} = \frac{\rho U D_h}{\mu}\sqrt{\frac{D_h}{2R}} \tag{3}$$

where R is the radius of the curvature of the spiral channel.

Particles of different sizes focus along separate streamlines in a spiral channel based on the above theory and then are typically collected at the outlet of the spiral channel for further downstream analyses. A common approach is to locate two or more outlet channels separated from each other considering the loci of the upcoming particles (Figure 1a). However, as the difference between the sizes of the particles becomes smaller, the separation distance between the loci of the particles (d in Figure 1a) also decreases. In such cases, the outlet design becomes as important as the spiral design itself. Mihandoust et al. showed that the performance of a slanted spiral could be improved by properly modifying the outlet geometry [23]. Here, we propose to gently widen the spiral channel at the outlet section (Figure 1b) and pose successive streamlined pillars in the shape of asymmetric hydrofoils upstream of the separation tip, as illustrated in Figure 1c.

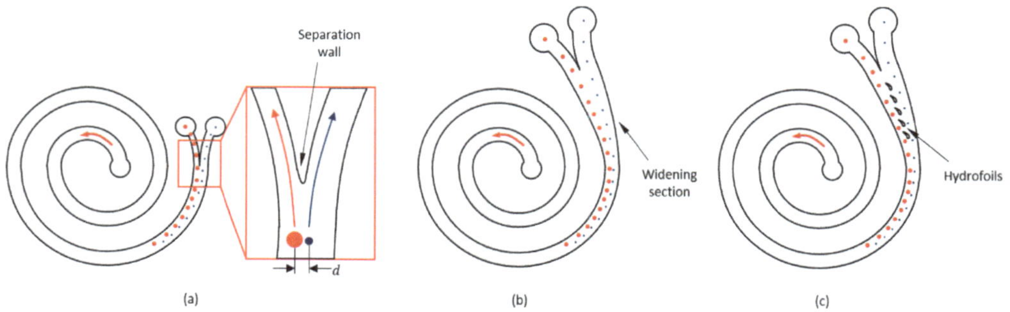

Figure 1. (**a**) Typical Archimedean spiral microfluidic channel with separation wall at the outlet section. (**b**) Spiral microfluidic channel with widening outlet section. (**c**) Spiral microfluidic channel with widening outlet section and hydrofoils.

To design the spiral channel and the widening outlet section with the hydrofoils, we utilized a hybrid approach combining analytical, numerical, and empirical solutions. Considering the limitations imposed by the fabrication method and the operating conditions explained in Section 2 (the Materials and Methods section), we started by setting the depth of the channel to 80 μm and the flow rate to 1500 μL/min. For the given depth and the flow rate, we searched for a set of feasible channel width and radius of curvature combinations that would allow 10 μm diameter and 14 μm diameter target particles (representing the white blood cells (WBCs) and circulating tumor cells (CTCs), respectively) to focus on definite loci along the spiral. In choosing the feasible set, the confinement ratio, λ, which is defined as the ratio between the particle diameter and the hydraulic diameter ($\lambda = a/D_h$), and the Dean number were set as the constraints. For curved channels, the minimum threshold for the confinement ratio for inertial focusing to take place was stated by Martel and Toner to be 0.07 ($\lambda > 0.07$) [24]. For $a_1 = 10$ μm and $a_2 = 14$ μm diameter particles, we aimed both λ_1 and λ_2 to be greater than 0.07 to ensure the focusing of both sizes of particles. Additionally, the Dean number should be sufficiently large for the Dean vortices across the channel to be effective to focus the particles within a feasible length of the spiral. On the other hand, Nivedita et al. showed that there is a critical Dean number, above which the primary Dean vortices are perturbed and secondary vortices form, which affects the focusing of the particles [25]. This critical Dean number is dependent on the aspect ratio of the channel, and for low-aspect ratio channels (0.2–0.6), it was reported as ranging between ~30 and 40 [25]. Accordingly, we set the feasible Dean number region to $10 < De < 30$. Based on these constraints ($\lambda_1 > 0.07$, $\lambda_2 > 0.07$, $10 < De < 30$), we found a set of channel-width and radius-of-curvature combinations. Within this set, we arbitrarily selected the channel width (w) and the average radius of curvature (R) as 250 μm and 4.5 mm, respectively (R is 4 mm at the inlet and 5 mm at the outlet), which resulted in $De = 16.7$, $\lambda_1 = 0.0825$, and $\lambda_2 = 0.1155$. In the following stages of the design, we further justified this selection by utilizing empirical and numerical models.

To determine the length of the spiral channel, we considered the time passed for the particles to migrate to their respective loci. The particles entering the spiral channel at an arbitrary location at the inlet section drift along the Dean vortices across the channel at average Dean velocity ($U_D = 1.8 \times 10^{-4}\mathrm{De}^{1.63}$) until they eventually reach their equilibrium positions. The maximum distance that a particle drifts across the channel before reaching its equilibrium position can be estimated as $2(w + h/2)$ (Figure 2a). Noting that the particle simultaneously streams along the channel at a flow velocity of $U = Q/(wh)$, the minimum length of the spiral channel that allows the particles to reach their equilibrium positions can be estimated as

$$L_{sp,min} = \frac{Q(2w + h)}{U_D wh} \quad (4)$$

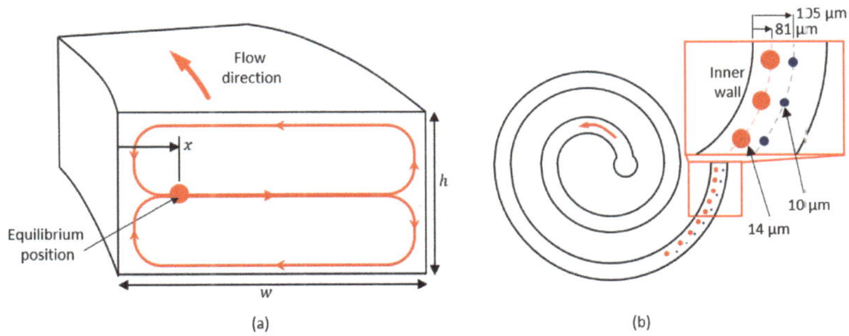

Figure 2. (**a**) Dean vortex across the channel and the equilibrium position of a particle. (**b**) Lateral equilibrium positions of 10 μm and 14 μm diameter particles at the exit of the spiral channel.

For selected channel dimensions and the flow rate, this length corresponds to a 2-loop Archimedean spiral with radius of curvature of 4.5 mm.

After deciding the geometry of the spiral channel, we estimated the loci of 10 μm and 14 μm diameter particles, respectively. For this purpose, we referred to the experimental data available in the literature [24], where the normalized lateral position of the particles along the width of the channel (x/w) are presented with respect to non-dimensional parameters, namely the Reynolds number ($\mathrm{Re} = \rho U_{max} D_h/\mu$, where U_{max} is 3/2 of the average flow velocity U), confinement ratio (λ), and the curvature ratio ($\delta = D_h/2R$). We interpolated the empirical data available in [24] to determine the lateral positions of 10 μm and 14 μm diameter particles at the exit of a 2-loop spiral microchannel with 80 μm height, 250 μm width, and radius of curvature of 4.5 mm. As a result, we calculated that under 1500 μL/min flow of aqueous solution, 14 μm diameter particles representing the CTCs and 10 μm diameter particles representing the WBCs exit the spiral microchannel at lateral positions of 81 μm and 105 μm (Figure 2b), respectively, with respect to the inner wall, making the separation distance $d = 24$ μm, which justifies the design of the spiral microfluidic channel.

To analyze the outlet geometry, we utilized a 2-dimensional numerical model by using COMSOL Multiphysics 5.2a. The outlet section was formed by slightly widening the upstream spiral channel. In our previous work [26], we showed that asymmetric hydrofoil-shaped pillars could be used to improve the separation. Accordingly, we located five successive asymmetric hydrofoils (NACA 9730 profile) in the widening section to enhance the separation distance between different-sized particles by keeping the velocity higher along the outer wall of the channel than along the inner wall of the channel, to keep the larger particles near the inner wall at a slower pace. Downstream of the hydrofoils, a separation wall was posed to direct different-sized particles to their respective outlets (product outlet to collect 14 μm diameter particles representing CTCs and waste outlet to collect 10 μm diameter particles representing WBCs).

Laminar flow and particle tracing modules were used to compute the velocity field and pressure distribution, and the location of the particles at the exit of the outlet section. To include the effect of the upstream channel, not only the outlet geometry but also the spiral microchannel was modeled. On the other hand, utilizing the abovementioned empirical model eliminated the need for a 3-dimensional model that would be required to determine the particle locations. Instead, we computed the particle trajectories by releasing 10 μm and 14 μm diameter particles at lateral positions of 81 μm and 105 μm, respectively, at the exit of the spiral section. Figure 3 illustrates the model and the boundary conditions.

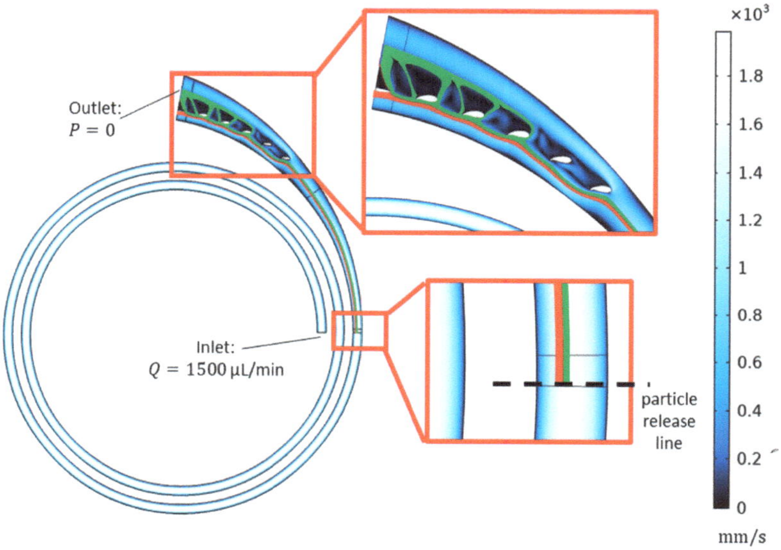

Figure 3. Two-dimensional model for analysis of the widening outlet section with hydrofoils.

2.2. Microfluidic Chip Fabrication

The microfluidic chips were fabricated with a MEMS-based silicon glass stack process where the microfluidic channels were formed on the silicon side. In the beginning, the microfluidic channel pattern was formed on the active side of the silicon wafer through DRIE etching with a target depth of 80 μm, through a photoresist mask. After the channel formation, the silicon wafers were cleaned and coated with thermal oxide, and further with PECVD oxide at the active side as a protection layer. The microfluidic inlet and outlet ports were then formed with a second DRIE process, which was applied from the backside of the wafer along its thickness using the thermally deposited SiO_2 as a masking layer as well as the photoresist used for patterning the SiO_2. This was followed by cleaning and oxide stripping processes. Then, 0.5 μm of sacrificial thermal oxide was successively deposited and stripped to remove possible residues and scallops formed on the channel and port walls during the DRIE processes. Afterward, the silicon wafer was coated with 0.3 μm thick thermal oxide and was anodically bonded with a blank glass wafer, closing the microfluidic channel and forming a transparent window. The process ended with the dicing of the wafers into chips. Channel sections were analyzed with SEM at 1.0 kV, x30 (METU MEMS Center).

2.3. Cell Culture, Blood Collection, and Sample Preparation

Cultured human breast cancer MCF-7 and BT-474 cell lines, the human non-small-cell lung cancer A549 cell line, and the human ovary adenocarcinoma SKOV-3 cell line were obtained from American Type Culture Collection (ATCC, Manassas, VA, USA). All cell lines were cultured at 37 °C in 5% CO_2. The MCF-7 and SKOV-3 cell lines were cultured in a growth medium containing Dulbecco's Modified Eagle's Medium—High Glucose (DMEM-HG) (Biological Industries, Kibbutz Beit HaEmek, Israel), 10% fetal bovine serum (FBS) (Biological Industries, Kibbutz Beit HaEmek, Israel), 1% minimum essential medium (MEM), non-essential amino acids (Biological Industries, Kibbutz Beit HaEmek, Israel) and 1% penicillin–streptomycin (Biological Industries, Kibbutz Beit HaEmek, Israel). The A549 cell line was cultured with DMEM-HG, 10% FBS, and 1% penicillin–streptomycin. The BT-474 cell line was cultured with RPMI-1640 medium (Biological Industries, Kibbutz Beit HaEmek, Israel), 10% FBS, 1% penicillin–streptomycin, 1% MEM non-essential amino acids, and 0.01% insulin (Lonza, Verviers, Belgium).

Cells were passaged until 70–80% confluency, and then they were detached from the culture flask with 0.25% trypsin-EDTA (Biological Industries, Kibbutz Beit HaEmek, Israel) and resuspended in phosphate-buffered saline (PBS) solution. For analytical performance evaluation, spiking experiments were carried out with cancer cells fluorescently labeled with Cell Tracker Red CMTPX Dye (Invitrogen, Waltham, MA, USA). The cell concentration was measured with a TC20 automated cell counter (BioRAD, Hercules, CA, USA). Trypan blue was used to measure cell viability. To achieve the desired cell number in spiking experiments for analytical performance evaluation studies, cells were diluted with the serial dilution and spiked (20–400 cells/0.5 mL PBS) into whole blood (5 mL) collected from healthy donors.

Ethical approval for blood collection was taken from the Ethical Committee of Zekai Tahir Burak Women's Health Research and Education Hospital, Ankara, Turkey, and the studies performed were in accordance with the ethical regulations under ethical-committee-approved protocols (Protocol No: MBS-CTC-HEU-AV-02, Date: 27 January 2019). An informed consent form, which was approved by the ethical committee, was obtained from each donor before participating in the study. Blood samples were collected in K_2EDTA blood collection tubes and processed on the same day.

The spiked whole blood was processed with the density-gradient centrifugation method using Ficoll Paque Plus (Cytiva, Uppsala, Sweden) to eliminate red blood cells (RBCs). The buffy coat containing peripheral blood mononuclear cells (PBMCs) and CTCs was collected and washed with PBS containing 1% FBS (F-PBS). The supernatant was discarded, and the cell pellet was resuspended in F-PBS (10 mL). Before the microfluidic CTC enrichment process, the cell suspension was filtered through a 30 μm cell strainer (Miltenyi Biotec, Bergisch Gladbach, Germany) to remove impurities and prevent clogging.

2.4. Experimental Setup

The experimental setup for microfluidic sample processing is represented in Figure 4. The fluid flow at a specified flow rate was supplied through a pressure-driven microfluidic setup supplied by a compressed N_2 line and consisting of a pressure controller (Fluigent Flow EZ 7000, Paris, France) to regulate the pressure to push the liquids at a specified flow rate, a thermal flow sensor (Fluigent Flow Unit XL, Paris, France) for real-time measurement of the flow rate, a custom-design chip holder that enables the connections of the chip to external microfluidic components and to inlet/outlet fluid reservoirs, and an inverted microscope (DMi8, Leica Microsystems, Wetzlar, Germany).

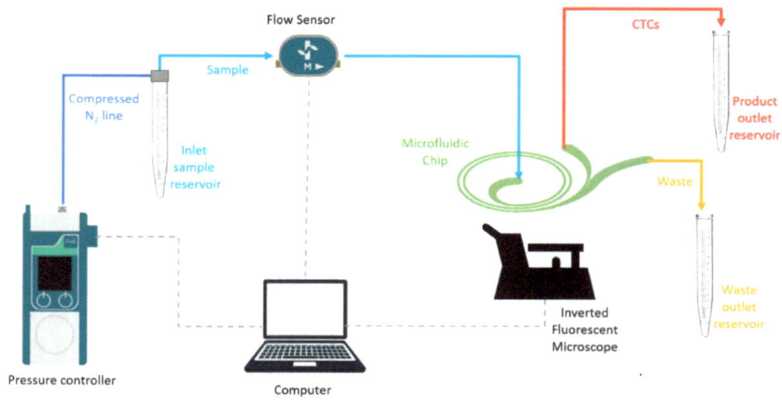

Figure 4. Experimental setup for microfluidic sample processing. The fluid flow is regulated with a flow controller supplied by compressed N_2 line. Flow rate is adjusted with the feedback control based on the readings of a thermal flow sensor. Channel is observed using an inverted microscope.

2.5. Chip Conditioning and Sample Processing

To prevent biofouling, the inner surfaces of the microfluidic chip and external fluidic components were coated with a random-graft co-polymer with a poly(L-lysine) backbone and poly(ethylene glycol) side chains (PLL-g-PEG Polymer—SuSoS, Zürich, Switzerland). Briefly, the channel was first conditioned with ethanol and washed with deionized (DI) water. Afterward, the channel was filled with PLL-g-PEG (1% in DI water) and incubated for 30 min. Then, the channel was washed with DI water and phosphate-buffered saline (PBS) solution before processing the cell suspension.

The cell suspension was fed to the chip with a constant volumetric flow rate, measured continuously with the flow sensor. A pressure controller regulated the applied pressure to the system with feedback from the flow sensor to maintain a steady flow rate. After the whole sample was processed on the system, the remaining sample in the system's dead volume (0.68 mL) was flushed with 1 mL of PBS with the same flow rate. The processed sample was collected into two outlets: the product outlet containing the enriched CTC sample and the waste outlet containing blood cells.

2.6. Chip Characterization Using Fluorescent Microbeads and Cells

The microfluidic chip design (namely 5H-50) and the two benchmark designs without the widening channel and without any hydrofoil structure, namely BARE and BARE-W, respectively, were characterized to identify their optimum operating conditions and preliminary performance characteristics. BARE (Figure 5a) was designed as a standard Archimedean spiral with a constant channel width. BARE-W (Figure 5b) has a similar channel geometry to that of 5H-50 but does not contain any hydrofoil structures in the widening channel, while 5H-50 (Figure 5c) contains five consecutive hydrofoils in the widening channel section specifically positioned to increase the performance characteristics, namely the CTC recovery rate and WBC depletion rate.

The CTC recovery rate was calculated by quantifying the number of CTCs at the product and waste outlets through triplicate measurements with the automated cell counter. The outlet suspension volumes and the average CTC concentrations in the waste and product outlet cell suspensions were used to calculate the recovery rate according to the following equation:

$$\text{CTC Recovery rate } (\%) = \frac{\text{\# of CTCs at product outlet}}{\text{\# of total CTCs at product and waste outlets}} \times 100 \quad (5)$$

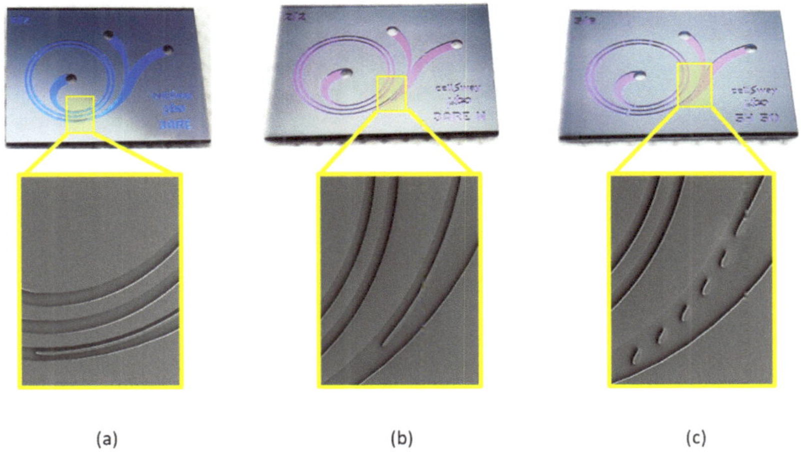

Figure 5. Microfabricated silicon–glass microfluidic chips. (a): BARE design, (b): BARE-W design, (c): 5H-50 design. Insets show the SEM images of separation regions for each chip design.

To estimate the WBC depletion rate, total cell concentrations were measured at the inlet and product outlet using the automated cell counter in triplicates. The average WBC concentrations in the inlet and product outlet cell suspensions were used to calculate WBC depletion rate according to the following equation:

$$\text{WBC Depletion rate } (\%) = \left(1 - \frac{\#\text{ of WBCs at product outlet}}{\#\text{ of WBCs at inlet sample}}\right) \times 100 \quad (6)$$

The chip characterization experiments were carried out initially using fluorescent polystyrene microbeads with 10.0 ± 0.6 µm and 18.7 ± 0.7 µm diameters (Polysciences Europe GmbH, Heidelberg, Germany) to represent WBCs and CTCs, respectively, and also using separate suspensions of MCF-7 breast cancer cell lines and WBCs. To identify the optimum flow rate, the bead or cell suspensions (1×10^5 particles/mL) were separately processed on the channel at flow rates varying between 500 µL/min and 2100 µL/min at an increment of 100 µL/min. At least 3 mL of particle suspension was processed at each selected flow rate to ensure that the set flow rate was stabilized. Video recordings were taken under an inverted fluorescent microscope equipped with a high-resolution camera (Hamamatsu ORCA Flash 4.0, Hamamatsu, Japan). The screenshots of the recordings were captured with VLC Video Player software (3.0.12) with a 30 fps capture rate and recording ratio of 2. The fluorescent intensity generated by microbeads across the channel width was analyzed by using the ImageJ software (1.50i) to find the focusing point of the microbeads across the channel. At least three independent experiments per bead and cell type were carried out with each chip design and the results were compared. These studies resulted in the preliminary identification of optimum volumetric flow rates of the chip designs and the initial performance characterization in terms of CTC recovery rates and WBC depletion rates.

2.7. Analytical Studies for Design Validation and Performance Characterization

Design validation studies were carried out using MCF-7 breast cancer cells (100 cells) spiked into healthy blood samples. We compared the performances of the proposed 5H-50 design with that of the BARE-W design at the optimum volumetric flow rate determined through initial characterization experiments. At least five independent experiments were carried out with each chip design, and the results were compared in terms of CTC recovery rate, WBC depletion rate, and cell viability at the outlet.

The CTC recovery rate was calculated by quantifying the number of Cell Tracker Red (CTR)-labeled MCF-7 cells. After processing the sample on the microfluidic chip, the cells collected at the product and waste outlets were centrifuged for 5 min and resuspended in 1 mL of PBS, and then seeded in a 96-well plate (100 μL per well) for optical examination. Then, images of the wells were acquired using the inverted fluorescence microscope equipped with a programmable motorized stage. All images were analyzed and MCF-7 cells in each well were automatically enumerated using a custom-designed software (Aurvis, Ankara, Turkey). The recovery rate was calculated according to Equation (5) after determining the total number of CTR-labeled CTCs in each suspension.

Further performance characterization studies were carried out in order to determine the recovery rate of the 5H-50 chip at different MCF-7-cell-spiking rates to show the linearity of the CTC recovery. MCF-7 cells were spiked to the whole blood with serial dilution at a spiking rate covering the range of 10^1–10^2 cells, and the blood sample was processed on the microfluidic chip after PMBC isolation. At each spiking rate, at least five experiments were conducted to show reproducibility and linearity between the spiked CTC number and the collected CTC number at the product outlet of the chip.

The applicability of the technology for different cancer types was demonstrated by testing the performance of the 5H-50 design with cancer cell lines derived from different cancer types, including non-small-cell lung cancer (A549, epithelial adenocarcinoma), ovarian cancer (SKOV-3, serous cystadenocarcinoma), and breast cancer (BT-474, ductal adenocarcinoma). The A549, SKOV-3, and BT-474 cell lines were spiked separately into whole blood (5 mL) at a spiking rate of 100 cells per 5 mL of blood. After PMBC isolation, the cell suspension was processed on the chip. At least five experiments were conducted with each cell line and their average recovery and depletion rates were compared with each other.

To investigate the effect of the microfluidic spiral chip process on CTC viability, cultured MCF-7 cancer cells (5 mL, 1×10^6 cells/mL) were processed on the chip at a flow rate of 1500 μL/min. Cell viability at the inlet and outlet cell suspensions was estimated via a Trypan blue (Sigma-Aldrich, Münich, Germany) exclusion assay and analyzed via an automated cell counter. Three independent experiments were carried out and all measurements were conducted in triplicate. The percent viability is defined as the number of live cells over the number of total cells in the inlet and product outlet suspensions, as in the equation given below.

$$\text{Viability}(\%) = \frac{\#\text{of viable cells}}{\#\text{of total cells}} \times 100 \quad (7)$$

3. Results

This section may be divided into subheadings. It should provide a concise and precise description of the experimental results and their interpretation, as well as the experimental conclusions that can be drawn.

3.1. Chip Characterization Using Fluorescent Microbeads and Cells

To corroborate the design principle and determine the flow conditions, different flow rates were systematically compared with a set of experiments. Firstly, fluorescent polystyrene beads mimicking WBCs and CTCs with 10 ± 0.6 μm and 18.7 ± 0.7 μm diameters, respectively, were tested separately through different microfluidic channel designs to observe the particle streams' behaviors between 500 and 2100 μL/min. Composite images of particle streams in each flow rate were captured and analyzed using ImageJ software. Figure 6 shows experimental data illustrating the distribution of 10 μm and 18.7 μm beads across 5H-50, BARE-W, and BARE microchannels for the chosen flow rates (Figure S1 for details). For both 10 μm and 18.7 μm particles, the variation in migration pattern with the flow rate were similar, regardless of the channel design. The results showed that 18.7 μm particles were not focused at low flow rates (<1200 μL/min). As the flow rate increased, especially around the design flow rate (1500 μL/min), larger particles tended towards the inner wall, as expected, while at flow rates higher than 1800 µL/min, the focusing was disturbed. On the contrary, 10 µm particles were directed toward the inner wall at low flow rates without a complete focusing. As the flow rate increased, they were directed to the waste outlet near the outer wall, and the focusing was still observed even at flow rates higher than 1800 µL/min. These studies have proven that the designed channels work as expected and the different-sized particles could be collected at the desired outlets by focusing them at different points across the width of the spiral channel. The results also showed that the flow rate range can be adjusted between 1200 µL/min and 1800 µL/min for 5H-50, BARE-W, and BARE channel designs to achieve the separation of particles with specific sizes. However, further increases in flow rate resulted in the dispersion of focus across the channel.

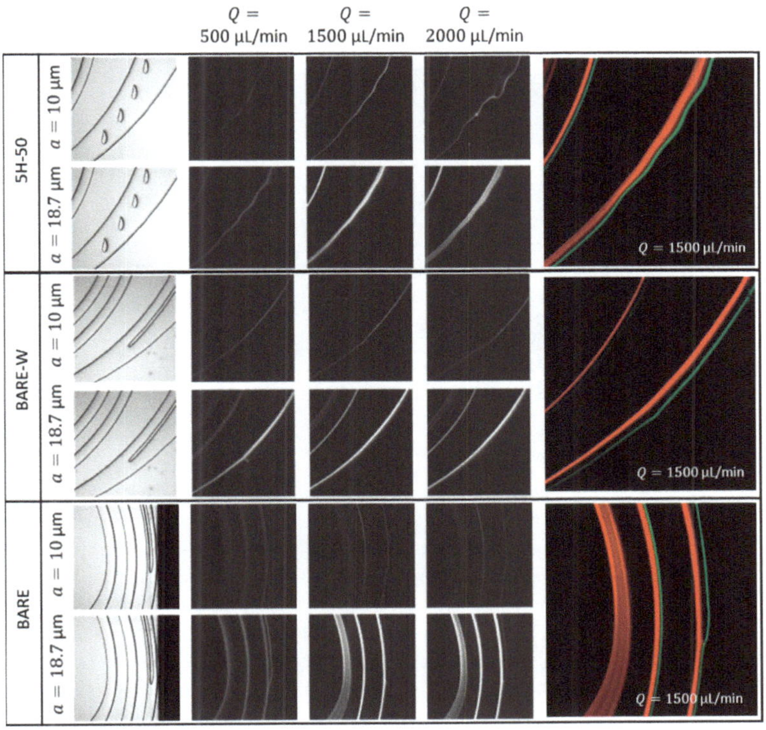

Figure 6. Bright-field and fluorescent images showing the focusing lines for 10 µm beads and 18.7 µm beads along the different channel designs at different flow rates. Fluorescent images were generated by overlaying the pseudocolored bright-field images for 18.7 µm beads (red) and 10 µm beads (green) at a 1500 µL/min flow rate.

After characterizing the channels for approximate operating flow rates and achieving proof of focusing, the channels were investigated for the focusing behavior of the MCF-7 cells and WBCs for different flow rates. Figure 7a shows the focusing lines for fluorescent stained MCF-7 cells and WBCs for three different channel designs at 1500 µL/min. Focusing for WBCs near the outer wall and focusing for MCF-7 cells near the inner wall were both observed for all the designs at this flow rate. Red-dotted traces marked on the bright-field images illustrate the focusing point identification lines across the channel width. Figure 7b was generated using the fluorescence intensity profiles of the particle distributions across the channel width for the 5H-50 channel at different flow rates. Focusing points on the plots were normalized to channel cross-sections where the fluorescence readings were taken so that 0.0 marked the inner wall and 1.0 marked the outer wall positions. The focusing point was defined as the distance from the highest fluorescence intensity to the inner wall. Fluorescence for WBCs was observed along a single line and the width of the line was thinner as the focusing occurred around 1500 µL/min. However, two or three focusing lines were observed for MCF-7 cells (Figure 7b), depending on the flow rate. One of these lines—the primary focusing line—had a more intense fluorescence. The other lines—the secondary and the tertiary focusing lines—exhibited a less intense fluorescence (note the line seen near the outer wall of 5H-50 on Figure 7a). These fainter lines may stem from the formation of the secondary Dean vortices at higher flow rates, i.e., at higher *Re*, as explained by Nivedita et al. in 2017 [25]. These secondary Dean vortices may cause the particles to migrate further and become trapped on an additional focusing line shifted from the inner wall to the outer wall of the channel.

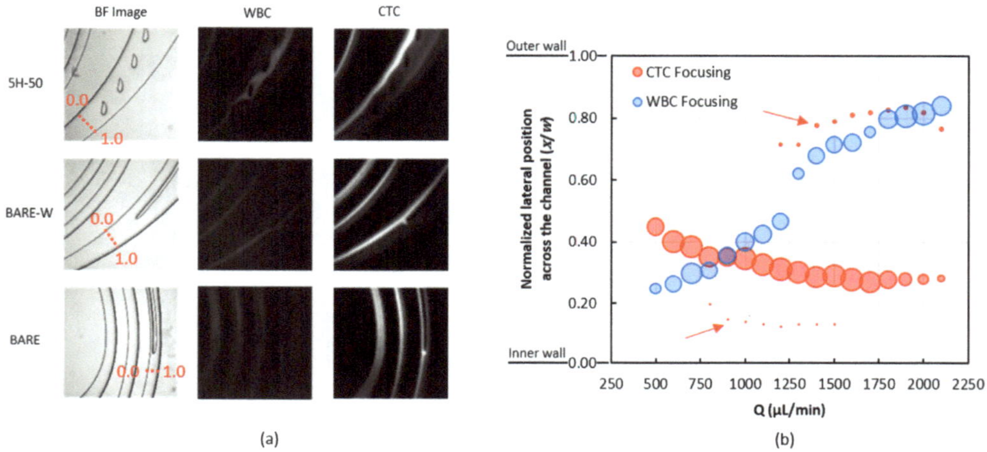

Figure 7. (**a**): Different focusing lines of fluorescently stained MCF-7 cells and WBCs inside 5H-50, BARE-W, and BARE chip designs, together with their corresponding bright-field images at 1500 µL/min flow rate. (**b**): Normalized lateral position (x/w) of MCF-7 cells and WBCs across the channel of 5H-50. The diameter of the data markers represents the normalized intensity. It is noted that MCF-7 cells were focused at secondary and tertiary positions (indicated by arrows) in addition to the primary equilibrium position. The data for fluorescent intensity distribution along the channel width were extracted on the red dotted lines drawn in the separation region as presented in (**a**).

Further experiments were carried out with separate suspensions of WBCs and MCF-7 cells to determine the optimum flow rate in the 1200–1800 µL/min range. Figure 8 shows the depletion and recovery rate performances for WBCs and MCF-7 cells, respectively, at the selected flow rate range for the 5H-50, BARE, and BARE-W channels. According to the results for the 5H-50 design, above 1300 µL/min, the effect of the flow rate on the depletion rate was insignificant. However, the recovery rate of the MCF-7 cells peaked at 1500 µL/min (Figure 8a). For this reason, the optimal flow rate was decided to be 1500 µL/min, which was also the design flow rate used in numerical calculations. At this flow rate, 80% of MCF-7 cells were recovered and 98% of the WBCs were depleted at the CTC outlet. In the BARE-W design, a similar trend was observed in terms of recovery rate. The highest WBC depletion rate was observed at 1300 µL/min, while the highest recovery rate was observed at 1500 µL/min (Figure 8b). Therefore, 1500 µL/min was chosen as the optimum flow rate for the BARE-W design. When Figure 8c was analyzed, it was seen that the benchmark Archimedean spiral had a high WBC depletion rate at all the tested flow rates; however, the maximum MCF-7 recovery at the optimum flow rate of 1400 µL/min could only be around 50%, which is significantly lower than that of the BARE-W and 5H-50 designs.

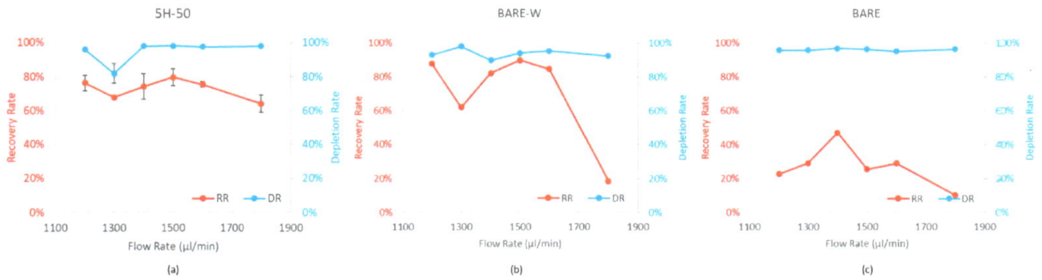

Figure 8. Variation in MCF-7 recovery and WBC depletion rates against flow rate when 1 × 10^5 cells/mL was used for (**a**) 5H-50, (**b**) BARE-W, and (**c**) BARE (**b**) chips. Optimal results were obtained at the design flow rate of 1500 µL/min in both 5H-50 and BARE-W.

To investigate the effect of the microfluidic chip process on cell viability, we carried out a viability analysis through a Trypan blue exclusion assay, utilizing MCF-7 cell suspension. The cell viability levels measured at the inlet and outlet cell suspensions were compared together with the total cell count in each suspension to see if there was any cell loss during the process. Cell viability data (obtained using Equation (7)) revealed that the inlet and product outlet cell suspensions have very close total cell viability values (82 ± 6% and 88 ± 5%, respectively), while the waste cell suspension has a slightly lower total cell viability (55 ± 15%). This can be interpreted as meaning that the collection of dead cells was mostly at the waste outlet, which is an expected result as dead cells are typically smaller in size. Total cell counts at inlet cell suspension (6.5 × 10^6) and outlet cell suspensions (6.2 × 10^6) revealed that the chip process does not cause significant cell loss (<5%).

In order to further validate the design, we compared the performances of the 5H-50 chip and the benchmark BARE-W and BARE chips with experiments carried out using MCF-7 breast cancer cell lines spiked into healthy blood samples (100 cells/5 mL blood) at 1500 µL/min. The average WBC depletion and MCF-7 recovery rates are presented in Figure 9. The BARE-W and 5H-50 designs had similar WBC depletion rates of 94.2 ± 2.4% and 94.2 ± 2.2%, respectively. The BARE chip had the highest depletion rate (98.4 ± 0.6%) but the lowest recovery rate (23.6 ± 2.1%). The average CTC recovery rate for the 5H-50 design (67.9 ± 5.2%) was higher than that obtained with the BARE-W design (56.7 ± 17.6%). Additionally, the reproducibility of the results was much better with the 5H-50 chip, as evident from the standard deviations of both data sets. These data show the experimental validation of the numerical design, confirming that the inclusion of hydrofoil structures into a widening spiral microfluidic channel section improves the enrichment of the circulating tumor cells. During the design of the microfluidic channel, the aim of adding hydrofoil structures was to keep the velocity higher along the outer wall of the channel than along the inner wall of the channel to keep the larger particles, in this study, MCF-7 cells, near the inner wall with a slower pace. We confirmed this behavior by reaching a higher recovery rate for the 5H-50 design. It should be noted that the recovery rate and depletion rate values obtained during the spiking experiments for all of the designs were lower than the ones obtained using separate 1 × 10^5 cells/mL suspensions. This might be attributed to particle–particle interaction being much more effective for the sample used in spiking experiments due to the existence of a high concentration of WBCs inside the solution (>2 × 10^6). It is possible that the focused flow generated by the much higher number of WBCs may have prevented the CTCs from reaching their equilibrium points.

In light of these results demonstrating the design validation, further analytical studies were carried out only on the 5H-50 design.

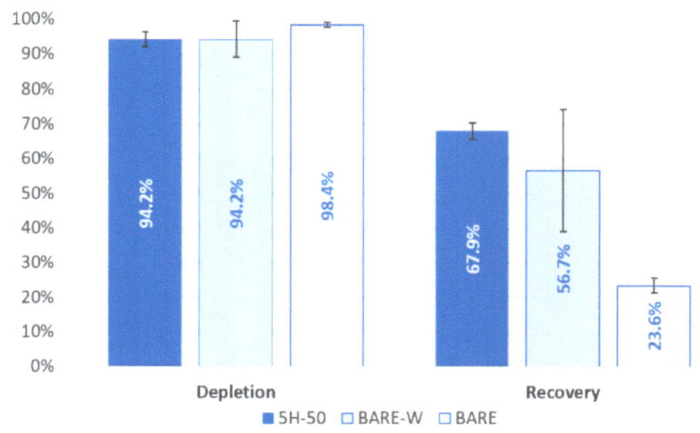

Figure 9. WBC depletion and MCF-7 recovery rates obtained from 5H-50, BARE-W, and BARE chip designs in spiking experiments. Data were collected by spiking 100 MCF-7 cells in 5 mL of whole blood at 1500 µL/min. Data show the average of at least three experiments with standard deviation.

3.2. Analytical Studies for Design Validation and Performance Characterization

Analytical validation of the 5H-50 chip design was performed initially with MCF-7 cell lines spiked into healthy blood samples at spiking rates varying between 20 and 400 per 5 mL of whole blood. Figure 10 shows the linear regression between the total number of cells spiked and the cells collected at the product outlet over a total of 27 experiments. The R^2 value of the regression model was calculated to be 0.9856, showing the high linearity of the MCF-7 cells' recovery over the experimented spiking rate range. The slope of the line was calculated to be 0.6831, which can be interpreted as an average 68.3% recovery rate of MCF-7s. The average depletion rate obtained from these experiments was 94.8 ± 2.1%. Both the recovery and depletion rates were in good agreement with the ones obtained during the 100 MCF-7 cell spiking experiments. Although the literature reports the purity (recovered CTC amount per contaminating WBC amount) as a metric for the cell separation systems, here, we prefer to report depletion rate since (i) it is more informative as a characterization metric, especially for comparison purposes, and (ii) the purity in clinical samples varies drastically as the numbers of CTCs and WBCs significantly differ from patient to patient.

The performance of the 5H-50 chip design was also evaluated with cancer cell lines derived from different cancer types, including lung (A549) and ovarian cancer (SKOV-3), as well as with another breast cancer cell line (BT-474). At the spiking rate of 100 cells/5 mL of whole blood, recoveries above 60% were obtained for all cell lines, reaching a maximum of 82.9 ± 9.9% with BT-474 (Figure 11). The average depletion rate was calculated to be 94.0 ± 1.0%. Difference in the recovery rates mainly depends on the different average cell size of each cell line. Among the studied cell lines, the BT-474 cells (20.9 ± 4.2 µm) were the largest, followed by the SKOV-3 cells (19.9 ± 3.6 µm) and the A549 cells (19.1 ± 4.1 µm). The MCF-7 cells (16.5 ± 2.3 µm) were the smallest. Since the designed microchannel separates cells based on their size differences, the BT-474 cells had the highest recovery (82.9 ± 9.9%), as expected. Although the MCF-7 cells were the smallest, they had a higher recovery rate than the A549 cells. This can be explained by the differences in their deformability, i.e., the alteration in their morphology under stress. It is documented in the literature that A549 cells, which had the lowest recovery (62.3 ± 8.4%), are more deformable than the other three cell lines and BT-474 cells are the least deformable among them [27–29]. Thus, we may conclude that the other than particle size differences, the extent of their deformability is also important because shape of the particles affects their migration pattern [30]. A similar effect was also observed when the recovery rate for BT-474 was compared with that of SKOV-3 cells, which are more deformable than BT-474. Although the average cell sizes of these two cell lines are very close, the recovery rate observed for BT-474 cells (82.9 ± 9.9%) was significantly higher than that obtained for SKOV-3 cells (71.0 ± 6.5%). Figure S2 shows the immunofluorescent staining of enriched BT-474 cell suspension, which exemplifies the intactness of the cells processed inside the spiral channel.

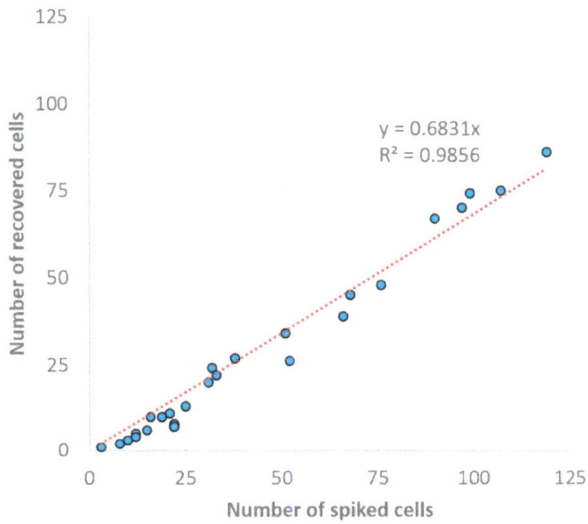

Figure 10. The relationship between spiked and collected MCF-7 cells. Data were generated through 27 independent experiments carried out at varying MCF-7-spiking rates. The linear regression was calculated to be R^2 = 0.9856.

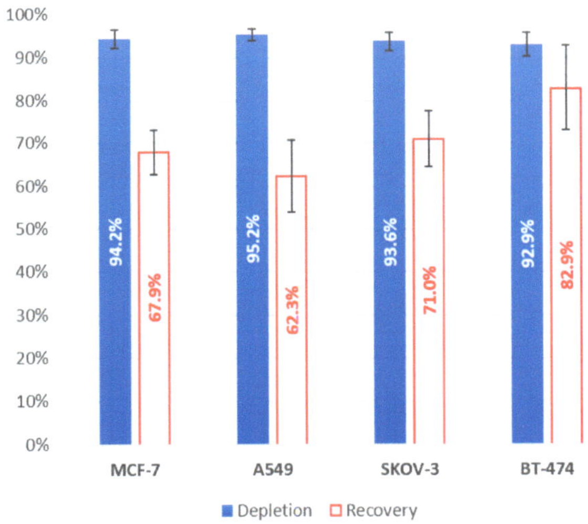

Figure 11. Recovery rates obtained with MCF-7 (n = 8), A549 (n = 5), SKOV-3 (n = 7), and BT-474 (n = 5) cancer cell lines spiked into healthy blood sample (5 mL).

4. Conclusions

CTCs are considered some of the most promising liquid biopsy biomarkers for obtaining real-time information on tumor evolution during therapy monitoring and cancer management. However, their current use in clinical practice is mainly restricted, used for prognostic stratification and monitoring as opposed to their vast potential. The isolation of extremely rare CTCs from peripheral blood cells with high efficiency is still the main challenge that impedes their extensive use in routine clinical practice as this challenge limits the development of sensitive identification and characterization assays on CTCs.

This paper addresses this challenge with a microfluidic channel design to enhance the size-based enrichment efficiency of viable CTCs from blood. For this purpose, we first developed a design methodology for the channel design by combining analytical, numerical, and empirical solutions. The front end of the channel was designed as a two-turn Archimedean spiral, and a gently widening outlet posing successive streamlined pillars in the shape of asymmetric hydrofoils was designed using 2-D numerical modelling. The single-inlet and two-outlet microfluidic channel operates at high throughput (1500 μL/min) and enables a simpler operation without the need to use sheath flow at the inlet. The performance of the design (namely 5H-50) was analytically demonstrated and benchmarked with two individually optimized designs comprising (i) only an Archimedean spiral (BARE) and (ii) an Archimedean spiral followed by a widening outlet (BARE-W). The optimized flow rate was determined for each design through fluorescent intensity data generated using microbeads and depletion and recovery rate values obtained with WBCs and MCF-7 cell lines. The designs were analytically validated with 100 MCF-7 cells spiked in 5 mL healthy whole-blood samples. It was demonstrated that the 5H-50 design outperformed the benchmark designs with a 67.9 ± 5.2% recovery rate, with a slightly lower depletion rate of 94.2 ± 2.2%. The R^2 value of the linear fit was measured to be 0.9856 within the MCF-7 spiking range of 10^1–10^2 cells per 5 mL. The viability of the cancer cell lines collected at the product outlet was measured to be 88 ± 5%, which is higher than the viability at the inlet. The design was further validated with the spiking of the A549, SKOV-3, and BT-474 cell lines yielding recovery rates above 60%, the highest being recorded as 82.9 ± 9.9% with BT-474 cells. The results indicate a high CTC recovery rate, independent of the type of spiked cancer cells, with no damage to cells in terms of integrity and viability. The method allows the rapid and precise phenotypical characterization of CTCs at DNA, protein, and gene-expression levels through the application of NGS, immunofluorescence, and FISH/RNA-ISH techniques, at a single-cell level. However, for the bulk molecular characterization of isolated cells, the number of WBCs at the product output should be reduced.

Future work will include conducting validation studies with patient blood samples mainly on breast and non-small-cell lung cancer. The studies can also be expanded to cover different cancer types, as the enrichment technique is suitable for use for almost all cancer types except hematological cancers. Moreover, the CTC enrichment workflow will be automated to reduce the hands-on time and to better adapt the developed workflow for use in clinical laboratories. The validation studies will explore both the identification and characterization of enriched CTCs from patient blood samples; hence, they require the integration of immunofluorescent microscopy techniques into the workflow. In our opinion, with the demonstration of the capability of identifying relevant biomarkers on CTCs in a standardized way, this workflow will be a step forward in revealing the clinical utility of CTCs through the generated clinically actionable data to be used for patient follow-up, therapy guidance, and precision medicine.

Supplementary Materials: The following supporting information can be downloaded at: https://www.mdpi.com/article/10.3390/bios13100938/s1, Figure S1: Bright-field and fluorescent images showing the focusing lines for 10 μm beads and 18.7 μm beads along the different channel designs at different flow rates; Figure S2: IF staining of BT-474 cells and white blood cells. BT-474 cells are ER/PR+, HER2+, CK+, DAPI+, and CD45-. Channels; Green: ER/PR, BF: Bright Field, Orange: HER2, Cy5: CD45, Cy7: pan Cytokeratin and DAPI.

Author Contributions: Conceptualization, B.S.-D., E.Y., E.O. and O.Z.; Data curation, B.S.-D., E.O. and O.Z.; Formal analysis, B.S.-D., M.A.D., B.S., E.Y., E.M., E.O. and O.Z.; Funding acquisition, E.O., O.Z. and H.K.; Investigation, B.S.-D., M.A.D., B.S., G.K., S.S. and E.M.; Methodology, B.S.-D., E.Y., E.M., T.B.T., E.O. and O.Z.; Project administration, B.S.-D., E.O. and O.Z.; Resources, B.S.-D., T.B.T., E.O. and O.Z.; Supervision, B.S.-D., E.Y., E.O., O.Z. and H.K.; Validation, B.S.-D., M.A.D., B.S., E.Y., E.M., E.O. and O.Z.; Visualization, B.S.-D., M.A.D., B.S. and E.Y.; Writing—original draft, B.S.-D.,

M.A.D., B.S., E.Y., G.K., S.S., E.O. and O.Z.; Writing—review and editing, B.S.-D., E.Y., E.O., O.Z. and H.K. All authors have read and agreed to the published version of the manuscript.

Funding: This research was funded by TÜBİTAK, grant number 3189298.

Institutional Review Board Statement: The study was conducted according to the guidelines of the Declaration of Helsinki, and approved by the Ethics Committee of Zekai Tahir Burak Women's Health Research and Education Hospital, Ankara, Turkey (Protocol No: MBS-CTC-HEU-AV-02, Date: 27 January 2019).

Informed Consent Statement: Informed consent was obtained from all subjects involved in the study.

Data Availability Statement: Available upon request.

Conflicts of Interest: Ender Yildirim and Ozge Zorlu have a patent titled "Method of microfluidic particle separation enhancement and the device thereof" issued to Mikro Biyosistemler Elektronik Sanayi ve Ticaret A.Ş. The funders had no role in the design of the study; in the collection, analyses, or interpretation of data; in the writing of the manuscript; or in the decision to publish the results.

References

1. Micalizzi, D.S.; Maheswaran, S.; Haber, D.A. A Conduit to Metastasis: Circulating Tumor Cell Biology. *Genes Dev.* **2017**, *31*, 1827–1840. [CrossRef]
2. Agashe, R.; Kurzrock, R. Circulating Tumor Cells: From the Laboratory to the Cancer Clinic. *Cancers* **2020**, *12*, 2361. [CrossRef] [PubMed]
3. Bardia, A.; Haber, D.A. Solidifying Liquid Biopsies: Can Circulating Tumor Cell Monitoring Guide Treatment Selection in Breast Cancer? *J. Clin. Oncol.* **2014**, *32*, 3470–3471. [CrossRef] [PubMed]
4. Masuda, T.; Hayashi, N.; Iguchi, T.; Ito, S.; Eguchi, H.; Mimori, K. Clinical and Biological Significance of Circulating Tumor Cells in Cancer. *Mol. Oncol.* **2016**, *10*, 408–417. [CrossRef] [PubMed]
5. Keller, L.; Pantel, K. Unravelling Tumour Heterogeneity by Single-Cell Profiling of Circulating Tumour Cells. *Nat Rev. Cancer* **2019**, *19*, 553–567. [CrossRef]
6. Powell, A.A.; Talasaz, A.A.H.; Zhang, H.; Coram, M.A.; Reddy, A.; Deng, G.; Telli, M.L.; Advani, R.H.; Carlson, R.W.; Mollick, J.A.; et al. Single Cell Profiling of Circulating Tumor Cells: Transcriptional Heterogeneity and Diversity from Breast Cancer Cell Lines. *PLoS ONE* **2012**, *7*, e33788. [CrossRef]
7. Bin Lim, S.; di Lee, W.; Vasudevan, J.; Lim, W.T.; Lim, C.T. Liquid Biopsy: One Cell at a Time. *NPJ Precis. Oncol.* **2019**, *3*, 23.
8. CTC Studies That Are Actively Recruiting. Available online: https://clinicaltrials.gov/ct2/results?term=circulating+tumor+cells&cond=cancer&Search=Apply&recrs=a&age_v=&gndr=&type=&rslt= (accessed on 30 November 2022).
9. Ju, S.; Chen, C.; Zhang, J.; Xu, L.; Zhang, X.; Li, Z.; Chen, Y.; Zhou, J.; Ji, F.; Wang, L. Detection of Circulating Tumor Cells: Opportunities and Challenges. *Biomark. Res.* **2022**, *10*, 58. [CrossRef]
10. Cho, H.; Kim, J.; Song, H.; Sohn, K.Y.; Jeon, M.; Han, K.H. Microfluidic Technologies for Circulating Tumor Cell Isolation. *Analyst* **2018**, *143*, 2936–2970. [CrossRef]
11. Hao, S.J.; Wan, Y.; Xia, Y.Q.; Zou, X.; Zheng, S.Y. Size-Based Separation Methods of Circulating Tumor Cells. *Adv. Drug Deliv. Rev.* **2018**, *125*, 3–20. [CrossRef]
12. Bankó, P.; Lee, S.Y.; Nagygyörgy, V.; Zrínyi, M.; Chae, C.H.; Cho, D.H.; Telekes, A. Technologies for Circulating Tumor Cell Separation from Whole Blood. *J. Hematol. Oncol.* **2019**, *12*, 48. [CrossRef] [PubMed]
13. Pødenphant, M.; Ashley, N.; Koprowska, K.; Mir, K.U.; Zalkovskij, M.; Bilenberg, B.; Bodmer, W.; Kristensen, A.; Marie, R. Separation of Cancer Cells from White Blood Cells by Pinched Flow Fractionation. *Lab Chip* **2015**, *15*, 4598–4606. [CrossRef] [PubMed]
14. Karabacak, N.M.; Spuhler, P.S.; Fachin, F.; Lim, E.J.; Pai, V.; Ozkumur, E.; Martel, J.M.; Kojic, N.; Smith, K.; Chen, P.I.; et al. Microfluidic, Marker-Free Isolation of Circulating Tumor Cells from Blood Samples. *Nat. Protoc.* **2014**, *9*, 694–710. [CrossRef] [PubMed]
15. Fachin, F.; Spuhler, P.; Martel-Foley, J.M.; Edd, J.F.; Barber, T.A.; Walsh, J.; Karabacak, M.; Pai, V.; Yu, M.; Smith, K.; et al. Monolithic Chip for High-Throughput Blood Cell Depletion to Sort Rare Circulating Tumor Cells. *Sci. Rep.* **2017**, *7*, 10936. [CrossRef]
16. Bhagat, A.A.S.; Bow, H.; Hou, H.W.; Tan, S.J.; Han, J.; Lim, C.T. Microfluidics for Cell Separation. *Med. Biol. Eng. Comput.* **2010**, *48*, 999–1014. [CrossRef]
17. di Carlo, D.; Irimia, D.; Tompkins, R.G.; Toner, M. Continuous Inertial Focusing, Ordering, and Separation of Particles in Microchannels. *Proc. Natl. Acad. Sci. USA* **2007**, *104*, 18892–18897. [CrossRef]
18. Nivedita, N.; Papautsky, I. Continuous Separation of Blood Cells in Spiral Microfluidic Devices. *Biomicrofluidics* **2013**, *7*, 054101. [CrossRef]
19. Warkiani, M.E.; Khoo, B.L.; Wu, L.; Tay, A.K.P.; Bhagat, A.A.S.; Han, J.; Lim, C.T. Ultra-Fast, Label-Free Isolation of Circulating Tumor Cells from Blood Using Spiral Microfluidics. *Nat. Protoc.* **2016**, *11*, 134–148. [CrossRef]

20. Warkiani, M.E.; Guan, G.; Luan, K.B.; Lee, W.C.; Bhagat, A.A.S.; Kant Chaudhuri, P.; Tan, D.S.W.; Lim, W.T.; Lee, S.C.; Chen, P.C.Y.; et al. Slanted Spiral Microfluidics for the Ultra-Fast, Label-Free Isolation of Circulating Tumor Cells. *Lab Chip* **2014**, *14*, 128–137. [CrossRef]
21. Sun, J.; Liu, C.; Li, M.; Wang, J.; Xianyu, Y.; Hu, G.; Jiang, X. Size-Based Hydrodynamic Rare Tumor Cell Separation in Curved Microfluidic Channels. *Biomicrofluidics* **2013**, *7*, 011802. [CrossRef]
22. Ookawara, S.; Higashi, R.; Street, D.; Ogawa, K. Feasibility Study on Concentration of Slurry and Classification of Contained Particles by Microchannel. *Chem. Eng. J.* **2004**, *101*, 171–178. [CrossRef]
23. Mihandoust, A.; Maleki-Jirsaraei, N.; Rouhani, S.; Safi, S.; Alizadeh, M. Improvement of Size-Based Particle Separation Throughput in Slanted Spiral Microchannel by Modifying Outlet Geometry. *Electrophoresis* **2020**, *41*, 353–359. [CrossRef] [PubMed]
24. Martel, J.M.; Toner, M. Particle Focusing in Curved Microfluidic Channels. *Sci. Rep.* **2013**, *3*, 3340. [CrossRef]
25. Nivedita, N.; Ligrani, P.; Papautsky, I. Dean Flow Dynamics in Low-Aspect Ratio Spiral Microchannels. *Sci. Rep.* **2017**, *7*, 44072. [CrossRef] [PubMed]
26. Özkayar, G.; Mutlu, E.; Şahin, Ş.; Demircan Yalçın, Y.; Töral, T.; Külah, H.; Yıldırım, E.; Zorlu, Ö.; Özgür, E. A Novel Microfluidic Method Utilizing a Hydrofoil Structure to Improve Circulating Tumor Cell Enrichment: Design and Analytical Validation. *Micromachines* **2020**, *11*, 981. [CrossRef]
27. Mousapour, B. *Measuring Biomechanical Properties of Cancer Cells Using a High-Throughput Microfluidic Platform and Their Correlation with Surface Rheology and Internal Elasticity*; University of California: Irvine, CA, USA, 2016.
28. Hwang, J.Y.; Kim, J.; Park, J.M.; Lee, C.; Jung, H.; Lee, J.; Shung, K.K. Cell Deformation by Single-Beam Acoustic Trapping: A Promising Tool for Measurements of Cell Mechanics. *Sci. Rep.* **2016**, *6*, 27238. [CrossRef]
29. Jhaveri, D. *Study of Cell Mechanics and Surface Rheology of Cancer Cells Using High Throughput Microfluidic Device*; University of California: Irvine, CA, USA, 2015.
30. Masaeli, M.; Sollier, E.; Amini, H.; Mao, W.; Camacho, K.; Doshi, N.; Mitragotri, S.; Alexeev, A.; di Carlo, D. Continuous Inertial Focusing and Separation of Particles by Shape. *Phys. Rev. X* **2012**, *2*, 031017. [CrossRef]

biosensors

MDPI

Article

Rapid Detection of Microparticles Using a Microfluidic Resistive Pulse Sensor Based on Bipolar Pulse-Width Multiplexing

Ruiting Xu [1], Leixin Ouyang [1], Rubia Shaik [2], Heyi Chen [1], Ge Zhang [2] and Jiang Zhe [1,*]

1 Department of Mechanical Engineering, University of Akron, Akron, OH 44325, USA; rx7@uakron.edu (R.X.); lo10@uakron.edu (L.O.); hc77@uakron.edu (H.C.)
2 Department of Biomedical Engineering, University of Akron, Akron, OH 44325, USA; rs169@uakron.edu (R.S.); ge10@uakron.edu (G.Z.)
* Correspondence: jzhe@uakron.edu; Tel.: +1-330-972-7737

Abstract: Rapid and accurate analysis of micro/nano bio-objects (e.g., cells, biomolecules) is crucial in clinical diagnostics and drug discovery. While a traditional resistive pulse sensor can provide multiple kinds of information (size, count, surface charge, etc.) about analytes, it has low throughput. We present a unique bipolar pulse-width, multiplexing-based resistive pulse sensor for high-throughput analysis of microparticles. Signal multiplexing is enabled by exposing the central electrode at different locations inside the parallel sensing channels. Together with two common electrodes, the central electrode encodes the electrical signal from each sensing channel, generating specific bipolar template waveforms with different pulse widths. Only one DC source is needed as input, and only one combined electrical output is collected. The combined signal can be demodulated using correlation analysis and a unique iterative cancellation scheme. The accuracy of particle counting and sizing was validated using mixtures of various sized microparticles. Results showed errors of 2.6% and 6.1% in sizing and counting, respectively. We further demonstrated its accuracy for cell analysis using HeLa cells.

Keywords: resistive pulse sensor; signal multiplexing; bipolar pulse; high throughput; particle counting; microfluidics; iterative cancellation

Citation: Xu, R.; Ouyang, L.; Shaik, R.; Chen, H.; Zhang, G.; Zhe, J. Rapid Detection of Microparticles Using a Microfluidic Resistive Pulse Sensor Based on Bipolar Pulse-Width Multiplexing. *Biosensors* **2023**, *13*, 721. https://doi.org/10.3390/bios13070721

Received: 21 June 2023
Revised: 1 July 2023
Accepted: 6 July 2023
Published: 9 July 2023

1. Introduction

Bioanalysis can benefit significantly from the utilization of various types of micro/nano particles, which possess unique properties such as a large surface-to-volume ratio and excellent biocompatibility [1,2]. Detection of these particles in a solution can provide valuable information that reflects the biotarget situations across a broad spectrum of applications, including biomedical research, public health, and food safety [3–5]. Therefore, development of portable, cost-effective, and efficient devices for particle detection is crucial for bioanalysis.

Resistive pulse sensing (RPS) is an established technology used to rapidly detect nano/micro-scaled particles [6–8]. A typical RPS system comprises two electrodes placed on each side of a sensing channel, filled with a conducting electrolyte solution. The passage of a single particle causes a temporary change in the electrical resistance of the sensing channel, which generates a current/voltage pulse picked up by the pair of electrodes [9]. The magnitude and duration of the pulse are dependent on the particle size and shape, allowing for the determination of both sizes and concentrations of particles [10,11]. This technique is commonly used in biomedical and nanotechnology research for applications including cell counting [12,13], nanoparticle characterization [14,15], and biomolecular detection [16,17]. It is a versatile and non-destructive method that enables real-time monitoring of individual particles as they traverse the sensing channel, making it a valuable tool for particle analysis with a single-particle resolution. To increase the throughput of detection, RPS devices with multiple channels using space division multiplexing [18], CODES multiplexing [19,20], geometry multiplexing [21], and frequency division multiplexing [22] have been developed. The space multiplexing method [18] has demonstrated high-throughput parallel analysis with eight sensing channels, each equipped with individual measurement electronics. However, as the number of channels increases, it becomes impractical to implement individual detection electronics for each channel. CODES division [20] and geometry multiplexing [21] method have been proposed, but they require complex, unique patterns of microelectrodes or microchannel geometry for each sensing channel to generate specific waveform patterns; both methods are difficult to apply in nanoscale particle detection. Additionally, the frequency-division multiplexing method [22] requires the device be operated in the resistance-dominant frequency region; only a limited number of sensing channels can be used, resulting in limited scalability.

Here, we introduce a bipolar pulse-width multiplexing microfluidic sensor that provides a simple and scalable solution for high-throughput micro/nano particle counting. The sensor does not need complex electrodes or geometry; specific waveforms can be generated by three electrodes operated in tandem using only one DC power source. Microparticle presence and sizes can be determined through correlation analysis and an iterative cancellation scheme, even when multiple particles are present in the sensing channels.

2. Materials and Methods

2.1. Materials

Polystyrene microparticles (15 μm, product #74964 and 20 μm, product #74491, Sigma-Aldrich, St. Louis, MO, USA) were used for demonstrating the multiplexed sensor, while 3-aminopropyl)triethoxysilane (APTES, 99%, product #440140, from Sigma-Aldrich) was used to enhance the bonding between photoresist (SU-8 6002) and polydimethylsiloxane (PDMS). Dulbecco's phosphate-buffered saline (DPBS, 1×, product# MT21031CV, Thermo Fisher Scientific) was used to prepare the particle solution. Silver foil (0.5 mm thick, 99.9% metals basis, 25 × 25 mm, Catalog #AA39181FF) was purchased from Fisher Scientific, serving as the central electrode.

Human negroid cervix epithelioid carcinoma cells (HeLa, product# 93021013), minimum essential medium eagle, with ear (EMEM, product# M2279), L-glutamine solution Bioxtra (product# G7513), MEM non-essential amino acid (NEAA, product# M7145), fetal bovine serum (FBS, product# F0926), and 0.25% trypsin-EDTA solution (product# T4049) were purchased from Sigma-Aldrich. Dulbecco's phosphate-buffered salt solution 1× (DPBS, Cat. No: MT21031CV), penicillin–streptomycin (Cat. No: 15-140-122), trypan blue solution 0.4% (Cat. No: BW17-942E), and saline solution (Cat. No. L97815) were obtained from Fisher Scientific. Briefly, HeLa cells were cultured in complete EMEM medium containing 2 mM L-glutamine solution, 1% NEAA, 10% FBS, and 1% penicillin. The cells were seeded at a density of 3500 cells/cm^2 in a T75 flask and incubated at 37 °C. The following day, the media was changed and then changed every other day until the flask reached about 90% confluency. To harvest the cells, they were washed once with 1× DPBS, and then 4 mL of 0.25% trypsin–EDTA solution was added and incubated at 37 °C for 5 min. Additional 8 mL of media was added, and the cells were pelleted in a 4 °C centrifuge at 220 g for 5 min. The cells were counted after staining with 0.4% trypan blue solution and then resuspended in 1× DPBS to a final working concentration of 10^5 cells/mL followed by high-throughput counting.

2.2. Sensing Principle

A multichannel RPS device with four parallel sensing channels was designed to demonstrate the bipolar pulse-width multiplexing method for high-throughput microparticle counting, as shown in Figure 1. The microfluidic device comprised (1) four microfluidic channels made of PDMS; (2) one Ag/AgCl central electrode embedded on the bottom of each microchannel separating each sensing channel to two sections at different locations;

(3) two Ag/AgCl common electrodes positioned on each side of the sensing channels; and (4) a pair of inlet and outlet reservoirs. The central electrode was formed through the following steps: (a) bonding a 0.5 mm silver layer to the glass substrate; (b) coating an insulating SU-8 layer on top of the Ag layer; (c) patterning four openings at different positions on the bottom of the four sensing channels, to expose the Ag electrode to the electrolyte; (d) treating the Ag layer with an electrochemical reaction to convert the exposed Ag to an Ag/AgCl electrode (see details in Section 2.3).

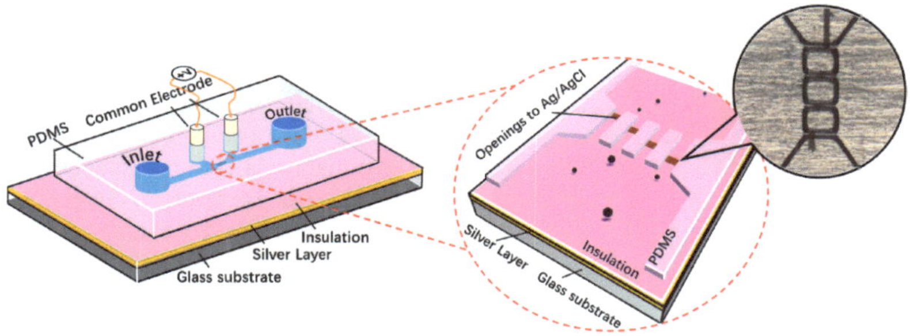

Figure 1. Left: illustration of the bipolar pulse-width multiplexing RPS device with four parallel sensing channels. Right: an enlarged image of the four parallel sensing channels.

A bridge circuit is used to monitor the resistance changes of sensing channels, as shown in Figure 2a,b. R_{eq_left} and R_{eq_right} represent the total resistance of the left sensing section and the right sensing section separated by the central electrode. R_1 and R_2 are the external adjustable resistors that form a Wheatstone bridge with R_{eq_left} and R_{eq_right}. A 1 V DC input voltage (V_{in}) is applied to the circuit. When a particle passes through the sensing channels (left section and right section), it induces a resistance change in R_{eq_left} or R_{eq_right}, causing a differential voltage between A and B. The voltage change is detected as the electrical output signal (V_{out}). When the particle passes through the left sensing section, R_{eq_left} increases while R_{eq_right} remains the same, causing a drop in voltage output. Similarly, when the particle passes the right sensing section, R_{eq_right} increases, causing a rise in voltage output. Hence, when a particle passes through each sensing channel consisting of two consecutive sensing sections, it generates a bipolar voltage pulse. Because the central electrode separates each sensing channel at different positions, the resistance/voltage drop and rise in each sensing channel occur differently when a particle passes through each sensing channel. Hence, the bipolar output from each sensing channel has a unique waveform (with different pulse widths t_1, t_2), as illustrated in Figure 2c. If multiple particles are present in different sensing channels, the overall voltage output measured across the common electrodes would be a combination of individual signals from each sensing channel. From the correlation analysis with the standard waveforms of each channel (generated by a fixed-size particle) and iterative cancellation scheme, we can identify in which channels the particles are transiting, as well as the count and size of particles passing each sensing channel. The magnitudes of the standard waveforms can be used to determine the sizes of the particles.

2.3. Device Fabrication

To fabricate the microfluidic device, the standard soft lithography method was utilized. First, the SU-8 2025 (MicroChem, Newton, MA, USA) mold was created, which included the four parallel sensing channels, two detecting reservoirs for the common electrodes, and one inlet/outlet reservoir. A PDMS (polydimethylsiloxane, Sylgard 184, Dow Corning) slab was then made by pouring the PDMS onto the top of the SU-8 mold, followed by degassing and curing the PDMS at 70 °C for 2 h. To fabricate the substrate with embedded Ag/AgCl, the 0.5 mm silver foil was bonded to a glass slide and was subsequently coated with SU-8

6002 (MicroChem, Newton, MA, USA). The insulating SU-8 layer was then subjected to a soft bake at 95 °C for 3 min and exposed to UV-light to pattern four openings for the Ag layer. After a post-bake (95 °C for 2 min) and development by SU-8 developer, four openings were created on the insulation layer. The insulated silver foil with four openings was immersed in AgCl solution, and DC power (5 V) was applied between the silver foil (positive electrode) and the silver rod (negative electrode) to facilitate the conversion of Ag to AgCl at the four openings. Next, the PDMS slab was punched to create inlet/outlet reservoirs and electrode holes for common electrodes. The PDMS slab was treated with air plasma (200 mTorr, 50 W, 50 s). The treated PDMS slab and SU-8 coated silver layer were immersed in 5% v/v aqueous APTES solution for 20 min, washed with DI water, and dried. Under a microscope, the two parts (PDMS slab and the substrate with four openings) were aligned and placed on a hotplate at 90 °C for over 30 min to form an irreversible bond. The nominal dimensions of the sensing channels were 120 μm (length), 40 μm (width), and 35 μm (height). The dimensions of the sensing channels were measured by a surface profilometer (Dektak 150, Veeco Instrument, Plainview, NY, USA): 122.4 ± 0.5 μm (length), 40.8 ± 0.3 μm (width), and 35.7 ± 0.2 μm (height).

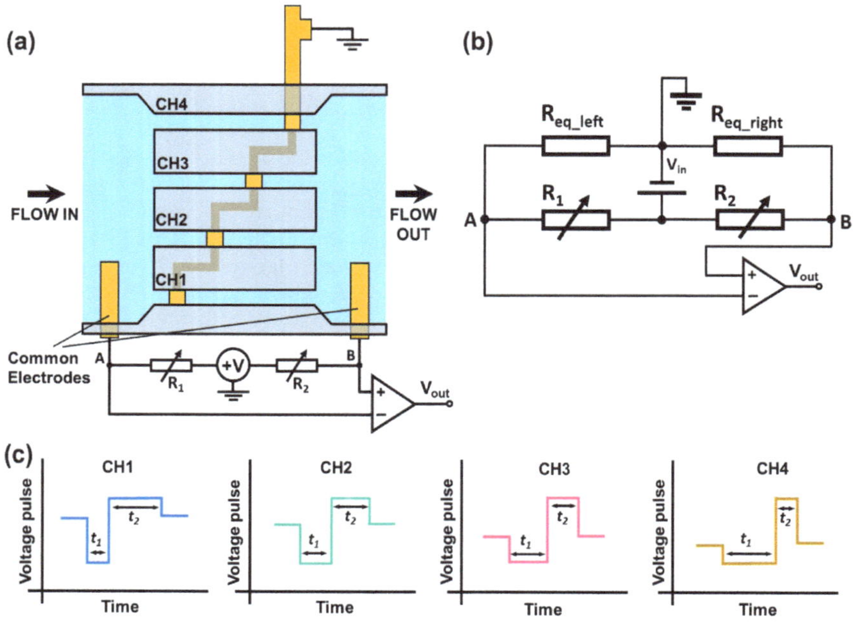

Figure 2. Schematic of the bipolar pulse-width multiplexing RPS device. (**a**) Illustration of the measurement set-up. (**b**) Diagram of a circuit that measures the resistive pulses generated by particles transiting through the sensing channels. (**c**) Illustration of bipolar voltage pulses/waveforms when one particle transits through different sensing channels (Blue: channel 1, Green: channel 2, Pink: channel 3, Yellow: channel 4). The bipolar pulses generated in different channels have different pulse widths t_1 and t_2.

3. Results

15 μm polystyrene particles suspended in DI water were used to generate the specific voltage template waveforms from each sensing channel. Typical waveforms generated in all sensing channels were recorded separately, as shown in Figure 3a. These waveforms were used as template waveforms to correlate with the combined output signal of the device. When a particle passes one specific sensing channel (e.g., channel 1), the combined signal must contain a component that is highly similar to the waveform of this channel

(e.g., Channel 1). Hence a large correlation coefficient (maximum value > 0.4) would be generated. In other words, a high correlation coefficient indicates a high similarity between the combined signal and the template waveform due to the passage of a particle through this sensing channel. Visa versa, a lower correlation coefficient (maximum value < 0.4) indicates low similarity between the two signals, or no particle passage through this specific sensing channel. Thus, by correlating the combined output signal with the four standard template waveforms, the presence of particles in each channel can be determined based on the maximum correlation coefficient (Figure 3b). The template waveform shows a higher correlation coefficient with itself than other template waveforms. Also, the starting and ending time of each pulse (representing the entry and exit of each particle) can be determined using the time lag [23]. Prior studies indicated that the correlation coefficient of less than 0.4 suggests a weak similarity between two signals [24,25]. Thus, here we set correlation coefficient of 0.4 as a threshold to judge whether the detected signal contains a specific waveform. In addition, a smaller particle passing through a sensing channel would generate a highly similar waveform but with a reduced magnitude.

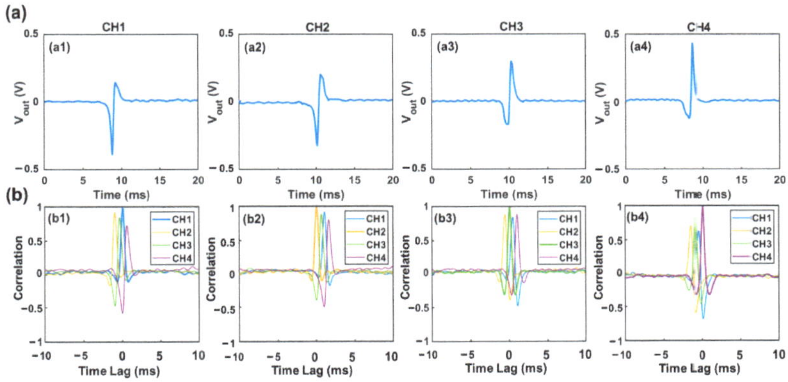

Figure 3. Illustration of correlation analysis (**a**) Voltage template waveforms generated by a 15 µm particle transiting through each sensing channel. (**b**) Correlation analysis between the combined voltage output and the four different template waveforms. The maximum correlation coefficient can be used to identify the presence of a particle in a specific sensing channel. The starting time of the signal can be determined based on the time lag it exhibits.

The correlation coefficient can be calculated by $r_{xy} = \frac{\sum (x_i - x_m)\cdot(y_i - y_m)}{\sqrt{(x_i - x_m)^2\cdot(y_i - y_m)^2}}$, where x_i, y_i represents each set of data (i.e., the combined output signal and a specific template waveform), and x_m, y_m is the average of respective data value [25,26]. A maximum correlation coefficient > 0.4 between two sets of data indicates the combined signal contains a specific waveform (or the presence of a particle in a specific channel where the waveform is generated). To extract the desired waveform from the combined signal, an iterative cancellation scheme is employed, as illustrated in Figure 4. The sequence is as follows: After the correlation analysis, if the maximum correlation coefficient is positive and larger than 0.4 (indicating the combined signal contains the corresponding waveform), a fraction (represented by 'A', e.g., 0.8×) of the template waveform is subtracted. The remaining signal is then correlated with all template waveforms in the next round. During the correlation-cancellation procedure, if the maximum correlation coefficient is larger than 0.4 but negative, it indicates the template signal was over-subtracted. Hence a smaller fraction of subtraction (e.g., subtraction of $\frac{1}{2}A\times$ waveform) should be used. The procedure is repeated until all correlation coefficients with all template waveforms are less than 0.4. The total magnitude of all subtractions can be used to calculate the size of particles, as the particle volume is proportional to the magnitude of the detected signal [9,27].

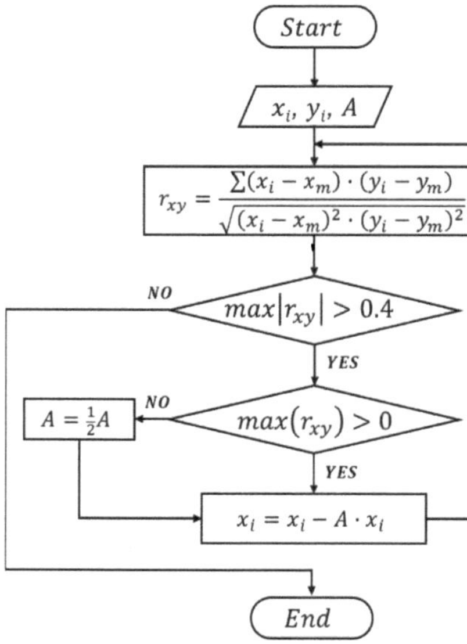

Figure 4. Flowchart of the iterative cancellation procedure to demodulate the detected signal (combined signal from all sensing channels).

3.1. Validation of Bipolar Pulse-Width Multiplexing Method

To demonstrate the principle of the bipolar pulse-width multiplexing method, we analyzed a case where multiple particles were present in different sensing channels, as shown in Figure 5. The particle solution consisted of two different sized polystyrene microparticles (15 ± 0.2 μm and 20 ± 0.3 μm). For validation purpose, a high-speed camera (MU043M-FL, United Scope LLC, Irvine, CA, USA) was employed to record the particle transits through the sensing channel. Template waveforms were generated by a 15 μm microparticle passing through each individual sensing channel, as depicted in Figure 3a. Correlation analysis was first performed between the combined signal and the template waveforms to identify the sensing channel through which the particle had passed.

Figure 5 illustrates the demodulation procedures when multiple particles were present in different sensing channels (Figure 5(a1)). The recorded electrical signal is shown in Figure 5(b1). First, the recorded signal was correlated with all template waveforms. The maximum correlation coefficient ($max(r_{xy}) > 0.4$) occurred in channel 3 (Figure 5(a2)), indicating the presence of one particle in this channel. Subsequently, a 0.8× waveform of channel 3 was subtracted from the combined signal, as depicted in Figure 5(b2). The correlation analysis was then repeated with all template waveforms, and the maximum correlation coefficient ($max(r_{xy}) > 0.4$) was found in channel 4 (Figure 5(a3)). Consequently, a 0.8× waveform of channel 4 was subtracted from the remaining signal (Figure 5(b3)). This interactive cancellation procedure was continued. An 0.8× subtraction of waveform 3 (Figure 5(b4)), an 0.8× subtraction of waveform 4 (Figure 5(b5)), an 0.8× subtraction of waveform 2 (Figure 5(b6)), an 0.8× subtraction of waveform 3 (Figure 5(b7)), and an 0.8× subtraction of waveform 4 (Figure 5(b8)) were performed one by one based on the occurrence and magnitude, respectively, of the maximum correlation coefficients (Figure 5(a4–a8)). After this step, the maximum correlation coefficient occurred in channel 3 once again (Figure 5(a9)). Hence, 0.8× and 0.4× subtractions of waveform 3 were tried (Figure 5(b9,b10)); however, both generated a large negative correlation coefficient between the remaining signal and the template waveforms, indicating an over-subtraction had occurred (Figure 5(a10,a11)). As a result, a 0.2× waveform 3 was subtracted (Figure 5(b11)). The remaining signal was then correlated with all template waveforms, and the maximum correlation coefficient ($max(r_{xy}) > 0.4$) was observed in channel 2 (Figure 5(a12)). Similar to before, 0.8×

and 0.4× subtractions were tried (Figure 5(b12,b13)); highly negative maximum correlation coefficients were observed (Figure 5(a13,a14)). Hence, an 0.2× waveform 2 was subtracted (Figure 5(b14)). After this, the maximum correlation coefficient became positive and still occurred in channel 2 (Figure 5(a15)). Therefore, a 0.1× waveform 2 was subtracted (Figure 5(b15)). After the subtraction, the maximum correlation coefficient (>0.4) still occurred in channel 2 (Figure 5(a16)). An 0.05× waveform 2 was subtracted (Figure 5(b16)). Afterwards, the maximum correlation coefficient was positive but below 0.4 (Figure 5(a17)); the interactive cancellation was completed. From the overall subtracted magnitudes of channel 2 (1.15×), channel 3 (2.6×), and channel 4 (2.4×), the particles' sizes were estimated to be 15.7 µm, 20.6 µm, and 20.1 µm, respectively. The entire interactive cancellation procedure was automatic with a MatLab code and took about 1 ms. It is important to note that although an initial 0.8× subtraction was employed for the magnitude of the iterative cancellation process, aiming to minimize the interaction steps and attain a decent resolution for particle sizing, we also conducted the interactive cancellations with both 0.5× and 0.3× initial subtractions. The obtained particle counts and sizes were nearly identical to those obtained with the 0.8× subtraction.

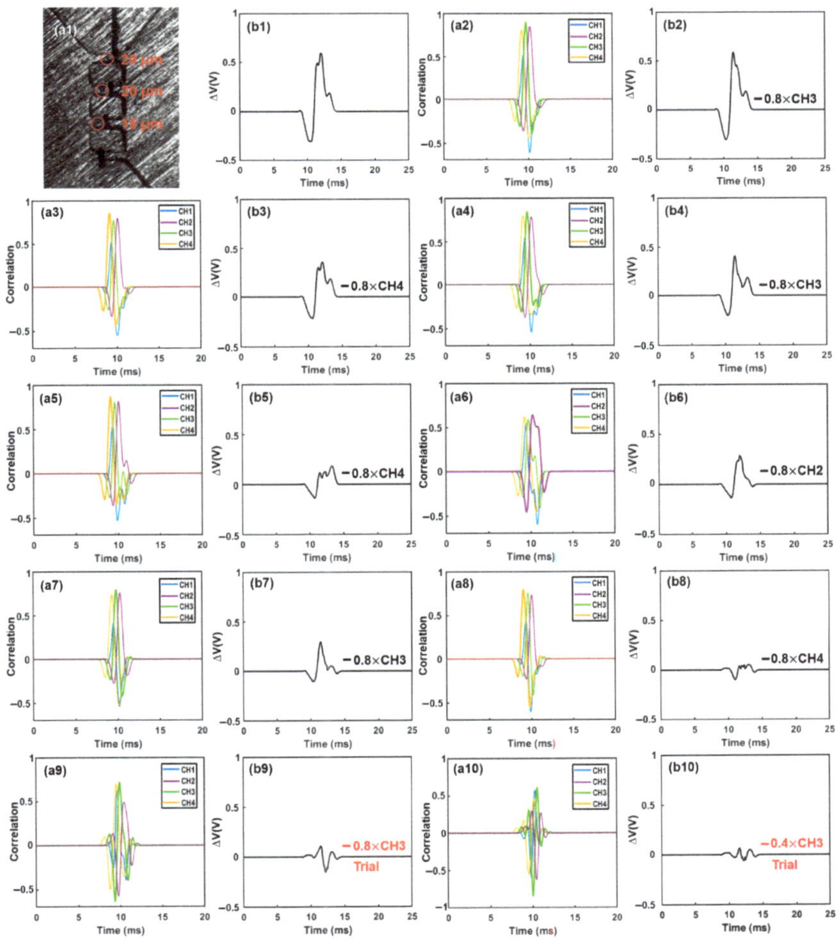

Figure 5. *Cont.*

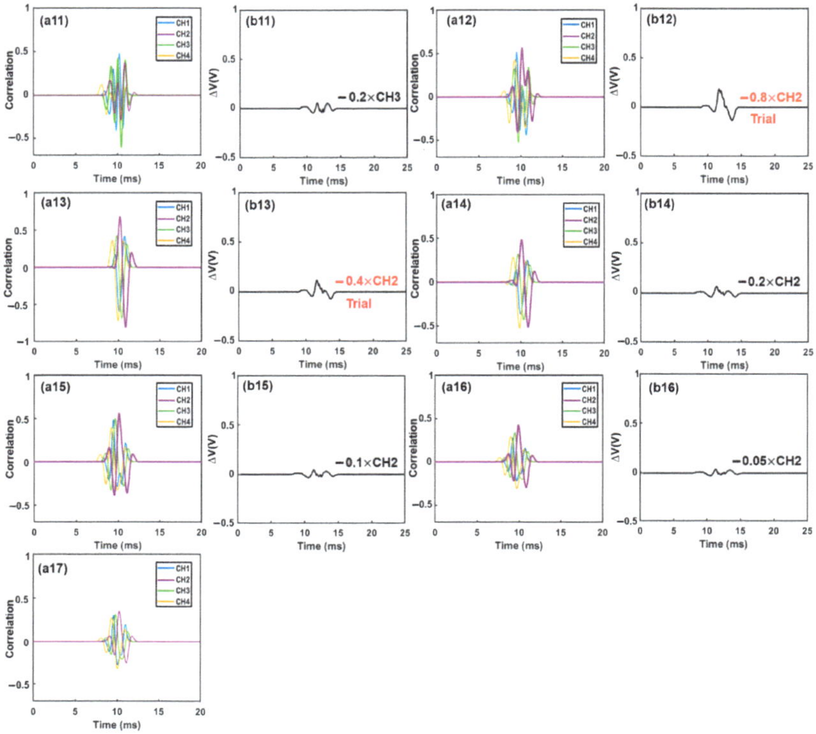

Figure 5. Iterative cancellation procedures for demodulation of the combined electrical signal. The combined signal was generated by three differently sized microparticles. The desired waveform was determined based on the correlation coefficient and then extracted from the remaining signal with a specific amplitude (0.8×, 0.4×, 0.2×, 0.1×, etc.). Initially, subtraction of a waveform with an amplitude of 0.8× was performed. If the correlation coefficient reached a highly negative value (indicating an over-subtraction occurred), subsequent subtractions of smaller amplitudes (e.g., 0.4×, 0.2×, 0.1×, 0.05×) were used to replace the 0.8× subtraction. Subtractions marked in red were trial subtractions that resulted in over-subtraction (highly negative correlation coefficient) and were not executed. The actual subtractions, marked in black, were performed on the combined signal ((**a1**)–(**a17**): correlation analysis between the combined signal or remaining signal and the four template waveforms. (**b1**): combined signal. (**b2**)–(**b16**): remaining signals after each subtraction).

3.2. Demonstration of Sizing and Counting Accuracy

To evaluate the sizing and counting accuracy of the microfluidic sensor, mixtures of particles with different sizes were used. Specifically, solutions containing two differently sized polystyrene microparticles (15 μm and 20 μm) at varying concentrations were prepared by dilution and loaded to the device. While the particle solution flowed through the device, the output signals were recorded across the pair of common electrodes. The output signal was demodulated using the interactive cancellation procedure shown in Figure 4.

The resulting particle concentrations and sizes of measurement are depicted in Figure 6 (indicated by vertical lines). For sample 1, the measured concentrations were 27.5/mL ± 1.5/mL (for 20 μm) and 50.5/mL ± 3.1/mL (for 15 μm particles); for sample 2, the measured concentrations were 92.4/mL ± 5.3/mL (for 20 μm particles) and 149.6/mL ± 8.5/mL (for 15 μm particles); for sample 3, the measured concentrations were 163.7/mL ± 9.6/mL (for 20 μm particles) and 284.6/mL ± 13.2/mL (for 15 μm particles). For comparison, actual concentrations of the three samples were measured using an AccuSizer^{TM} 780 (optical particle sizer) and are shown in Figure 6a. Encouragingly, the measured concentrations aligned well with the actual concentrations. Figure 6b shows a quantitative comparison between the measurement results and the estimated concentrations. The maximum difference is 5.9% (for 20 μm particles) from sample 3 and 6.1% (for 15 μm particles) from sample 1. The two sets of concentrations matched reasonably well with each other. The small errors in counting may have been caused by the settlement of a small number of cells, considering the polystyrene particles have a slightly higher density than the medium solution.

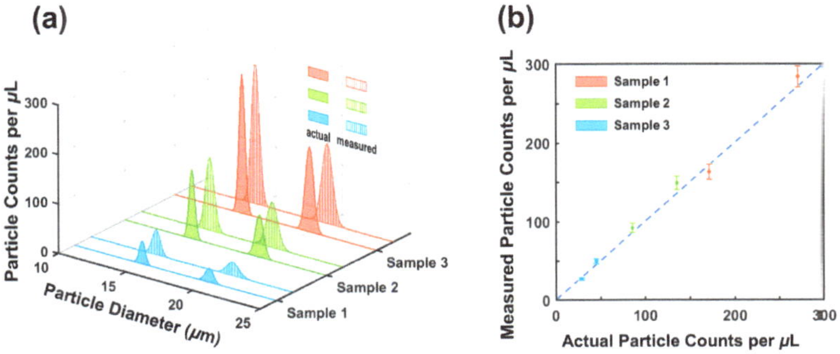

Figure 6. (**a**) Comparison of measured and actual concentrations and diameters of microparticles in the mixed solutions. (**b**) Comparison of measured particle concentrations vs. actual particle concentrations. RPS measurements are represented by small solid circles with error bars. Actual particle concentrations and sizes were measured by an AccuSizer^{TM} 780 (optical particle sizer).

The sizes of particles were also obtained from the interactive cancellation. In sample 1, the measured particle sizes were 20.5 μm ± 0.4 μm (for 20 μm particles) and 14.8 μm ± 0.5 μm (for 15 μm polystyrene particles); in sample 2, the measured particle sizes were 19.7 μm ± 0.6 μm and 15.1 μm ± 0.4 μm, respectively; in sample 3, the measured sizes were 20.1 μm ± 0.7 μm and 14.7 μm ± 0.5 μm, respectively. The average of the three sizes of measurements was 20.1± 0.6 μm. We also measured the particle sizes using a particle analyzer (AccuSizer^{TM} 780). The particle sizes were measured to be 14.95 μm ± 0.57 μm and 19.6 μm ± 0.74 μm. The RPS measurement and the AccuSizer measurement were in good agreement. The difference was 2.6%. Electrical noise originating from flow fluctuations may have caused errors in the size measurements, as the particle sizes were correlated to the magnitude of the output signal.

The above results demonstrated the capability of bipolar pulse-width multiplexed RPS to accurately measure the sizes and counts of microparticles in a mixed solution. The sizing of particles was accomplished through the interactive cancellation process, which relied on correlation analysis (maximum correlation coefficient). The correlation analysis indicated the time the maximum correlation coefficient occurred. This information makes it possible to determine the start and ending points of each pulse [28], even if several particles are present in different sensing channels at the same time.

Finally, we used the sensor to count HeLa cells. HeLa cells are human negroid cervix epithelioid carcinoma cells. These cells play a crucial role in studying the propagation status of cells and advancing our understanding of cancer and viral infections [29,30]. HeLa cells were cultured in DPBS solution (see details in Section 2.1). One HeLa cell sample was loaded to the RPS sensor. The concentration and the size of the HeLa cells were measured to be 0.98×10^5 mL^{-1} and 15.9 ± 3.9 μm (sizes ranging from 8.5 μm to 25.4 μm) using an AccuSizer^{TM} 780. Output signals were collected, which were subsequently demultiplexed by interactive cancellation. Figure 7b shows the cell analysis results in the four sensing channels. The measured concentration of the HeLa cells was approximately $(0.93 \pm 0.12) \times 10^5$ mL^{-1} (obtained from three measurements), which matched with the actual concentration reasonably well. The difference is likely because a small portion of cells settled on the substrate of the channels or attached to the channel walls. This method achieved an accuracy of 94.9% for HeLa cell counting. Compared to the counting accuracies of other electrical multiplexing methods, including the frequency division multiplexing method (i.e., 88%) [22] and CODES method (i.e., 96.15%) [20], this method exhibited a higher or comparable counting accuracy. In terms of particle sizing, this method had a 97.4% accuracy, which is higher than that of the frequency division multiplexing method (i.e., 94.8%). The sizes of the HeLa cells were measured to be 16.6 $\pm$ 4.6 μm (ranging from 8.9 μm to 24.5 μm), which were in good agreement with the measured values from the AccuSizer^{TM} 780.

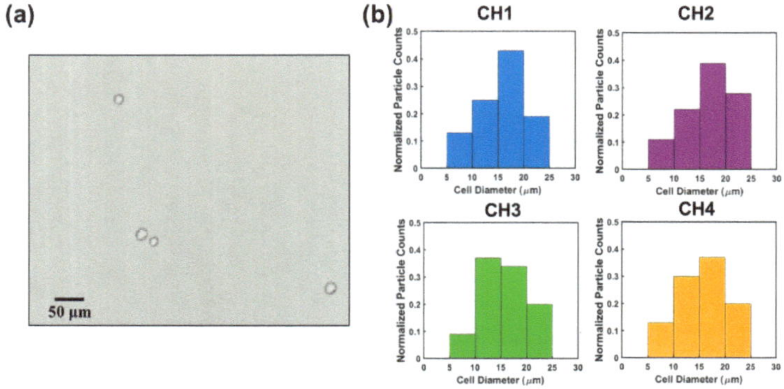

Figure 7. (**a**) HeLa cell image under a microscope. (**b**) Histograms of cell diameter distribution measured across channel 1, channel 2, channel 3, and channel 4.

The bipolar pulse-width multiplexing method enables high-throughput microparticle counting and sizing without the need for complex electrode designs or intricate geometry configurations of sensing channels. Only one DC power source, one pair of common electrodes, and one central electrode are required to encode the detected signal. This not only reduces the complexity of the detection electronics but also decreases the data size. While a four-channel device is presented for demonstration of high-throughput particle analysis, the throughput can be further improved by incorporation of additional sensing channels. This method can be extended to nanoscale particle counting, including proteins, nucleic acids, and viruses, by fabricating openings at different positions to expose the central electrode inside the sensing channels. Compared to other multiplexing methods, this fabrication process is simple. It can be extended to measure nanoscale particles by fabricating smaller sensing channels with openings at different locations. Note that the silver coil in the multiplexed RPS can be replaced by sputtering a thin silver film. In addition, this multiplexed RPS can be combined with many well-established antibody- or aptamer-based bio-recognition methods to detect various of micro and nano bio-objects including cells [31,32] and biomolecules [33–35] with high throughput and high specificity.

4. Conclusions

We developed a bipolar pulse-width multiplexing-based resistive pulse sensor capable of high-throughput counting and sizing of microparticles through the utilization of multiple parallel sensing channels. The unique bipolar pulse-width multiplexing facilitated the encoding of electrical signals generated by the passage of microparticles, while only requiring one DC source. The combined signal can be demodulated using correlation analysis and an iterative cancellation scheme. This multiplexed RPS was demonstrated using mixtures of differently sized microparticles with varying concentrations. The RPS sensor can predict the sizes and concentrations of standard polystyrene particles accurately with errors of 2.6% and 6.1% in sizing and counting, respectively. The sensor was demonstrated to accurately count HeLa cells, with an error of 5.6% in concentration. Due to the simple fabrication process, smaller sensing channels can be fabricated for detection and analysis of nanoscale analytes. With its high throughput and accuracy, this RPS sensor holds promise for the rapid analysis of micro and nano objects including biomolecules, viruses, and bacteria, especially in resource-limited environments.

Author Contributions: Writing, original draft preparation, R.X.; experimental planning, J.Z. and G.Z.; testing and analysis, R.X., L.O. and H.C.; visualization, L.O.; writing, review and editing, J.Z. and H.C.; cell culture, R.S.; supervision and funding acquisition, J.Z. and G.Z. All authors have read and agreed to the published version of the manuscript.

Funding: This work was supported by the National Science Foundation of the USA, under award numbers DBI 1911526 and ECCS 1905786, and ECCS 2232940.

Institutional Review Board Statement: Not applicable.

Informed Consent Statement: Not applicable.

Data Availability Statement: Data will be made available on request.

Conflicts of Interest: The authors declare no conflict of interest.

References

1. Holzinger, M.; Le Goff, A.; Cosnier, S. Nanomaterials for biosensing applications: A review. *Front. Chem.* **2014**, *2*, 63. [CrossRef] [PubMed]
2. Tansil, N.C.; Gao, Z. Nanoparticles in biomolecular detection. *Nano Today* **2006**, *1*, 28–37. [CrossRef]
3. Ventura-Aguilar, R.I.; Bautista-Baños, S.; Mendoza-Acevedo, S.; Bosquez-Molina, E. Nanomaterials for designing biosensors to detect fungi and bacteria related to food safety of agricultural products. *Postharvest Biol. Technol.* **2023**, *195*, 112116. [CrossRef]
4. Pividori, M.I.; Alegret, S. Micro and nanoparticles in biosensing systems for food safety and environmental monitoring. An example of converging technologies. *Microchim. Acta* **2010**, *170*, 227–242. [CrossRef]
5. Xianyu, Y.; Wang, Q.; Chen, Y. Magnetic particles-enabled biosensors for point-of-care testing. *TrAC Trends Anal. Chem.* **2018**, *106*, 213–224. [CrossRef]
6. Kozak, D.; Anderson, W.; Vogel, R.; Trau, M. Advances in resistive pulse sensors: Devices bridging the void between molecular and microscopic detection. *Nano Today* **2011**, *6*, 531–545. [CrossRef]
7. Weatherall, E.; Willmott, G.R. Applications of tunable resistive pulse sensing. *Analyst* **2015**, *140*, 3318–3334. [CrossRef]
8. DeBlois, R.W.; Bean, C.P. Counting and Sizing of Submicron Particles by the Resistive Pulse Technique. *Rev. Sci. Instrum.* **1970**, *41*, 909–916. [CrossRef]
9. Song, Y.; Zhang, J.; Li, D. Microfluidic and Nanofluidic Resistive Pulse Sensing: A Review. *Micromachines* **2017**, *8*, 204. [CrossRef]
10. Yang, L.; Yamamoto, T. Quantification of Virus Particles Using Nanopore-Based Resistive-Pulse Sensing Techniques. *Front. Microbiol.* **2016**, *7*, 1500. [CrossRef]
11. Vaclavek, T.; Prikryl, J.; Foret, F. Resistive pulse sensing as particle counting and sizing method in microfluidic systems: Designs and applications review. *J. Sep. Sci.* **2019**, *42*, 445–457. [CrossRef] [PubMed]
12. Zhou, T.; Song, Y.; Yuan, Y.; Li, D. A novel microfluidic resistive pulse sensor with multiple voltage input channels and a side sensing gate for particle and cell detection. *Anal. Chim. Acta* **2019**, *1052*, 113–123. [CrossRef] [PubMed]
13. Pan, R.; Hu, K.; Jia, R.; Rotenberg, S.A.; Jiang, D.; Mirkin, M.V. Resistive-Pulse Sensing Inside Single Living Cells. *J. Am. Chem. Soc.* **2020**, *142*, 5778–5784. [CrossRef] [PubMed]
14. Luo, L.; German, S.R.; Lan, W.-J.; Holden, D.A.; Mega, T.L.; White, H.S. Resistive-Pulse Analysis of Nanoparticles. *Annu. Rev. Anal. Chem.* **2014**, *7*, 513–535. [CrossRef]
15. Sikora, A.; Shard, A.G.; Minelli, C. Size and Zeta-Potential Measurement of Silica Nanoparticles in Serum Using Tunable Resistive Pulse Sensing. *Langmuir* **2016**, *32*, 2216–2224. [CrossRef]
16. Blundell, E.L.C.J.; Mayne, L.J.; Billinge, E.R.; Platt, M. Emergence of tunable resistive pulse sensing as a biosensor. *Anal. Methods* **2015**, *7*, 7055–7066. [CrossRef]
17. Sivakumaran, M.; Platt, M. Tunable resistive pulse sensing: Potential applications in nanomedicine. *Nanomedicine* **2016**, *11*, 2197–2214. [CrossRef]
18. Song, Y.; Yang, J.; Pan, X.; Li, D. High-throughput and sensitive particle counting by a novel microfluidic differential resistive pulse sensor with multidetecting channels and a common reference channel. *Electrophoresis* **2015**, *36*, 495–501. [CrossRef]
19. Liu, R.; Waheed, W.; Wang, N.; Civelekoglu, O.; Boya, M.; Chu, C.-H.; Sarioglu, A.F. Design and modeling of electrode networks for code-division multiplexed resistive pulse sensing in microfluidic devices. *Lab Chip* **2017**, *17*, 2650–2666. [CrossRef]
20. Liu, R.; Wang, N.; Kamili, F.; Sarioglu, A.F. Microfluidic CODES: A scalable multiplexed electronic sensor for orthogonal detection of particles in microfluidic channels. *Lab Chip* **2016**, *16*, 1350–1357. [CrossRef]
21. Xu, R.; Ouyang, L.; Shaik, R.; Zhang, G.; Zhe, J. Multiplexed resistive pulse sensor based on geometry modulation for high-throughput microparticle counting. *Sens. Actuators Rep.* **2023**, *5*, 100140. [CrossRef]

22. Jagtiani, A.V.; Carletta, J.; Zhe, J. A microfluidic multichannel resistive pulse sensor using frequency division multiplexing for high throughput counting of micro particles. *J. Micromech. Microeng.* **2011**, *21*, 065004. [CrossRef]
23. Dinan, E.; Jabbari, B. Spreading codes for direct sequence CDMA and wideband CDMA cellular networks. *IEEE Commun. Mag.* **1998**, *36*, 48–54. [CrossRef]
24. Schober, P.; Boer, C.; Schwarte, L.A. Correlation Coefficients: Appropriate Use and Interpretation. *Anesth. Analg.* **2018**, *126*, 1763–1768. [CrossRef] [PubMed]
25. Akoglu, H. User's guide to correlation coefficients. *Turk. J. Emerg. Med.* **2018**, *18*, 91–93. [CrossRef]
26. Mukaka, M.M. Statistics corner: A guide to appropriate use of correlation coefficient in medical research. *Malawi Med. J.* **2012**, *24*, 69–71.
27. Wu, X.; Chon, C.H.; Wang, Y.-N.; Kang, Y.; Li, D. Simultaneous particle counting and detecting on a chip. *Lab A Chip* **2008**, *8*, 1943–1949. [CrossRef]
28. Shen, C. Analysis of detrended time-lagged cross-correlation between two nonstationary time series. *Phys. Lett. A* **2015**, *379*, 680–687. [CrossRef]
29. Strauss, N.; Hendee, E.D. The effect of diphtheria toxin on the metabolism of Hela cells. *J. Exp. Med.* **1959**, *109*, 145–163. [CrossRef]
30. Girardi, A.J.; Mc Michael, H., Jr.; Henle, W. The use of HeLa cells in suspension for the quantitative study of virus propagation. *Virology* **1956**, *2*, 532–544. [CrossRef]
31. Saleem, A.; Husheem, M.; Härkönen, P.; Pihlaja, K. Inhibition of cancer cell growth by crude extract and the phenolics of *Terminalia chebula* retz. fruit. *J. Ethnopharmacol.* **2002**, *81*, 327–336. [CrossRef] [PubMed]
32. Xu, R.; Ouyang, L.; Chen, H.; Zhang, G.; Zhe, J. Recent Advances in Biomolecular Detection Based on Aptamers and Nanoparticles. *Biosensors* **2023**, *13*, 474. [CrossRef] [PubMed]
33. Xu, R.; Abune, L.; Davis, B.; Ouyang, L.; Zhang, G.; Wang, Y.; Zhe, J. Ultrasensitive detection of small biomolecules using aptamer-based molecular recognition and nanoparticle counting. *Biosens. Bioelectron.* **2022**, *203*, 114023. [CrossRef] [PubMed]
34. Billinge, E.R.; Broom, M.; Platt, M. Monitoring Aptamer–Protein Interactions Using Tunable Resistive Pulse Sensing. *Anal. Chem.* **2014**, *86*, 1030–1037. [CrossRef] [PubMed]
35. Billinge, E.R.; Platt, M. Multiplexed, label-free detection of biomarkers using aptamers and Tunable Resistive Pulse Sensing (AptaTRPS). *Biosens. Bioelectron.* **2015**, *68*, 741–748. [CrossRef] [PubMed]

biosensors

Article

Detection of Glutamate Decarboxylase Antibodies and Simultaneous Multi-Molecular Translocation Exploration by Glass Nanopores

Chongxin Tao [1,†], Yun Bai [1,†], Jiang Chen [1], Jing Lu [2], Yan Bi [2,*] and Jian Li [1,*]

1 Key Laboratory of DGHD, MOE, School of Life Science and Technology, Southeast University, Nanjing 210096, China
2 Department of Endocrinology, Nanjing Drum Tower Hospital, Affiliated Hospital of Medical School, Nanjing University, Nanjing 210008, China
* Correspondence: biyan@nju.edu.cn (Y.B.); jianli2014@seu.edu.cn (J.L.)
† These authors contributed equally to this work.

Abstract: Glutamic acid decarboxylase antibody (GADAb) has emerged as a significant biomarker for clinical diagnosis and prognosis in type 1 diabetes (T1D). In this study, we investigated the potential utilization of glass capillary solid-state nanopores as a cost-effective and easily preparable platform for the detection of individual antigens, antibodies, and antigen-antibody complexes without necessitating any modifications to the nanopores. Our findings revealed notable characteristic variations in the translocation events of glutamic acid decarboxylase (GAD65) through nanopores under different voltage conditions, discovered that anomalous phenomenon of protein translocation events increasing with voltage may potentially be caused by the crowding of multiple proteins in the nanopores, and demonstrated that there are multiple components in the polyclonal antibodies (GADAb-poly). Furthermore, we achieved successful differentiation between GAD65, GADAb, and GADAb-GAD65 complexes. These results offer promising prospects for the development of a rapid and reliable GADAb detection method, which holds the potential to be applied in patient serum samples, thereby facilitating a label-free, cost-effective, and early diagnosis of type I diabetes.

Keywords: glass nanopore; single molecule detection; glutamate decarboxylase antibody; multi-molecular translocation; type I diabetes

Citation: Tao, C.; Bai, Y.; Chen, J.; Lu, J.; Bi, Y.; Li, J. Detection of Glutamate Decarboxylase Antibodies and Simultaneous Multi-Molecular Translocation Exploration by Glass Nanopores. *Biosensors* **2024**, *14*, 255. https://doi.org/10.3390/bios14050255

Received: 21 March 2024
Revised: 9 May 2024
Accepted: 14 May 2024
Published: 17 May 2024

1. Introduction

Diabetes mellitus is a very common endocrine disease in clinical practice, which is mainly divided into type I diabetes (T1DM), characterized by pancreatic β-cell damage, and type II diabetes (T2DM), characterized by insulin resistance [1]. Due to the large differences in the disease process and treatment methods between type I and II diabetes, early diagnosis and classification are of great significance for their treatment and prognosis. As an autoimmune disease, the diagnosis and classification of type I diabetes depend on a series of autoimmune antibodies against the secreted proteins of pancreatic β-cells. Among them, the autoimmune antibodies of glutamic acid decarboxylase (GAD65), are one of the most sensitive and specific clinical markers, which have been used in combination with several other autoimmune antibodies (like ICA and IAA) as the diagnostic criteria for type I diabetes in clinical practice [2,3]. Currently, the detection of GADAb in clinical practice mainly relies on several traditional methods, including ELISA [4,5], RIA [6], and immunofluorescence [7], which exhibit excellent reliability and are widely used for the quantitative detection of specific antigens and antibodies. Apart from these gold standard methods, single-molecule detection technology has gradually become a hotspot in the field of biological analysis due to its high sensitivity, as well as high-throughput advantages. However, the reported single-molecule immunofluorescence detection and single-molecule electrochemical immunoassay methods still rely on various biomarkers [8,9], which limits the promotion of single-molecule immunoassay technology in applications.

Nanopore technology provides a label-free, high-throughput single-molecule detection method at the nanometer scale [10–12]. Its detection principle is based on the concept of the Coulter counter [13]. When a stable external electric field is applied on both sides of the nanopore, the ions in the electrolyte on both sides of the nanopore move directionally and produce a weak but relatively stable ion current. When charged analyte biomolecules pass through the nanopore under the driving force of the electric field, they will interfere with the ion current and produce detectable pulse signals. According to the characteristics of signal amplitude, duration, and event frequency, it can be inferred that the detected biomolecules have certain structural, size, and charge states [14–17].

In previous studies, biological nanopores have been widely used in DNA sequencing [18], microRNA detection [19], and analysis of oligopeptides and proteins [20]. Research on antigen-antibody interactions has also been reported [21], but the lower stability and fixed pore size limit the development of biological nanopores in practical immunoassay applications. At the same time, solid-state nanopores have more stable physicochemical properties and adjustable sizes compared to biological nanopores, which greatly expand the application potential of solid-state nanopores in the field of single-molecule detection [22,23]. Glass capillary solid-state nanopores, as members of solid-state nanopores, have advantages of high stiffness, low cost, and reproducible size and shape. Their related applications in protein recognition have been reported many times [24–26]. In these applications, the differences in protein size and charge state are the keys to distinguishing them.

Therefore, in this study, quartz glass nanopores were used to investigate the translocation behavior of glutamate decarboxylase (GAD65), glutamate decarboxylase antibodies (GADAb), and the antigen-antibody complex (GAD65-GADAb complex). Due to the differences in charge and molecular diameter among the three molecules, there are differences in the characteristics of translocation signals. By distinguishing the translocation characteristics of the immune complexes, it provides a potential application for the development of new single-molecule detection methods for GAD65 and GADAb in clinical practice, which is cheap, sensitive, and high-throughput.

2. Materials and Methods

2.1. Chemical and Reagents

GAD65 was purchased from Abcam (Cambridge, UK) and GADAb (both monoclonal and polyclonal) were purchased from Sigma-Aldrich (Shanghai, China). Tris-HCl (pH = 8.0) was produced by Solarbio (Beijing, China). Both KCl and ethanol were purchased from Sinopharm (Shanghai, China). Electrolyte for translocation through nanopore was 1 M KCl solution contain 10 mM Tris-HCl (pH = 8.0). All solutions were prepared with Milli-Q water (18 MΩ·cm resistivity) from a Millipore system (Merck, Darmstadt, Germany).

2.2. Device Construction

Two Ag/AgCl electrodes were inserted into the electrochemical cell filled with electrolyte (1 M KCl, 10 mM Tris-HCl, pH = 8.0). During the detection of protein molecules, the protein molecules were added to the electrolyte solution, with the glass nanopore fixed to the negative electrode and the positive electrode added outside the nanopore (as shown in Figure 1a). The other ends of the electrodes were connected to the Axopatch 700B patch clamp amplifier (Molecular Devices, San Jose, CA, USA). The signals were digitized using the Axon Digidata 1550A digital-to-analog converter (Molecular Devices, San Jose, CA, USA) and viewed using Clampfit software (Version 10.5.2.6).

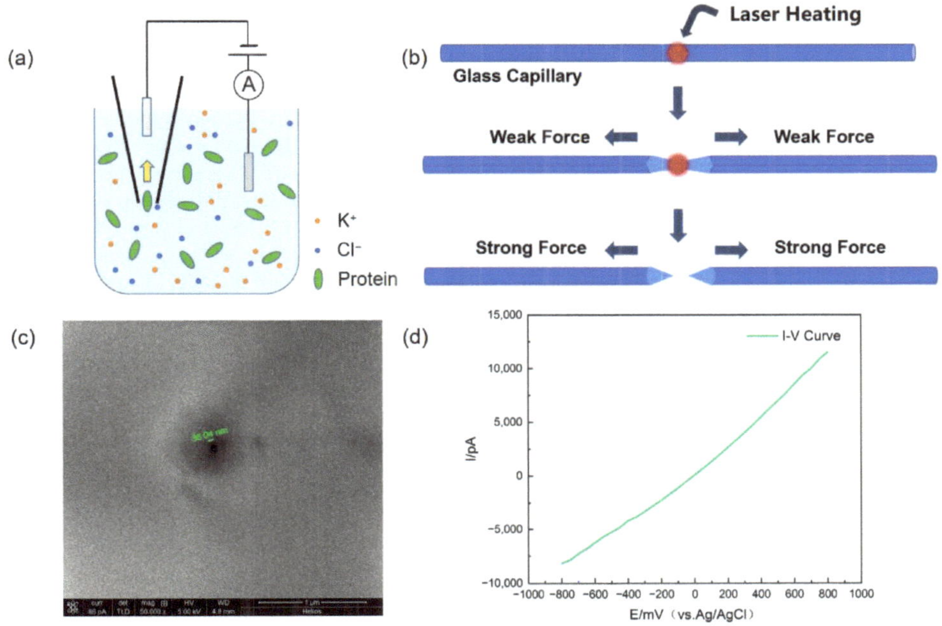

Figure 1. Fabrication of glass nanopores and device structure. (**a**) Schematic diagram of glass nanopore device for detecting proteins in ionic solutions. (**b**) Schematic of glass nanopore fabrication. (**c**) Characterization of glass nanopore size using scanning electron microscopy (SEM). (**d**) Characterization of glass nanopore size using I-V curve approximation.

2.3. Fabrication of Glass Nanopores and Characterization

Quartz glass capillaries used in the experiments (QF100-70-10, Sutter Instrument Co., Novato, CA, USA) have an outer diameter of 1 mm, an inner diameter of 0.7 mm, and a length of 7.5 cm. Before use, the glass capillaries were ultrasonically cleaned in ethanol and pure water for 15 min, and the liquid residue on the tube walls was removed using a nitrogen gas stream. The capillaries were then pulled into nanopores using a CO_2 laser capillary puller (model P-2000, Sutter Instruments Co., Novato, CA, USA) with the parameters shown in Table 1:

Table 1. Fabrication parameters of glass nanopore.

Heat	Filament	Velocity	Delay	Pull
760	4	29	140	168

The glass nanopore was electrochemically characterized in electrolyte (1 M KCl, 10 mM Tris-HCl, pH = 8.0) using cyclic voltammetry. The ion current was measured at intervals of 100 mV in the range of −500 mV to +500 mV, and an I-V scan curve was plotted accordingly. The conductance value of the nanopore was determined using the slope of the curve, and the diameter of the glass nanopore was calculated. Subsequently, the morphological characteristics and diameter of the glass nanopore were characterized using scanning electron microscopy (SEM) to verify the accuracy of the nanopore diameter estimation obtained using electrochemical methods.

2.4. Protein Translocation

The glutamate decarboxylase (GAD65, Abcam ab206646) and glutamate decarboxylase monoclonal antibody (GADAb-mono, Sigma-Aldrich G1166, derived from the GAD-6 hybridoma produced by the fusion of mouse myeloma cells and splenocytes from a mouse immunized with purified rat brain GAD) were diluted to 1540 pmol/L and 667 pmol/L, respectively, in electrolyte buffer (1 M KCl, 10 mM Tris-HCl, pH = 8.0). The glutamate decarboxylase polyclonal antibody (GADAb-poly, Sigma-Aldrich G5163, produced in rabbits using a synthetic peptide KDIDFLIEEIERLGQDL corresponding to the C-terminal region of GAD 67 of human origin as immunogen) was diluted to 10,000 times of its original concentration using the buffer. The diluted samples were injected into the glass nanopore using a microinjector, while the electrochemical cell was filled with electrolyte.

The ionic currents were measured with Ag/AgCl electrodes inserted in buffer solution and recorded with the amplifier Axopatch 200B (Molecular Devices, San Jose, CA, USA) in voltage-clamp mode using a low-pass Bessel filter of 5 kHz. The signals were digitized with DigiData 1550s digitizer at 100 kHz and viewed with Clampfit 10.2 software. The protein translocation events were recorded, and the event features were extracted using the Transalyzer analysis package (Version RC1b) based on Matlab [27]. The amplitude, duration time, and baseline data of each translocation event were imported into Origin software (Version 2023b) to generate frequency distribution histograms and fitted curves for the basic characteristics of the translocation events under various conditions. The different characteristics of the protein translocation events were investigated by varying the applied voltage.

2.5. Detection of Immune Complex

The antigen and antibody (monoclonal) proteins mentioned in the previous steps were mixed in equal volumes at their respective concentrations and incubated at room temperature for 30 min. The resulting immune complexes were injected into the glass nanopore using the microinjector. The translocation currents generated by the immune complexes passing through the nanopore were measured and recorded using the device described earlier. The amplitude size, duration time, and baseline data of each translocation event were plotted as translocation time and $\Delta I/I_0$ frequency distribution histograms. The average effective molecular diameter of the immune complexes was calculated based on the statistical results and compared with the molecular diameter of the monomer antigen or antibody to distinguish between antigens, antibodies, and their immune complexes formed by binding.

2.6. Statistical Analysis Methods

The general statistical analysis methods for translocation signals include plotting scatter plots of duration time versus signal amplitude, histograms of duration time–frequency distribution, and histograms of signal amplitude of blockage ($\Delta I/I_0$) frequency distribution. Specifically, the raw data are processed using the Transalyzer analysis package (Version RC1b) and then imported into the Origin software (Version 2023b) to generate these three types of statistical analysis graphs. In the histogram of frequency distribution, the average duration time and the amplitude of blockage ($\Delta I/I_0$) can be obtained by fitting a Gaussian curve and determining its mean value, representing the average state of all signals. The equation for fitting the Gaussian curve is as follows:

$$y = y_0 + \frac{A}{w\sqrt{\pi/2}} e^{-2\frac{(x-x_c)^2}{w^2}} \quad (1)$$

3. Results and Discussion

Through the laser-pulling instrument, glass nanopores with a diameter of 30 to 50 nm were obtained (Figure 1b). As the diameter affects the signal-to-noise ratio of the glass nanopore detection [28], the pore diameter was calculated before each detection by measuring the I-V curve of the nanopore electrolyte and calculating the resistance in the glass nanopore (Figure 1d). Due to the geometric and charge asymmetry of glass nanopores, a significant ion rectification phenomenon can be observed through the I-V curve [29], manifesting as a smaller current magnitude in the negative voltage region compared to the positive voltage region at the same voltage level. This phenomenon may be related to ion accumulation/dissipation [30,31] or electrophoretic capture of mobile ions [32]. The pore diameter of the glass nanopore can be approximately calculated using the following formula [33,34]:

$$R = \frac{\gamma \cot \frac{\theta}{2}}{\pi a} \tag{2}$$

In this case, R is the resistance value of the nanopore calculated from the measured I-V curve, θ is the tip angle of the conical nanopore (approximately 15° ± 3°), γ is the resistivity of the electrolyte, and a represents the radius of the nanopore. In theory, during the pulling process, a glass capillary with an inner diameter of 0.7 mm and a length of 7.5 cm will be pulled into a pair of glass nanopores with the same pore diameter. The size of the nanopore is characterized using scanning electron microscopy (SEM), and the exact nanopore diameter can be obtained (Figure 1c). By comparing the calculated and measured values of the glass nanopores pulled from the same glass capillary using both computational and scanning electron microscopy methods, it can be observed that the estimated values were in good agreement with the SEM imaging results, indicating that the computational method can provide accurate measurements of the nanopore diameter.

The glass nanopore was filled with electrolyte solution (1 M KCl, 10 mM Tris-HCl, pH = 8.0) in both the nanopore and the electrochemical cell. A constant voltage was applied to detect the signals generated by the passage of GAD65 and GADAb through the nanopore. When GAD65 or GADAb were added to the electrolyte solution, due to the negative charge carried by the antigen or antibody molecules, the electrophoretic force acting on the protein molecules was opposite to the direction of the electric field. On the other hand, the negatively charged inner wall of the glass nanopore attracts cations in the solution, forming a cation layer on the nanopore wall, with the direction of the electroosmotic flow opposite to the electrophoretic force. In this case, the electrophoretic force is stronger than the electroosmotic flow, resulting in the protein molecules moving in the same direction as the electrophoresis. When the direction of the electric field is from the negative electrode to the positive electrode (represented by a negative voltage), the proteins flow from outside the nanopore into the nanopore (Figure 1a).

The three-dimensional structure of GAD65 shows that the dimension of GAD65 is about 12.0 × 9.9 × 7.8 nm (Figure 2a) [35]. When different voltages were applied across the two ends of the nanopore, proteins translocated through the nanopore, causing changes in the current and generating translocation signals. The translocation signals under different voltages showed different shapes (Figure 2b). As can be seen from the current graphs, when the voltage applied to both ends of the nanopore was increased from −300 mV to −500 mV, both the baseline current and the amplitude of current changes due to molecular translocation increased. It is known that the translocation time of a molecule through a nanopore is related to the length of the molecule, while the ratio of blocked current to baseline current is related to the cross-sectional area of the molecule [36]. In the three voltage conditions tested, all typical translocation signals within 9 s were selected. It was found that the duration time increased with voltage (Figure 2c), which is different from the characteristics of faster translocation speed and shorter translocation time at higher voltages within a low voltage range [37]. This suggests that the effective length of molecules passing through the pore increases. The ratio of blocked current to baseline current $\Delta I/I_0$ also increased with voltage (Figure 2d), indicating a slight increase in the effective diameter of the molecules passing through the nanopore. One possible reason for this is that at different voltages, molecules pass through the nanopore at different orientations [38]. However, it is also possible that as the voltage increases, the translocation speed of molecules further increases, leading to multiple molecules translocating through the nanopore simultaneously, which causes a temporarily crowded in the nanopore (Figure 3a). This results in some translocation events showing longer duration times and larger blocked currents.

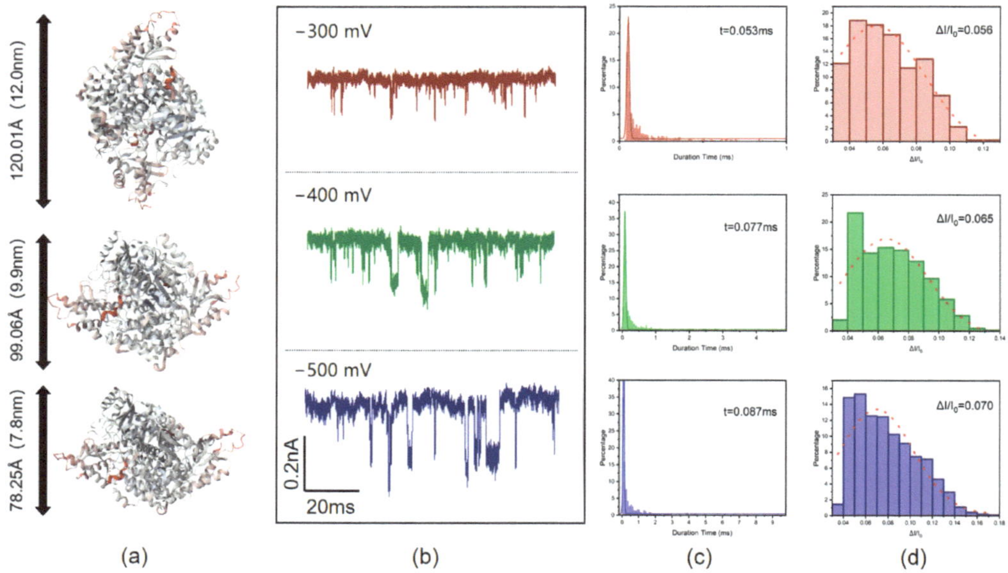

Figure 2. Detection of glutamate decarboxylase molecules using glass nanopores. (**a**) Three-dimensional image of GAD65 protein molecules [35]. (**b**) Current signals are generated by GAD65 molecules passing through glass nanopores under voltages of −300 mV to −500 mV; the signal is rotated 180° for easy observation. (**c**) GAD65 molecule translocation time under voltages of −300 mV to −500 mV. (**d**) Blockade current amplitude ratio $\Delta I/I_0$ of GAD65 molecules under voltages of −300 mV to −500 mV.

To further validate the above conjecture, taking the potential of −400 mV as an example, the current characteristics corresponding to different duration times are shown in Figure 3b. At −400 mV voltage, events with a duration time of less than 0.2 ms are considered as single-molecule translocation events, while events with a duration time greater than 0.2 ms are considered as multiple-molecule translocation events. The scatter plot of duration time and blocking current amplitude (Figure 3c) shows that events with duration time less than 0.2 ms account for 60% of the total events (the blue-shaded area). To ensure statistical significance, under different voltage conditions, the first 60% and the last 40% of the events are analyzed according to the order of translocation time. The results are summarized in Table 2. It can be seen that when the voltage is continuously increased, the duration time and $\Delta I/I_0$ of the 0 to 60% events do not change significantly, indicating that the molecules pass through the nanopore as single molecules influenced by the combined effects of electrophoresis and electroosmosis in a rapid manner. However, with further increase in voltage, multiple molecules pass through the nanopore simultaneously and accumulate momentarily at the nanopore, resulting in an increase in the effective molecular diameter inside the nanopore. Therefore, the duration time and $\Delta I/I_0$ of the 60% to 100% events increase significantly with increasing voltage, which subsequently affects the overall (0 to 100%) values. When the voltage rises to a high voltage region (greater than −600 mV), the increase in voltage has a more significant effect on molecular motion speed than on accumulation at the nanopore, so the translocation time and blocking current ratio decrease with further increase in current.

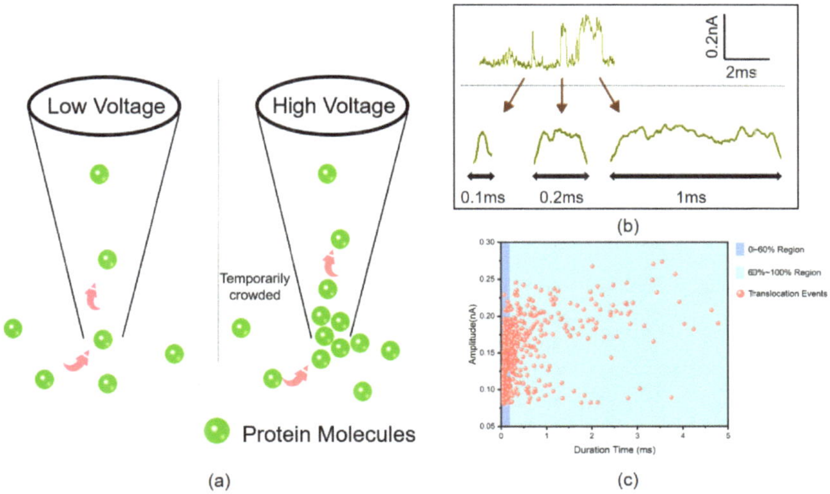

Figure 3. Transient molecular congestion phenomena in nanopores. (**a**) Simultaneous translocation of multiple glutamate decarboxylase molecules through a nanopore. (**b**) Current signatures corresponding to different translocation time events of GAD65 at −400 mV voltage. (**c**) Scatter plot of translocation time versus blockage current amplitude of GAD65 at −400 mV voltage.

Table 2. Fitting results of the normal distribution mean for Duration time and $\Delta I/I_0$ corresponding to GAD65 translocation events with translocation times in the 0–60% and 60–100% ranges, which are ranked by translocation time from shortest to longest.

Grouping of Events	Analysis Type	−300 mV	−400 mV	−500 mV	−600 mV
0–60%	Duration Time	0.052 ms	0.056 ms	0.053 ms	0.056 ms
	$\Delta I/I_0$	0.051	0.055	0.060	0.060
60–100%	Duration Time	0.12 ms	0.28 ms	0.66 ms	0.29 ms
	$\Delta I/I_0$	0.081	0.089	0.10	0.090
0–100%	Duration Time	0.053 ms	0.077 ms	0.087 ms	0.086 ms
	$\Delta I/I_0$	0.056	0.065	0.070	0.066

The antibody produced by a single B cell clone with high homogeneity and only targeting a specific epitope is called a monoclonal antibody. The mixture of antibodies produced by different antibody B cells in the body under the stimulation of antigenic determinants is referred to as polyclonal antibodies [39]. In animals, GADAb may exist in the form of polyclonal antibodies. The glutamate decarboxylase antibody (GADAb) analyzed in the experiment has a Y-type structure similar to other IgG antibodies, but the differences in specific amino acid fragments lead to different surface charges, which further affect the characteristics of specific monoclonal antibodies passing through nanopores. In this case, the glutamate decarboxylase polyclonal antibody (GADAb-poly) was passed through the nanopore (Figure 4a), and the resulting current signals at different voltages were studied (Figure 4b). Under a low voltage of −200 mV, the frequency distribution histogram of translocation times and $\Delta I/I_0$ in the low voltage region both show two distinct event peaks at 2 ms and 9 ms, 0.12 and 0.17 suggesting the distinct statistical characteristics of polyclonal antibodies in nanopore detection compared to monoclonal antibodies. This provides a way for the practical detection of GADAb in the human body.

In other studies that use glass nanopores and rely on the specific binding of antigen-antibody interactions for antibody detection, the typical approach involves modifying the antigen molecules on the inner surface of the glass nanopore and then placing the nanopore in an electrolyte solution containing antibodies for detection, the specific binding between antigen-antibody molecules is confirmed through characteristics such as extended translocation time [21,40]. In contrast, our study attempted to premix the antigen-antibody molecules in an electrolyte solution (1 M KCl, 10 mM Tris-HCl, pH = 8.0) and directly detect the mixed solution using unmodified glass nanopores. Using the optimal concentration of antigen or antibody solutions that produce significant current signals in their respective single solutions, GAD65 and GADAb were 1540 pmol/L and 667 pmol/L, respectively. During this part, for clearer results, a monoclonal antibody of GADAb (GADAb-mono) was used, and the two were premixed in a 1:1 volume ratio for 30 min before detection. It was found that at a voltage of 300 mV, the relative current amplitude ratios ($\Delta I/I_0$) of GAD65 and GADAb solutions were 0.08 and 0.016, respectively (Figure 5c,d). Tests conducted on a mixed solution of GAD65-GADAb also observed these two peaks, while a peak with an amplitude of approximately 0.23, which is approximately equal to the sum of the abscissas of the two peaks, was also observed (Figure 5e). This peak is suspected to correspond to the specific binding of antigen-antibody complexes. Furthermore, a minor peak can be observed at $\Delta I/I_0 \approx 0.32$, which is likely representative of the binding of an antibody to a bimolecular antigen complex, demonstrating the bivalent nature of the IgG antibody. Additionally, the fitted peak height in Figure 5e exhibits a good correlation with the antigen and antibody concentrations before mixing. This study achieved simultaneous identification of glutamate decarboxylase (GAD65), glutamate decarboxylase antibody (GADAb), and GAD65-GADAb antigen-antibody complexes in a complex system, providing a new method for early screening of type I diabetes.

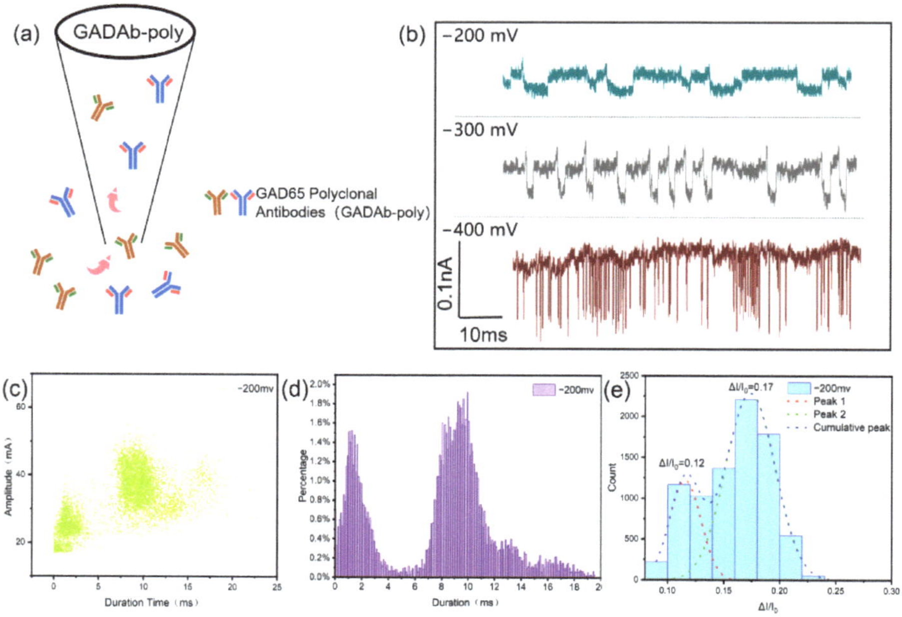

Figure 4. Detection of glutamate decarboxylase polyclonal antibody (GADAb-poly) using glass nanopores. (**a**) Schematic of GADAb-poly translocation through the nanopore. (**b**) Current signals are generated by GADAb-poly passing through glass nanopores under voltages of −200 mV to −400 mV; the signal is rotated 180° for easy observation. (**c**) Scatter plot of translocation time versus blockage current amplitude of GADAb-poly at −200 mV voltage. (**d**) Histograms of translocation time–frequency distribution of GADAb-poly at −200 mV voltage. (**e**) Histograms of $\Delta I/I_0$ distribution of GADAb-poly at −200 mV voltage.

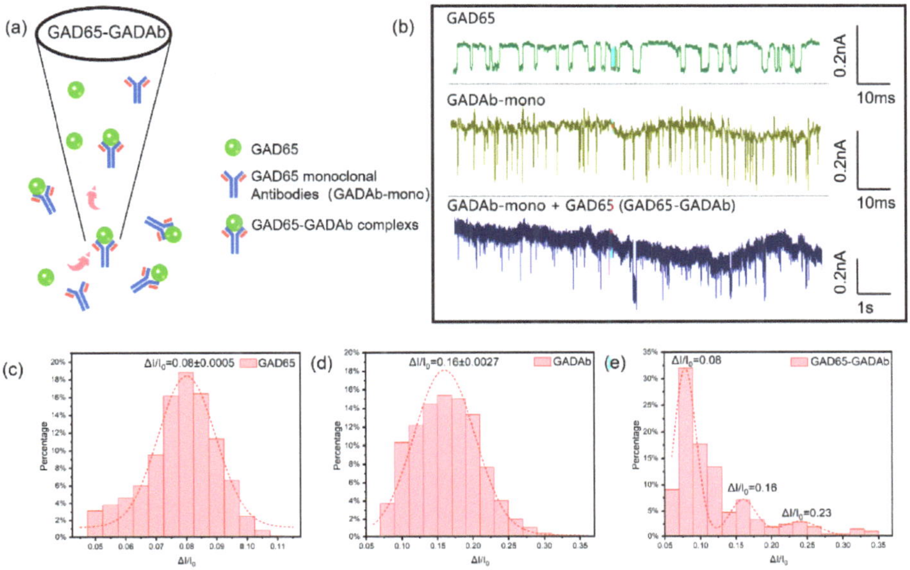

Figure 5. Detection of glutamate decarboxylase (GAD65) and glutamate decarboxylase antibody (GADAb-mono) mixed solutions using glass nanopores. (**a**) Schematic of mixed solution translocation through the nanopore. (**b**) Current signals generated by GAD65, GADAb-mono and their mixture passing through glass nanopores under voltage of 300 mV. (**c**–**e**) Histograms and fitted normal distributions of $\Delta I/I_0$ at 300 mV voltage for GAD65, GADAb-mono, and the mixed solution of GAD65 and GADAb-mono.

4. Conclusions

The glass capillary nanopore has the characteristics of low cost, simple manufacturing, high hardness, easy control of size and shape, and easy to be mass-produced. This article has completed the preparation of glass capillary solid-state nanopores with a pore size of 30 nm to 50 nm and applied glass nanopores to the detection and identification of protein molecules and antigen-antibody complexes. The focus is on the detection of glutamate decarboxylase (GAD65) protein, glutamate decarboxylase polyclonal antibodies (GADAb), and their antigen-antibody complexes (GAD65-GADAb). Using GAD65 as a typical protein molecule, this study investigates the significant differences in the translocation time and amplitude ratio of obstructed current between higher voltage ranges (−300 mV to −500 mV) and low voltage ranges (less than −300 mV). Through a proportional division approach, the characteristic events of individual molecules translocating and multiple molecules translocating simultaneously are approximately studied. It is revealed that within this voltage range, as the voltage increases, the temporary accumulation phenomenon inside the nanopore becomes the main factor affecting the overall characteristics of the events. When the voltage is further increased, the effect of voltage on the acceleration of molecular movement is significantly greater than that of pore accumulation, thus reducing the temporary accumulation phenomenon inside the nanopore. In the low voltage range (−200 mV), preliminary identification of the number of antibody species in glutamate decarboxylase polyclonal antibodies (GADAb-poly) is completed, indicating that there are at least two or more monoclonal antibodies. Finally, GAD65 is directly mixed with glutamate decarboxylase monoclonal antibodies (GADAb-mono) and passed through the nanopore without modification or labeling. The identification of three protein binding states in solution is completed, providing a new method for rapid and low-cost detection of GADAb, which is a preliminary screening indicator for autoimmune type I diabetes (immune-mediated).

Author Contributions: Conceptualization and Supervision, J.L. (Jian Li) and Y.B. (Yan Bi); Experimental work and investigation, C.T., Y.B. (Yun Bai), J.C. and J.L. (Jing Lu); Data analysis, C.T.; Funding acquisition, J.L. (Jian Li); Resources, J.L. (Jing Lu) and Y.B. (Yan Bi); Writing—original draft, C.T. and Y.B. (Yun Bai); Writing—review & editing, C.T. All authors have read and agreed to the published version of the manuscript.

Funding: This research was funded by the general project of the National Natural Science Foundation of China, grant number 32270607. This work was supported by Natural Science Foundation of Jiangsu Province Major Project (BK20222008).

Institutional Review Board Statement: Not applicable.

Informed Consent Statement: Not applicable.

Data Availability Statement: The raw data supporting the conclusions of this article will be made available by the authors on request.

Acknowledgments: Special thanks are given to Jingjie Sha from Jiangsu Key Laboratory for Design and Manufacture of Micro-nano Biomedical Instruments for providing experimental platform support for this research.

Conflicts of Interest: The authors declare no conflicts of interest.

References

1. American Diabetes Association. 2. Classification and Diagnosis of Diabetes: Standards of Medical Care in Diabetes—2021. *Diabetes Care* **2020**, *44* (Suppl. S1), S15–S33. [CrossRef] [PubMed]
2. Singla, R.; Homko, C.; Schey, R.; Parkman, H.P. Diabetes-Related Autoantibodies in Diabetic Gastroparesis. *Dig. Dis. Sci.* **2015**, *60*, 1733–1737. [CrossRef] [PubMed]
3. Towns, R.; Pietropaolo, M. GAD-65 Autoantibodies and their role as biomarkers of type 1 diabetes and latent autoimmune diabetes in adults (LADA). *Drugs Future* **2011**, *36*, 847–854. [CrossRef] [PubMed]
4. Amoroso, M.; Achenbach, P.; Powell, M.; Coles, R.; Chlebowska, M.; Carr, L.; Furmaniak, J.; Scholz, M.; Bonifacio, E.; Ziegler, A.-G.; et al. 3 Screen islet cell autoantibody ELISA: A sensitive and specific ELISA for the combined measurement of autoantibodies to GAD65, to IA-2 and to ZnT8. *Clin. Chim. Acta* **2016**, *462*, 60–64. [CrossRef] [PubMed]
5. Zhang, D.; Huang, J.; Hu, J. Improved diagnosis of type-1 diabetes mellitus using multiplexed autoantibodies ELISA array. *Anal. Biochem.* **2022**, *649*, 114722. [CrossRef] [PubMed]
6. Murayama, H.; Matsuura, N.; Kawamura, T.; Maruyama, T.; Kikuchi, N.; Kobayashi, T.; Nishibe, F.; Nagata, A. A sensitive radioimmunoassay of insulin autoantibody: Reduction of non-specific binding of [125I]insulin. *J. Autoimmun.* **2006**, *26*, 127–132. [CrossRef] [PubMed]
7. Ankelo, M.; Westerlund, A.; Blomberg, K.; Knip, M.; Ilonen, J.; Hinkkanen, A.E. Time-Resolved Immunofluorometric Dual-Label Assay for Simultaneous Detection of Autoantibodies to GAD65 and IA-2 in Children with Type 1 Diabetes. *Clin. Chem.* **2007**, *53*, 472–479. [CrossRef]
8. Zhong, Z.Y.; Peng, N.; Qing, Y.; Shan, J.L.; Li, M.X.; Guan, W.; Dai, N.; Gu, X.Q.; Wang, D. An electrochemical immunosensor for simultaneous multiplexed detection of neuron-specific enolase and pro-gastrin-releasing peptide using liposomes as enhancer. *Electrochim. Acta* **2011**, *56*, 5624–5629. [CrossRef]
9. Parker, D.M.; Winkenbach, L.P.; Parker, A.; Boyson, S.; Nishimura, E.O. Improved Methods for Single-Molecule Fluorescence In Situ Hybridization and Immunofluorescence in Caenorhabditis elegans Embryos. *Curr. Protoc.* **2021**, *1*, e299. [CrossRef]
10. Kasianowicz, J.J.; Brandin, E.; Branton, D.; Deamer, D.W. Characterization of individual polynucleotide molecules using a membrane channel. *Proc. Natl. Acad. Sci. USA* **1996**, *93*, 13770–13773. [CrossRef]
11. He, Y.; Tsutsui, M.; Zhou, Y.; Miao, X.-S. Solid-state nanopore systems: From materials to applications. *NPG Asia Mater.* **2021**, *13*, 48. [CrossRef]
12. Xue, L.; Yamazaki, H.; Ren, R.; Wanunu, M.; Ivanov, A.P.; Edel, J.B. Solid-state nanopore sensors. *Nat. Rev. Mater.* **2020**, *5*, 931–951. [CrossRef]
13. Coulter, W.H. Means for Counting Particles Suspended in a Fluid. US2656508A, 20 October 1953.
14. Ying, Y.L.; Cao, C.; Long, Y.T. Single molecule analysis by biological nanopore sensors. *Analyst* **2014**, *139*, 3826–3835. [CrossRef] [PubMed]
15. Arjmandi, N.; Van Roy, W.; Lagae, L.; Borghs, G. Measuring the Electric Charge and Zeta Potential of Nanometer-Sized Objects Using Pyramidal-Shaped Nanopores. *Anal. Chem.* **2012**, *84*, 8490–8496. [CrossRef] [PubMed]
16. Tsutsui, M.; Yoshida, T.; Yokota, K.; Yasaki, H.; Yasui, T.; Arima, A.; Tonomura, W.; Nagashima, K.; Yanagida, T.; Kaji, N.; et al. Discriminating single-bacterial shape using low-aspect-ratio pores. *Sci. Rep.* **2017**, *7*, 17371. [CrossRef] [PubMed]
17. Houghtaling, J.; Ying, C.; Eggenberger, O.M.; Fennouri, A.; Nandivada, S.; Acharjee, M.; Li, J.; Hall, A.R.; Mayer, M. Estimation of Shape, Volume, and Dipole Moment of Individual Proteins Freely Transiting a Synthetic Nanopore. *ACS Nano* **2019**, *13*, 5231–5242. [CrossRef] [PubMed]

18. Stoddart, D.; Heron, A.J.; Klingelhoefer, J.; Mikhailova, E.; Maglia, G.; Bayley, H. Nucleobase Recognition in ssDNA at the Central Constriction of the α-Hemolysin Pore. *Nano Lett.* **2010**, *10*, 3633–3637. [CrossRef] [PubMed]
19. Hiratani, M.; Kawano, R. DNA Logic Operation with Nanopore Decoding To Recognize MicroRNA Patterns in Small Cell Lung Cancer. *Anal. Chem.* **2018**, *90*, 8531–8537. [CrossRef] [PubMed]
20. Merstorf, C.; Cressiot, B.; Pastoriza-Gallego, M.; Oukhaled, A.; Betton, J.M.; Auvray, L.; Pelta, J. Wild Type, Mutant Protein Unfolding and Phase Transition Detected by Single-Nanopore Recording. *ACS Chem. Biol.* **2012**, *7*, 652–658. [CrossRef]
21. Ying, Y.L.; Yu, R.J.; Hu, Y.X.; Gao, R.; Long, Y.T. Single antibody-antigen interactions monitored *via* transient ionic current recording using nanopore sensors. *Chem. Commun.* **2017**, *53*, 8620–8623. [CrossRef]
22. Lin, K.B.; Li, Z.W.; Tao, Y.; Li, K.; Yang, H.J.; Ma, J.; Li, T.; Sha, J.J.; Chen, Y.F. Surface Charge Density Inside a Silicon Nitride Nanopore. *Langmuir* **2021**, *37*, 10521–10528. [CrossRef] [PubMed]
23. Venkatesan, B.M.; Dorvel, B.; Yemenicioglu, S.; Watkins, N.; Petrov, I.; Bashir, R. Highly Sensitive, Mechanically Stable Nanopore Sensors for DNA Analysis. *Adv. Mater.* **2009**, *21*, 2771–2776. [CrossRef] [PubMed]
24. Steinbock, L.J.; Krishnan, S.; Bulushev, R.D.; Borgeaud, S.; Blokesch, M.; Feletti, L.; Radenovic, A. Probing the size of proteins with glass nanopores. *Nanoscale* **2014**, *6*, 14380–14387. [CrossRef] [PubMed]
25. Bandara, Y.M.N.D.Y.; Freedman, K.J. Enhanced Signal to Noise Ratio Enables High Bandwidth Nanopore Recordings and Molecular Weight Profiling of Proteins. *ACS Nano* **2022**, *16*, 14111–14120. [CrossRef] [PubMed]
26. Wang, F.; Zhao, C.; Zhao, P.; Chen, F.; Qiao, D.; Feng, J. MoS_2 nanopore identifies single amino acids with sub-1 Dalton resolution. *Nat. Commun.* **2023**, *14*, 2895. [CrossRef]
27. Plesa, C.; Dekker, C. Data analysis methods for solid-state nanopores. *Nanotechnology* **2015**, *26*, 084003. [CrossRef]
28. Cai, H.; Wang, Y.; Yu, Y.; Mirkin, M.V.; Bhakta, S.; Bishop, G.W.; Joshi, A.A.; Rusling, J.F. Resistive-Pulse Measurements with Nanopipettes: Detection of Vascular Endothelial Growth Factor C (VEGF-C) Using Antibody-Decorated Nanoparticles. *Anal. Chem.* **2015**, *87*, 6403–6410. [CrossRef] [PubMed]
29. Kovarik, M.L.; Zhou, K.M.; Jacobson, S.C. Effect of Conical Nanopore Diameter on Ion Current Rectification. *J. Phys. Chem. B* **2009**, *113*, 15960–15966. [CrossRef] [PubMed]
30. Woermann, D. Electrochemical transport properties of a cone-shaped nanopore: High and low electrical conductivity states depending on the sign of an applied electrical potential difference. *Phys. Chem. Chem. Phys.* **2003**, *5*, 1853–1858. [CrossRef]
31. Liu, Q.; Wang, Y.; Guo, W.; Ji, H.; Xue, J.; Ouyang, Q. Asymmetric properties of ion transport in a charged conical nanopore. *Phys. Rev. E* **2007**, *75*, 6. [CrossRef]
32. Siwy, Z.; Fulinski, A. Fabrication of a synthetic nanopore ion pump. *Phys. Rev. Lett.* **2002**, *89*, 4. [CrossRef] [PubMed]
33. Wang, H.; Tang, H.; Yang, C.; Li, Y. Selective Single Molecule Nanopore Sensing of microRNA Using PNA Functionalized Magnetic Core-shell Fe_3O_4-Au Nanoparticles. *Anal. Chem.* **2019**, *91*, 7965–7970. [CrossRef] [PubMed]
34. Wang, D.; Qi, G.; Zhou, Y.; Zhang, Y.; Wang, B.; Hu, P.; Jin, Y. Single-cell ATP detection and content analyses in electrostimulus-induced apoptosis using functionalized glass nanopipettes. *Chem. Commun.* **2020**, *56*, 1561–1564. [CrossRef] [PubMed]
35. Fenalti, G.; Law, R.H.P.; Buckle, A.M.; Langendorf, C.; Tuck, K.; Rosado, C.J.; Faux, N.G.; Mahmood, K.; Hampe, C.S.; Banga, J.P.; et al. GABA production by glutamic acid decarboxylase is regulated by a dynamic catalytic loop. *Nat. Struct. Mol. Biol.* **2007**, *14*, 280–286. [CrossRef] [PubMed]
36. Sha, J.; Hasan, T.; Milana, S.; Bertulli, C.; Bell, N.A.W.; Privitera, G.; Ni, Z.; Chen, Y.; Bonaccorso, F.; Ferrari, A.C.; et al. Nanotubes Complexed with DNA and Proteins for Resistive-Pulse Sensing. *ACS Nano* **2013**, *7*, 8857–8869. [CrossRef] [PubMed]
37. Chen, H.; Lin, Y.; Long, Y.-T.; Minteer, S.D.; Ying, Y.-L. Nanopore-based measurement of the interaction of P450cam monooxygenase and putidaredoxin at the single-molecule level. *Faraday Discuss.* **2022**, *233*, 295–302. [CrossRef] [PubMed]
38. Sha, J.; Si, W.; Xu, B.; Zhang, S.; Li, K.; Lin, K.; Shi, H.; Chen, Y. Identification of Spherical and Nonspherical Proteins by a Solid-State Nanopore. *Anal. Chem.* **2018**, *90*, 13826–13831. [CrossRef] [PubMed]
39. Lipman, N.S.; Jackson, L.R.; Trudel, L.J.; Weis-Garcia, F. Monoclonal Versus Polyclonal Antibodies: Distinguishing Characteristics, Applications, and Information Resources. *ILAR J.* **2005**, *46*, 258–268. [CrossRef] [PubMed]
40. Fu, F.; Zhou, Z.Z.; Sun, Q.; Xu, B.; Sha, J. Label-free Detection of PD-1 Antibody and Antigen Immunoreaction Using Nano-Sensors. *Acta Chim.* **2019**, *77*, 287–292. [CrossRef]

biosensors

MDPI

Communication

Development of a Microfluidic Chip System with Giant Magnetoresistance Sensor for High-Sensitivity Detection of Magnetic Nanoparticles in Biomedical Applications

Tzong-Rong Ger [1,*], Pei-Sheng Wu [1], Wei-Jie Wang [1,2], Chiung-An Chen [3,*], Patricia Angela R. Abu [4] and Shih-Lun Chen [5]

1 Department of Biomedical Engineering, Chung Yuan Christian University, Chung-Li 320314, Taiwan
2 Division of Nephrology, Department of Internal Medicine, Taoyuan General Hospital, Ministry of Health and Welfare, Taoyuan 115204, Taiwan
3 Department of Electrical Engineering, Ming Chi University of Technology, New Taipei City 243303, Taiwan
4 Department of Information Systems and Computer Science, Ateneo de Manila University, Quezon City 1108, Philippines
5 Department of Electronic Engineering, Chung Yuan Christian University, Taoyuan City 320314, Taiwan; chrischen@cycu.edu.tw
* Correspondence: sunbow@cycu.org.tw (T.-R.G.); joannechen@mail.mcut.edu.tw (C.-A.C.); Tel.: +886-3-2654536 (T.-R.G.)

Abstract: Magnetic nanoparticles (MNPs) have been widely utilized in the biomedical field for numerous years, offering several advantages such as exceptional biocompatibility and diverse applications in biology. However, the existing methods for quantifying magnetic labeled sample assays are scarce. This research presents a novel approach by developing a microfluidic chip system embedded with a giant magnetoresistance (GMR) sensor. The system successfully detects low concentrations of MNPs with magnetic particle velocities of 20 mm/s. The stray field generated by the magnetic subject flowing through the microchannel above the GMR sensor causes variations in the signals. The sensor's output signals are appropriately amplified, filtered, and processed to provide valuable indications. The integration of the GMR microfluidic chip system demonstrates notable attributes, including affordability, speed, and user-friendly operation. Moreover, it exhibits a high detection sensitivity of 10 μg/μL for MNPs, achieved through optimizing the vertical magnetic field to 100 Oe and the horizontal magnetic field to 2 Oe. Additionally, the study examines magnetic labeled RAW264.7 cells. This quantitative detection of magnetic nanoparticles can have applications in DNA concentration detection, protein concentration detection, and other promising areas of research.

Keywords: magnetic particles; magnetoresistive sensors; microfluidics

Citation: Ger, T.-R.; Wu, P.-S.; Wang, W.-J.; Chen, C.-A.; Abu, P.A.R.; Chen, S.-L. Development of a Microfluidic Chip System with Giant Magnetoresistance Sensor for High-Sensitivity Detection of Magnetic Nanoparticles in Biomedical Applications. *Biosensors* **2023**, *13*, 807. https://doi.org/10.3390/bios13080807

Received: 10 July 2023
Revised: 5 August 2023
Accepted: 6 August 2023
Published: 11 August 2023

1. Introduction

Magnetic nanoparticles (MNPs) have found widespread applications in various fields, including MRI contrast agents, magnetic recording devices, bio-sensing, drug delivery, thermal therapy, and biomolecule separation [1–6]. However, the use of large quantities of MNPs for cell manipulation in these techniques can lead to nanotoxicity concerns. Current cell assay methods for evaluating target quantities often involve techniques such as Prussian blue staining [7], T2 relaxometry [8], and UV/VIS spectrometry [9,10]. However, these approaches are primarily limited to the static detection of stray fields from immobilized labels. To address this limitation, researchers have explored the use of magnetic sensors in microfluidic applications to detect small biological samples. One such approach involves the use of a magnetic immunoassay, a novel type of immunoassay that enables the quantitative detection of biomolecules. The number of biomolecular targets, such as DNA or cells, can be determined by measuring the magnetic subject, which can be achieved through techniques such as measuring the remnant magnetic flux [11], magnetization relaxation time of magnetic particle clusters [12], or reduction in alternating current (AC) magnetic susceptibility [13]. In the past decade, magnetoresistive (MR) sensors have been utilized in magnetic biosensing to estimate the amount of target biomolecules or cells. These MR-based biosensors measure variations in MR signals caused by magnetic microparticles or nanoparticles attached to the target sample [14–16]. The first significant contributions to the field of detecting magnetic markers using measurements of resistance/impedance change in magnetic field sensors, including measurements in a continuous flow, can be attributed to pioneering studies. David R. Baselt et al. developed a biosensor capable of measuring the forces that bind DNA–DNA, antibody–antigen, or ligand–receptor pairs at the single-molecule level. Known as the Bead Array Counter (BARC), this biosensor utilizes interaction forces to immobilize magnetic microbeads on a solid substrate. Microfabricated magnetoresistive transducers on the substrate then indicate whether the beads are displaced when subjected to magnetic forces [17]. G. V. Kurlyandskaya et al. employed a commercial Ferrofluid® liquid thin layer to cover the ribbon surface of their sensor. This innovative approach revealed that the magnetoimpedance response was significantly affected by the presence of the magnetic Ferro liquid, the applied magnetic field's intensity, and the driving current parameters. The proposed magnetoimpedance-based prototype demonstrated high sensitivity to the fringe field generated by magnetic nanoparticles, thereby offering great promise as a biosensor [18]. F. Blanc-Béguin et al. conducted research to determine the optimal conditions for producing cell samples suitable for imaging with the detection of modifications in the magnetic field caused by maghemite (Fe_2O_3) nanoparticles. These nanoparticles acted as a high-sensitivity magnetic biosensor based on the giant magnetoimpedance (GMI) effect. The preliminary results from this study provided valuable insights into the production of biological samples, laying the groundwork for further advancements in GMI biosensor technology [19]. A. García-Arribas et al. introduced a microfluidic device capable of determining the concentration of magnetic beads under a continuous flow of the carrier fluid, utilizing the giant magnetoimpedance effect (GMI) [20].

However, research has demonstrated that the sensor exhibited exceptional sensitivity to liquid and materials present in the microfluidic chamber. Although it successfully detected magnetic microparticles in a static regime and magnetic nanoparticles under a continuous flow, the measurements proved to be delicate and challenging to simple measurement and analysis devices. Further research and development are required to address these intricacies and enhance the sensor's reproducibility. In this research, the measurement systems for studying the stray fields from MNPs in MR-based biosensors rely on lock-in amplifier detection. We propose an integrated GMR microfluidic chip system as a cost-effective alternative to lock-in amplifiers. This system offers the advantages of being fast and easy to operate, as well as significantly reducing the sensing instrument costs.

2. Materials and Methods

The GMR general structure consisted of two ferromagnetic layers with a nonmagnetic layer sandwiched. The principle of GMR is shown in Figure 1 (red square) when the magnetization of two ferromagnetic layers parallel in the same direction and the spin electron passed through the layer with the same magnetization direction. The spin electron had a lower possibility of scattering, leading to lower magnetic resistance. On the other hand, when the two antiparallel ferromagnetic layers were magnetized, the spin electron had a higher possibility of scattering, leading to higher resistance [21]. Figure 1 is a schematic diagram of the integration of the GMR microfluidic chip. A Charge Coupled Device (CCD) camera was used to monitor the flow of MNPs, the microfluidic channel in which the MNPs flowed from the inlet to the outlet, and the permanent magnet for magnetizing the superparamagnetic MNPs. The upper blue square is the real image of the

GMR microfluidic chip; the lower blue square is the MNP subjected to the applied magnetic field (blue arrow), which produced the magnetization in the same direction (red arrow). The white arrow indicates the direction of the magnetization of GMR, and the green arrow indicates the direction of the microfluidic flow. On the right is a simple diagram of signal processing. This experiment process was divided into three parts: the microfluidic channel, the synthesis of dextran-coated MNPs, and cell culture.

The fabrication process of the microfluidic MNP detection chip shown in Figure 2a consisted of two main steps: the fabrication of the microfluidic channel and the integration of the GMR sensor and microfluidic chip. As for the microchannel fabrication, negative photoresist SU-8 was patterned and developed with (1) spin coating and soft bake on a glass substrate, (2) mask aligning, and (3) a UV lithography process (4) as a mother mold for the microchannel; then, (5) Polydimethylsiloxane (PDMS) was poured onto the mold and cured. The surface of the GMR sensor was covered by the demolding PDMS layer to form the sensor chip used in this study, and then oxygen plasma was used for the bonding between the sensor and the microchannel. Figure 2b is a schematic diagram of the microfluidic chip with a microchannel for magnetic nanoparticle flowing and the sensor chip, as well as a readout circuit for MNP detection. Figure 2c is an image of PDMS-based microfluidics with the inlet, sensing, and outlet areas. The external magnet was applied for the stable magnetization of MNPs.

In this study, dextran-coated magnetic nanoparticles (DEX-MNPs) with an average diameter of 10 nm were synthesized and utilized. Dextran, known for its good dispersibility in aqueous solutions, was chosen as the coating material due to its widely recognized surface modification capabilities. The synthesis of DEX-MNPs followed a previously published method [22,23]. In summary, 0.405 g of iron (III) was mixed with 0.694 g of dextran in 10 mL of deionized water. The solution was injected into a 3-necked flask containing 30 mL of preheated deionized water at 80 °C under a N_2 gas atmosphere. After 5 min of constant agitation, 0.833 mL of N_2H_4 was added, followed by the injection of 0.148 g of iron (II) in 10 mL of water after another 5 min. Subsequently, 8 mL of NaOH was added, and the solution was dialyzed for 24 h to remove unreacted compounds. The resulting DEX-MNPs were obtained through freeze drying. Figure 3 shows the scanning electron microscopy (SEM) images of the MNPs (Figure 3a) and DEX-MNPs (Figure 3b). The saturation magnetization of the MNPs and DEX-MNPs was measured to be 73.64 emu/g and 15.39 emu/g, respectively (Figure 3c). Figure 3d shows the magnetization curves taken in zero-field-cooling (ZFC) and field-cooling (FC) modes with an applied magnetic field of 100 Oe. The sample showed superparamagnetic behavior at room temperature, with blocking transition at TB = 71.82 K (H = 100 Oe), which is similar to results from other published papers [24,25]. For cell sample preparation, RAW 264.7 cells, from a murine macrophage cell line, were cultured in Dulbecco's Modified Eagle's Medium (DMEM) and supplemented with 1% penicillin and 10% fetal bovine serum (FBS) at 37 °C in a 5% CO_2 environment. Once the cells reached 80–90% confluence in 6-well culture plates, the medium was replaced with a medium containing DEX-MNPs at a concentration of 260 μg/mL of iron, and the cells were incubated for 12 h.

This study involved the establishment of two systems: the integration of the GMR microfluidic chip system and the lock-in MNP detection system. The GMR microfluidic chip system, illustrated in Figure 4a, comprised a solution pump, a detection area, and a data display device. The solution pump facilitated the flow of microfluidics through the channel, while the detection area consisted of microfluidic channels and GMR sensors (AAH002, NVE Corp., Eden Prairie, MN, USA) integrated with an external magnetic field for detecting the concentration of MNPs. As magnetically labeled particles passed through the GMR sensor, they induced variations in the magnetic field, leading to changes in the resistance and voltage of the sensor (ΩGMR). Thus, the concentration of MNPs could be calculated based on the signal difference. The data display device included a signal processing circuit, a microcontroller, and a display. The signal processing circuit incorporated a differential amplifier, a high-pass filter, and a low-pass filter, with high-pass and low-pass cutoff frequencies set at 5 Hz and 15 Hz, respectively, each tailored for different time spans and shapes. During the measurement of particle concentration in the flow, the sensor signals exhibited a range between 0.01 and 0.1 mVpp, with a total amplifier gain of 75 dB. The amplified output signal was captured by the MSP430f5529 microcontroller unit (Texas Instruments Incorporated, Dallas, TX, USA), which converted the analog signal from the processing circuit into a digital signal. The microprocessor then utilized a correlation equation between voltage and MNP concentration to convert the signal into concentration, which was presented on the LC display PVC160203P (Picvue Electronics CO., Hsinchu, Taiwan). The LabVIEW software enabled the synchronization of data from the microprocessor and displayed it on the computer, as depicted in Figure 4b. As the MNPs passed through the GMR microfluidic chip, signal vibrations were observed on the front panel of LabVIEW.

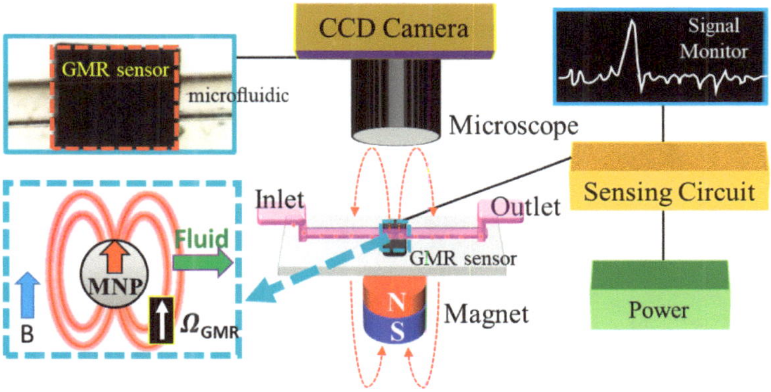

Figure 1. Schematic diagram of the integrated GMR microfluidic chip system, including the solution pump, detection area, and display device. The CCD camera monitors the flow of MNPs. The middle section represents the microfluidic channel. The lower part features a permanent magnet for magnetizing the superparamagnetic MNPs. The upper left inset shows the real image of the GMR microfluidic chip, while the lower left inset illustrates the MNP under the applied magnetic field (blue arrow), resulting in magnetization in the same direction (red arrow). The white arrow indicates the direction of the magnetization of the GMR, and the green arrow indicates the direction of the microfluidic flow.

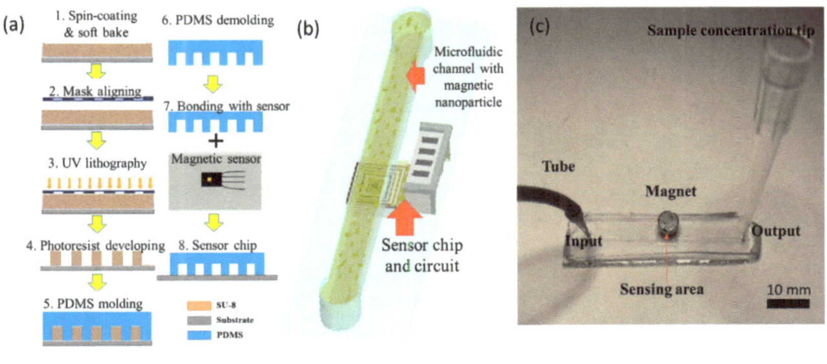

Figure 2. (**a**) The fabrication process of the microfluidic MNP detection chip. The lock-in MNP detection system containing the solution pump, detection area, and lock-in amplifier; (**b**) the schematic diagram of the microfluidic chip with a microchannel for magnetic nanoparticle flowing and the sensor chip and a readout circuit. (**c**) The external magnet was applied for alternative magnetization of MNPs.

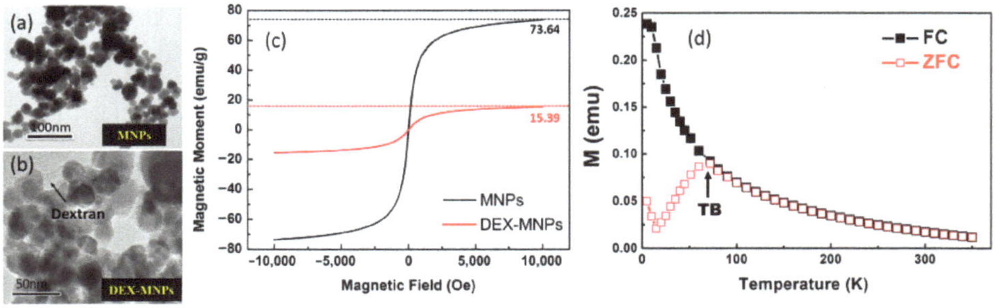

Figure 3. SEM image of (**a**) MNPs and (**b**) dextran-coated MNP particles. (**c**) SQUID magnetic hysteresis loops of magnetic nanoparticles (MNPs) and dextran-coated MNPs (DEX-MNPs). Magnetization ratio of DEX-MNPs and MNPs was about 0.21 (15.39/73.64). (**d**) ZFC/FC curves for MNP samples with an applied field of 100 Oe (TB = 71.82 K).

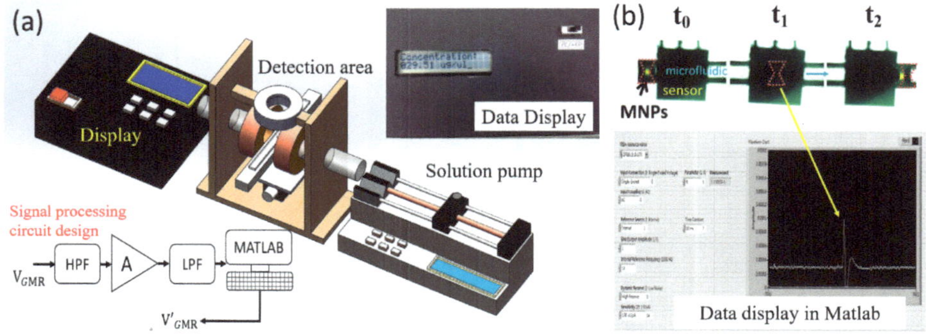

Figure 4. (**a**) Integration of the GMR microfluidic chip system comprising the solution pump, detection area, and data display device. The upper right inset shows the photo of the data display device. The signal processing circuit design (lower left) includes a differential amplifier, high-pass filter, low-pass filter, and amplifier with a total gain of 75 dB. (**b**) The LabVIEW software enables data synchronization from the microprocessor and display on the computer. Signal vibrations are observed on the front panel of LabVIEW when MNPs pass through the GMR microfluidic chip.

3. Results and Discussion

After confirming that the GMR sensor could detect MNPs, the GMR magnetization conditions were explored to optimize the measurement process. A preliminary detection of MNPs was conducted using an oscilloscope, which revealed that the MNPs passing through the GMR sensor caused voltage waveform vibrations, indicating the sensor's ability to detect a small number of MNPs (Figure 5). Simultaneously, the CCD camera screened the MNPs near and above the microfluidic chip (Figure 5a,b). Figure 5c shows MNPs of the same concentration passing through the GMR microfluidic chip, while Figure 5d presents the corresponding data from the oscilloscope, demonstrating the system's stability.

To achieve an optimal balance between the MR sensor and the MNPs' magnetization, an applied magnetic field was used. The vertical (Z-axis) magnetic field was employed to magnetize the MNPs, while the horizontal magnetic field was adjusted to find the optimal value. Experimental results indicated that a vertical magnetic field of 100 Oe and a horizontal magnetic field of 2 Oe provided the optimal measurement environment. Figure 6a displays the voltage signals of different MNP concentrations passing through the GMR microfluidic chip, illustrating a stronger signal with a higher MNP concentration. The relationship between MNP concentration and voltage showed linearity (Figure 6b) with a coefficient of determination (R^2) of 0.99984. When examining dextran-coated MNPs with a concentration of 50 μg/μL in the integrated GMR microfluidic chip system, the measured voltage was substituted into the MNP concentration and voltage correlation function. The result indicated an MNP concentration of approximately 12 μg/μL (not shown in the figure). The magnetization ratio of dextran-coated MNPs, measured using SQUID (Figure 3c), was approximately 0.21 (15.39/73.64). This means that only 21% (10.5 μg/μL) of the MNPs in DEX-MNPs with a concentration of 50 μg/μL were magnetized, which closely aligns with the examined data.

To examine the intracellular localizations of MNPs, cells treated with MNPs were washed three times with phosphate-buffered saline (PBS) and fixed in 4% paraformaldehyde, as depicted in Figure 7a,b. Next, a staining reagent of 2% potassium ferrocyanide with 6% HCl (1:1 *v*/*v*), known as Prussian blue stain, was added to the cells and incubated for 10 min. To quantify the MNPs, single-cell magnetophoresis was performed by subjecting the magnetic-labeled cells in suspension to a controlled magnetic field gradient. The magnetic cells suspended in the medium were exposed to the magnetic field. In the steady-state regime, the magnetic force $Fm = m_{bead}\ dB/dx$ (where m_{bead} represents the magnetic moment of the magnetic beads and dB/dx is the magnetic field gradient) was balanced with the viscous force $F_{vis} = 6\pi\eta Rv$ (where R is the radius of the cell, η is the viscosity of the carrier liquid, and v is the cell velocity). The total magnetic moment of the MNPs inside a cell could be expressed as $m_{bead} = NcM_s\pi D^3/6$, where N is the total number of MNPs per cell, D is the diameter of an MNP, and c is the ratio of the net magnetization of the MNPs to their saturation magnetization M_s (set as 0.8 in this case). By setting the cell radius R and the carrier liquid viscosity η as 0.013 Pas, the number of MNPs loaded by cells N could be calculated using the following equation: $N = 36\eta Rv/(cM_sD^3(dB/dX))$.

A total of 510 mobile cells moving at a constant velocity toward the magnet were tracked using video microscopy. The average velocity of all the tracked cells was 35.6 ± 5.33 μm/s, as shown in Figure 7c. Applying the same method as in reference [26], the number of MNPs internalized by RAW cells was estimated to be $(25.8 \pm 3.86) \times 10^6$. Each cell contained an average of 1.13 ng/μL of MNPs, as determined using Inductively Coupled Plasma Mass Spectrometry (ICP-MS) measurements. The magnetic-labeled RAW cells were examined using the microfluidic chip system. Figure 7d displays the measured responses corresponding to the magnetic-labeled RAW cell passing over the microfluidic chip system, indicating an MNP concentration of 30 μg/μL. This implies that each cell carried approximately 3 ng/μL, which is close to the value of 1.13 ng/μL measured using ICP-MS (not shown here). The slight difference between the values obtained from ICP-MS and our device can be attributed to operational errors and should be addressed.

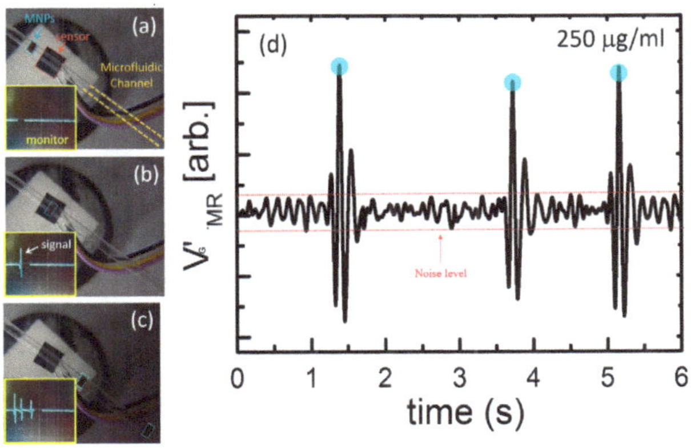

Figure 5. The picture of MNPs (**a**) close to the GMR sensor, (**b**) above the GMR sensor, and (**c**) pass the GMR sensor. The pictures in the lower left corner are the instant signal image. (**d**) The signal diagram of three consecutive 250 ug/mL MNPs passing through the sensor.

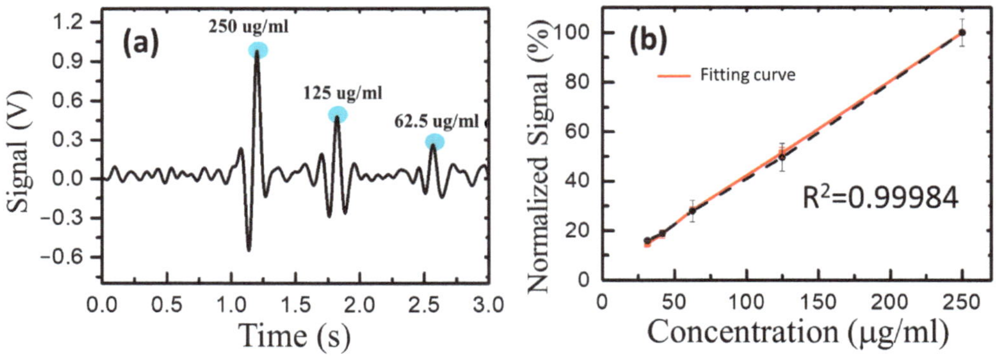

Figure 6. (**a**) The voltage signal of different MNP concentrations. It could be seen that, the higher the concentration of MNPs, the stronger the signal. (**b**) The linear relationship between the concentration of MNPs and the voltage. The coefficient of determination R^2 was 0.99984.

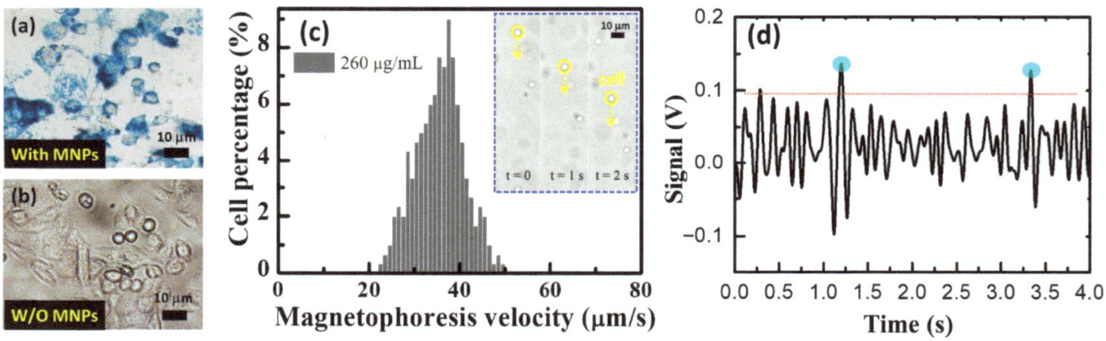

Figure 7. (**a**) Prussian blue staining results of RAW cells internalizing magnetic nanoparticles (MNPs) for 12 h. (**b**) Control corresponds to the RAW cells that were not treated with MPs in parallel to the treated group. (**c**) Cell velocity distributions of 510 magnetically labeled cells. Insets: consecutive optical micrographs of mobile cells at different time points. Scale bar represents 10 μm. (**d**) The measured signals responding to the magnetic-labeled RAW cells passing through the microfluidic chip system.

4. Conclusions

This study successfully developed an integrated GMR microfluidic chip system for the detection of MNPs. The optimal values for the horizontal and vertical magnetic fields, determined through the lock-in MNP detection system, were found to be 2 Oe and 100 Oe, respectively. The correlation between voltage and MNP concentration was examined for both the lock-in MNP detection system and the GMR microfluidic chip system. The accuracy of both systems was validated using dextran-coated MNPs. Furthermore, the magnetic-labeled RAW cell was tested in the GMR microfluidic chip system, yielding results that closely matched those obtained using ICP-MS. This confirms that the developed integration of the GMR microfluidic chip system for MNP detection in this study offers several advantages, including a relatively low cost, fast and easy operation, and accurate measurement of magnetic particle concentration.

Author Contributions: Conceptualization, T.-R.G. and P.-S.W.; methodology, T.-R.G. and W.-J.W.; validation, T.-R.G., P.-S.W., P.A.R.A. and C.-A.C.; investigation, T.-R.G., P.-S.W. and S.-L.C.; writing—original draft preparation, T.-R.G. and C.-A.C.; writing—review and editing, P.A.R.A. and S.-L.C.; visualization, T.-R.G., W.-J.W. and P.A.R.A.; supervision, W.-J.W.; project administration, T.-R.G.; funding acquisition, W.-J.W. All authors have read and agreed to the published version of the manuscript.

Funding: This research was funded by the Ministry of Science and Technology, Taiwan, under Grant MOST 111-2221-E-033-007-MY3, NSTC 111-2622-E-033-010, MOST-112-2410-H-033-014, MOST-111-2221-E-033-041, MOST-111-2823-8-033-001, MOST-111-2622-E-131-001. NSTC 112-2221-E-033-049-MY3, NSTC 112-2622-E-033-003.

Institutional Review Board Statement: Not applicable.

Informed Consent Statement: Not applicable.

Acknowledgments: This manuscript is dedicated to the memory of Shyh-Liang Lou. who sadly passed away on 12 September 2016, whose support of lab space and resource to the research process was invaluable.

Conflicts of Interest: The funders had no role in the design of the study, in the collection, analyses, or interpretation of data, in the writing of the manuscript, or in the decision to publish the results.

References

1. Pankhurst, Q.A.; Connolly, J.; Jones, S.K.; Dobson, J. Applications of magnetic nanoparticles in biomedicine. *J. Phys. D Appl. Phys.* **2003**, *36*, R167–R181. [CrossRef]
2. Figuerola, A.; Corato, R.D.; Manna, L.; Pellegrino, T. From iron oxide nanoparticles towards advanced iron-based inorganic materials designed for biomedical applications. *Pharmacol. Res.* **2010**, *62*, 126–143. [CrossRef] [PubMed]
3. Jong, W.H.D.; Borm, P.J. Drug delivery and nanoparticles: Applications and hazards. *Int. J. Nanomed.* **2008**, *2*, 133–149. [CrossRef] [PubMed]
4. Ger, T.R.; Huang, H.T.; Huang, C.Y.; Liu, W.C.; Lai, J.Y.; Liu, B.T.; Chen, J.Y.; Hong, C.W.; Chen, P.J.; Lai, M.F. Comparing the magnetic property of shell thickness controlled of Ag-Ni core-shell nanoparticles. *J. Appl. Phys.* **2014**, *115*, 17B528. [CrossRef]
5. Ger, T.R.; Huang, C.Y.; Lai, M.F. Cell Culture Arrangement Using Ferromagnetic Diamond-Shaped Thin Films. *MAG IEEE Trans* **2013**, *49*, 3453–3455. [CrossRef]
6. Akiyama, H.; Ito, A.; Kawabe, Y.; Kamihira, M. Fabrication of complex three-dimensional tissue architectures using a magnetic force-based cell patterning technique. *Biomed. Microdevices* **2009**, *11*, 713–721. [CrossRef] [PubMed]
7. Lu, S.S.; Liu, S.; Zu, Q.Q.; Xu, X.Q.; Yu, J.; Wang, J.W.; Zhang, Y.; Shi, H.B. In vivo MR imaging ofintraarterially delivered magnetically labeled mesenchymal stem cells in a canine stroke model. *PLoS ONE* **2013**, *8*, e54963.
8. Yocum, T.G.; Wilson, B.L.; Ashari, P.; Jordan, K.E.; Frank, A.J.; Arbab, A.S. Effect of human stem cells labeled with ferumoxides-poly-L-lysine on hematologic and biochemical measurements in rats. *Radiology* **2005**, *235*, 547–552. [CrossRef]
9. Riemer, J.; Hoepken, H.H.; Czerwinska, H.; Robinson, S.R.; Dringen, R. Colorimetric ferrozine-based assay for the quantitation of iron in cultured cells. *Anal. Biochem.* **2004**, *331*, 370–375. [CrossRef]
10. Rad, A.M.; Janic, B.; Iskander, A.S.; Soltanian-Zadeh, H.; Arbab, A.S. Measurement of quantity of iron in magnetically labeled cells: Comparison among different UV/VIS spectrometric methods. *Biotechniques* **2007**, *43*, 627–628. [CrossRef]
11. Enpuku, K.; Minotani, T.; Gima, T.; Kuroki, Y.; Itoh, Y.; Yamashita, M.; Katakura, Y.; Kuhara, S. Detection of magnetic nanoparticles with superconducting quantum interference device (SQUID) magnetometer and application to immunoassays. *Jpn J. Appl. Phys* **1999**, *38*, 1102–1105. [CrossRef]
12. Matz, H.; Hartwig, S.; Kerberger, T.; Atzpadin, H.; Trahms, L. PART I: Electronics Sessions-Applications of SQUID's II-Low Tc SQUID Measurement System for Magnetic Relaxation Immunoassays in Unshielded Environment. *IEEE Trans. Magn.* **2001**, *11*, 1371–1374.
13. Nikitin, P.I.; Vetoshko, P.M.; Ksenevich, T.I. New type of biosensor based on magnetic nanoparticle detection. *J. Magn. Magn. Mater.* **2007**, *311*, 445–449. [CrossRef]
14. Lee, C.P.; Lai, M.F.; Huang, H.T.; Lin, C.W.; Wei, Z.H. Wheatstone bridge giant-magnetoresistance based cell counter. *Biosens. Bioelectron.* **2014**, *57*, 48–53. [CrossRef] [PubMed]
15. Kokkinis, G.; Jamalieh, M.; Cardoso, F.; Cardoso, S.; Keplinger, F.; Giouroudi, I. Magnetic-based biomolecule detection using giant magnetoresistance sensors. *J. Appl. Phys.* **2015**, *117*, 17B731. [CrossRef]
16. Albisetti, E.; Petti, D.; Cantoni, M.; Damin, F.; Torti, A.; Chiari, M.; Bertacco, R. Conditions for efficient on-chip magnetic bead detection via magnetoresistive sensors. *Biosens. Bioelectron.* **2013**, *47*, 213–217. [CrossRef] [PubMed]
17. Baselt, D.R.; Lee, G.U.; Natesan, M.; Metzger, S.W.; Sheehan, P.E.; Colton, R.A. biosensor based on magnetoresistance technology. *Biosens. Bioelectron.* **1998**, *13*, 731–739. [CrossRef]
18. Kurlyandskaya, G.V.; Sánchez, M.L.; Hernando, B.; Prida, V.M.; Gorria, P.; Tejedor, M. Giant-magnetoimpedance-based sensitive element as a model for biosensors. *Appl. Phys. Lett.* **2003**, *82*, 3053–3056. [CrossRef]

19. Blanc-Béguin, F.; Nabily, S.; Gieraltowski, J.; Turzo, A.; Querellou, S.; Salaun, P.Y. Cytotoxicity and GMI bio-sensor detection of maghemite nanoparticles internalized into cells. *J. Magn. Magn. Mater.* **2009**, *321*, 192–197. [CrossRef]
20. García-Arribas, A.; Martínez, F.; Fernández, E.; Ozaeta, I.; Kurlyandskaya, G.V.; Svalov, A.V.; Berganzo, J.; Barandiaran, J.M. GMI detection of magnetic-particle concentration in continuous flow. *Sens. Actuators A Phys.* **2011**, *172*, 103–108. [CrossRef]
21. Hirota, E.; Sakakima, H.; Inomata, K. *Giant Magneto-Resistance Devices*; Springer Science & Business Media: Berlin/Heidelberg, Germany, 2013; Volume 40.
22. Matahum, J.; Su, C.M.; Wang, W.J.; Lou, S.L.; Ger, T.R. Effect of Surface Charge on the Uptake of Magnetic Nanoparticles on Mouse Fibroblast Cells. *IEEE Magn. Lett.* **2017**, *8*, 1–5. [CrossRef]
23. Wang, W.J.; Huang, Y.C.; Su, C.M.; Ger, T.R. Multi-functional drug carrier micelles with anti-inflammatory drug. *Front. Chem.-Polym. Chem.* **2019**, *7*, 93. [CrossRef] [PubMed]
24. Liu, X.-D.; Chen, H.; Liu, S.S.; Ye, L.Q.; Li, Y.P. Hydrothermal synthesis of superparamagnetic Fe3O4 nanoparticles with ionic liquids as stabilizer. *Mater. Res. Bull.* **2015**, *62*, 217–221. [CrossRef]
25. Arriortua, O.K.; Insausti, M.; Lezama, L.; de Muro, I.G.; Garaio, E.; de la Fuente, J.M.; Fratila, R.M.; Morales, M.P.; Costa, R.; Eceiza, M.; et al. RGD-Functionalized Fe3O4 nanoparticles for magnetic hyperthermia. *Colloids Surf. B Biointerfaces* **2018**, *165*, 315–324. [CrossRef] [PubMed]
26. Wilhelm, C.; Gazeau, F.; Bacri, J. Magnetophoresis and ferromagnetic resonance of magnetically labeled cells. *Eur. Biophys. J.* **2002**, *31*, 118–125. [CrossRef]

biosensors

MDPI

Article

Different Strategies for the Microfluidic Purification of Antibiotics from Food: A Comparative Study

Lorenzo Lunelli [1,2], Martina Germanis [1,3], Lia Vanzetti [1] and Cristina Potrich [1,2,*]

1 Bruno Kessler Foundation, Center for Sensors & Devices, Via Sommarive 18, 38123 Trento, Italy
2 National Research Council, Institute of Biophysics, Via alla Cascata 56/C, 38123 Trento, Italy
3 FTH Srl (Femtorays), Via Solteri 38, 38121 Trento, Italy
* Correspondence: cpotrich@fbk.eu

Abstract: The presence of residual antibiotics in food is increasingly emerging as a worrying risk for human health both for the possible direct toxicity and for the development of antibiotic-resistant bacteria. In the context of food safety, new methods based on microfluidics could offer better performance, providing improved rapidity, portability and sustainability, being more cost effective and easy to use. Here, a microfluidic method based on the use of magnetic microbeads specifically functionalized and inserted in polymeric microchambers is proposed. The microbeads are functionalized either with aptamers, antibodies or small functional groups able to interact with specific antibiotics. The setup of these different strategies as well as the performance of the different functionalizations are carefully evaluated and compared. The most promising results are obtained employing the functionalization with aptamers, which are able not only to capture and release almost all tetracycline present in the initial sample but also to deliver an enriched and simplified solution of antibiotic. These solutions of purified antibiotics are particularly suitable for further analyses, for example, with innovative methods, such as label-free detection. On the contrary, the on-chip process based on antibodies could capture only partially the antibiotics, as well as the protocol based on beads functionalized with small groups specific for sulfonamides. Therefore, the on-chip purification with aptamers combined with new portable detection systems opens new possibilities for the development of sensors in the field of food safety.

Keywords: antibiotic extraction; aptamer functionalization; antibody functionalization; microfluidic purification

Citation: Lunelli, L.; Germanis, M.; Vanzetti, L.; Potrich, C. Different Strategies for the Microfluidic Purification of Antibiotics from Food: A Comparative Study. *Biosensors* **2023**, *13*, 325. https://doi.org/10.3390/bios13030325

Received: 27 January 2023
Revised: 17 February 2023
Accepted: 24 February 2023
Published: 27 February 2023

1. Introduction

Antibiotics are widely used both to treat and prevent infections in humans and animals, and to obtain favorable effects on animal growth. This widespread use has unwanted side effects, i.e., the possible contamination of the environment and of the food chain [1,2], with, for example, milk and meat for human consumption that may contain antibiotics [3–5]. In addition, the overuse of antibiotics adversely impacts the onset of antibiotic-resistant bacterial strains [6,7]. To minimize these problems, legislative bodies limit the amount of antibiotics that may be found in food for human consumption [8–10]. Besides standard laboratory technologies, several kind of sensors have been developed and commercialized to comply with these limits, with the aim of their potential use in small laboratories. These sensors are mainly based on lateral flow technologies or take advantage of ELISA (enzyme-linked immunosorbent assay) based methods. All these methodologies, however, suffer from the drawback of assessing the compliance of the raw food after its collection, implying that contamination may be spread to several batches of food during its processing, before the contamination may be detected. Ideally, raw food should be tested in the field at the beginning of the collection chain, leading to the isolation of contaminated batches from the following processing chain.

In this scenario, the development of automated purification methods may lead to the establishment of in-field usable instruments [11,12], which also allow automated data communication with centralized facilities. Using purification methods, traces of antibiotics could be separated from complex food matrices, easing their further analysis with automated methods [13]. This requires a two-step process, i.e., a first phase, where antibiotics are bound and the food matrix is removed, and a second phase, where the release of the captured antibiotics is implemented, using a medium that is far more simple than the original matrix. Several different capture strategies can be exploited, based on molecular-imprinted polymers [14], antibodies [15,16], aptamers [17–19] and also taking advantage of specific interactions that the molecules of interest have with small functional groups [20].

The ideal biosensor for the on-site detection of food contaminations should be able to process raw food and precisely quantify the possible presence of antibiotics. These characteristics are difficult to obtain in a single system since modern detectors are often based on label-free methods, such as SPR and electrochemical biosensors. SPR is highly sensitive, but the equipment is costly and not portable. On the contrary, electrochemical biosensors are not sensitive enough to apply them to real samples, due to the problem of interface effect on the electrode surface, which produces an extremely high background that impedes the proper quantification of biomolecules [21,22]. Many attempts to overcome these problems and to produce biosensors usable in real settings are based on nanomaterials [23–25], which, however, need a critical implementation toward the desired application. A combination among nanomaterials, microfluidics and new sensors could, therefore, increase dramatically the success of developing antibiotic biosensors for real sample detection [12,26]. With the aim of setting up a microfluidic purification system able to deliver simplified solutions to an innovative detector, such as the label-free sensors, here, different functionalization strategies were applied to microbeads inserted in a microfluidic device.

In this work, we focus indeed on the evaluation and comparison of the performance in terms of the efficiency of antibiotics' capture and release, and of several binding strategies, namely aptamers, antibodies and small functional groups, using functionalized magnetic microbeads as a common platform. Two different kind of microbeads are evaluated, namely polystyrene based (Dynabeads™) and agarose based (PureCube), carrying different functional groups, with respect to their performance regarding the purification of the antibiotics tetracycline, sulfonamides and chloramphenicol.

2. Materials and Methods

2.1. Materials

The following reagents were purchased from Merck Life Science S.r.l. (Milan, Italy): methanol, acetonitrile, oxalic acid, ethanolamine, glycine, dithiothreitol (DTT), sulfanilic acid, hydrochloric acid, ammonium hydroxide, ammonium sulphate and all powders for buffer solutions.

The following microbeads were used: Dynabeads™ M-270 Epoxy (DynaEpoxy) and Dynabeads™ M-270 Streptavidin (DynaSA) from Thermo Scientific (Waltham, MA, USA); MAGAR-cN (MagarNA) from Immagina Biotechnology (Trento, Italy); PureCube NHS-Activated MagBeads (PureCubeNHS) and PureCube Maleimide-Activated MagBeads (PureCube Maleimide) from Cube Biotech (Monheim am Rhein, Germany).

All aptamers were obtained from IDT Integrated DNA Technologies (Leuven, Belgium). Aptamers are listed in Tables 1 and 2.

Anti-antibiotics antibodies were obtained from Fitzgerald (Fitzgerald Industries International; Acton, MA, USA): sheep polyclonal anti-tetracycline antibody (a-TC Ab); sheep polyclonal anti-chloramphenicol antibody (a-CAF Ab); and rabbit polyclonal anti-sulfamethazine antibody (a-SMZ Ab). The fluorescent antibody Ab-A568 (rabbit anti-goat IgG (H+L), Cross-Adsorbed Secondary Antibody, AlexaFluor™568 and protein G were purchased from Thermo Scientific (Waltham, MA, USA).

Table 1. List of aptamers tested. Aptamers were all DNA based, except for a-TCmu, which was RNA based. They were purchased without modifications or with one of the modifications listed in the last column. Some aptamers were also modified at the 3′ end with a fluorescent label (/36-TAMSp/). All modifications are referred to the standard nomenclature of the manufacturer. TC: tetracycline; SDM: sulfadimethoxine; SMZ: sulfamethazine.

Name	Sequence (5′-3′)	Target	5′ Modification
a-TC8	CGG TGG TG	TC	/5AmMC12/; /5ThioMC6-D/
a-TC40	GTT TGT GTA TTA CAG TTA TGT TAC CCT CAT TTT TCT GAA C	TC	/5AmMC12/
a-TC76	CGT ACG GAA TTC GCT AGC CCC CCG GCA GGC CAC GGC TTG GGT TGG TCC CAC TGC GCG TGG ATC CGA GCT CCA CGT G	TC	/5AmMC12/; /5ThioMC6-D/; /5Biosg/
a-TCmu [1]	GGG CCU AAA ACA UAC CAG AUC GCC ACC CGC GCU UUA AUC UGG AGA GGU GAA GAA UAC GAC CAC CUA GGC UC	TC	/5Biosg/
a-SDM	GAG GGC AAC GAG TGT TTA TAG A	SDM	/5Biosg/; /5ThioMC6-D/
a-SMZ	TTA GCT TAT GCG TTG GCC GGG ATA AGG ATC CAG CCG TTG TAG ATT TGC GTT CTA ACT CTC	SMZ	/5Biosg/; /5ThioMC6-D/

[1] RNA sequence.

Table 2. List of nonsense (NS) aptamers. All modifications are referred to the standard nomenclature of the manufacturer.

Name	Sequence (5′-3′)	5′ Modification
NS-NH_2	CCG TCG AGC AGA GTT CCG TCG AGC AGA	/5AmMC12//iSp9/
NS-Thio	CCG TCG AGC AGA GTT CCG TCG AGC AGA	/5ThioMC6-D//iSp9/
NS-b	GT TGG GCA CGT GTT GTC TCT CTG TGT CTC GTG CCC TTC GCT AGG CCC ACA	/5BiotinTEG/

The following buffers were used:

Binding Buffer (BB; used for TC and SMZ): 100 mM NaCl, 2 mM $MgCl_2$, 5 mM KCl, 1 mM $CaCl_2$, 20 mM Tris/HCl pH 7.6

Binding Buffer Muller (BBM): 100 mM NaCl, 10 mM $MgCl_2$, 20 mM potassium phosphate buffer pH 7.5

Binding and Washing Buffer (B&W): 1 M NaCl, 0.5 mM EDTA, 5 mM Tris/HCl pH 7.5

Coupling Buffer I (CBI): 150 mM NaCl, 100 mM sodium phosphate pH 7.4

Phosphate Buffered Saline (PBS): 138 mM NaCl, 2.7 mM KCl, 10 mM sodium phosphate pH 7.4

Binding Buffer SDM (BBSDM): 50 mM NaCl, 5 mM KCl, 5 mM $MgCl_2$, 20 mM Tris/HCl pH 8

The Istituto Zooprofilattico Sperimentale del Piemonte, Liguria e Valle d'Aosta (IZSPLV), provided raw milk and honey, all tested for the absence of antibiotics. IZSPLV also provided pure tetracycline, sulfonamides and chloramphenicol for spiking solutions.

Microfluidic chips, with four chambers of 100 µL volume each (Rhombic Chamber Chip eP1, Fluidic Design 221, PMMA, 600 µm depth), were purchased from microfluidic ChipShop GmbH (Jena, Germany).

2.2. *Interaction of Antibiotics with Their Aptamers or Antibodies*

The specific recognition of aptamers and antibodies (in this section generally designed as macromolecules) with their targets was evaluated in solution, using two different methods: the effect of binding on the intrinsic fluorescence of tetracycline was exploited for testing the a-TC aptamers [17] (Section 2.2.1), while an equilibrium filtration method [27,28] was utilized for all aptamers (apart from a-TC8, which is too small for this method), and for all the antibodies (Section 2.2.2).

2.2.1. Spectrofluorimetric Analysis

Different concentrations of TC were tested for their intrinsic fluorescence signal in order to optimize the concentration to be used as starting amount. Therefore, TC was dissolved at fixed nanomolar concentration in a total volume of 1 mL of buffer 100 mM NaCl, 2 mM $MgCl_2$, 5 mM KCl, 1 mM $CaCl_2$, 20 mM sodium phosphate pH 7.6 for a-TC8, a-TC40 and a-TC76 or BBM for a-TCmu and specific aptamers were added at increasing concentrations. The aptamers were titrated in such a way as to not exceed a total volume increase of 5%. The solution was stirred during each titration step and allowed to equilibrate for 5 min before data collection (longer incubation time were also tested, without significant differences in results). Fluorescence spectra were acquired for each titration point with a SPEX FluorMax spectrofluorimeter (Horiba Instruments Inc., Edison, NJ, USA) at 25 °C. An excitation wavelength of 370 nm was used to acquire the emission spectrum from 380 to 610 nm. The fluorescence emission signal from 500 to 535 nm was then integrated and plotted against the aptamer concentration.

2.2.2. Equilibrium Filtration Method

Vials with 100 µL solutions of macromolecules:antibiotic (1:1 volumes) were prepared at a fixed antibiotic concentration and increasing macromolecule concentration and left to reach equilibrium for 30 min. Afterwards, the solutions were transferred in the upper compartment of filters with a nominal cutoff of 10.000 Dalton (Microcon-10kDa Centrifugal Filter Unit with Ultracel-10 membrane, Merck Life Science S.r.l.; Milan, Italy), or 3.000 Dalton (Amicon Ultra-0.5 Centrifugal Filter Unit, Ultracel-3, Merck Life Science S.r.l.; Milan, Italy), depending on the MW of the tested aptamer. Then, the solutions were centrifuged for 8 min at 12.000× g, allowing around 50–60 µL of solution to flow to the lower filter compartment. The antibiotic concentration found in the flow through is a good estimate of the free antibiotic present in the original reaction vial. On the other hand, it is expected that the antibiotic bound to the aptamer is (mostly, see below for details) confined in the upper filter compartment.

2.2.3. Fit of the Antibiotic-Macromolecule Titration

The amount of bound antibiotic present in the solutions prepared as described above, was estimated from the saturation parameter $\nu \equiv L/M$, i.e., defined as the moles of bound ligand (antibiotic) per mole of macromolecule (aptamer or antibody), expressed using the equation from Cantor and Schimmel Biophysical chemistry (section 15-3, [29]). The parameters obtained from this equation were used to build two different models for TC fluorescence and equilibrium filtration experiments. The fitting of experimental data was implemented in Octave v. 6.1.0 [30], with the optim package v.1.6.0, using TeXmacs v. 2.1.1 [31] as the graphical front-end. The detailed description of the fitting method as well as the definition of the fitting parameters is detailed in the Supplementary Materials file. In brief, the equilibrium constant k is obtained in both models, while the fluorescence yields for the free (F_f) and bound (F_b) TC characterize the fit of fluorescence experiments.

Concerning the fit of equilibrium filtration experiments, the fitted parameters are (alongside k) the filter throughput losses λ and the aspecific antibiotic adhesion ($TC_{aspecific}$ —for TC).

2.3. *Functionalization of Microbeads*

The microbeads were functionalized by adapting the manufacturer's instructions, as detailed in the Supplementary Materials file.

DynaEpoxy and *PureCubeNHS*. These beads were conjugated either to amino-terminated aptamers or to the primary amines exposed by the basic amino acids in antibodies or in protein G, used as an intermediate step to better orient the antibody molecules. Finally, DynaEpoxy was conjugated also with sulfanilic acid for the capture of sulfonamides, adapting the protocol described by Hirsch et al. [32]. All fractions were collected and measured at the spectrophotometer in order to quantify the bound aptamer/antibody/sulphone groups (Section 2.4.3).

DynaSA and *MagarNA*. These beads expose streptavidin on their surface and were made to react with biotinilated aptamers. Additionally, in this case, the supernatant collected after reaction was measured, and the amount of bound aptamer was quantified (Section 2.4.3).

PureCube Maleimide. Thiolated aptamers, reduced just before use, were conjugated to these beads, as detailed in the Supplementary Materials.

2.4. Characterization of the Functionalized Microbeads

The binding of aptamers or antibodies to the beads was monitored by binding a fluorescent aptamer or antibody to the beads and checking the resulting fluorescence via confocal microscopy. Moreover, all the solutions of aptamer/antibody used for the functionalization of the microbeads were collected and quantified with the spectrophotometer to indirectly estimate the amount of bound aptamer/antibody. When PureCubeNHS beads are functionalized, this spectrophotometric determination of the bound aptamer/antibody is impossible because during the procedure, the NHS group is released, whose absorbance maximum is close to that of aptamers and antibodies.

In addition, the XPS analysis was employed for monitoring the functionalization of beads.

2.4.1. Confocal Analysis

The microbeads functionalized with the fluorescent aptamer or antibody were resuspended in PBS, deposited on a microscope slide and covered with a coverslip. A Leica SP5-II confocal microscope (Leica Instruments, Wetzlar, Germany), equipped with a helium/neon laser (543 nm), was employed for imaging the beads, acquiring data with a 40× objective in air. Images were also acquired using the additional channel of the transmitted light photomultiplier to image all the microbeads, independently of their functionalization. Spectral data of functionalized and untreated microbeads were acquired to confirm the correspondence of the detected light with the fluorescence spectra of the used fluorophores.

2.4.2. XPS Analysis

The XPS analysis is particularly suitable for the chemical characterization of samples carrying biomolecule layers, as already shown [33]. Samples for XPS measurements were deposited as 15 µL solutions of microbeads resuspended in pure water, deposited on 1 cm × 1 cm substrates of thermally grown silicon oxide, after improving the beads adhesion with argon plasma treatment (10.5 W, 2 mbar, 1 min). Samples were carefully dried at room temperature before their introduction in the chamber. XPS analyses were performed using a Kratos Axis Ultra^{DLD} instrument equipped with a hemispherical analyzer and a monochromatic AlKα (1486.6 eV) X-ray source, in spectroscopy mode.For beads functionalized with sulfanilic acid, O 1s, C 1s, N 1s, S 2p core lines were acquired, while for beads functionalized with a-TC Ab, O 1s, C 1s, N 1s, and Si 2p core lines were measured. XPS quantification was performed using the instrument sensitivity factors and the high-resolution spectra. Charge compensation was achieved using a charge neutralizer located at the bottom of the electrostatic input lens system. The quantification, reported as a relative elemental percentage, was performed by using the integrated area of the fitted core lines and by correcting for the atomic sensitivity factors. XPS data analysis was performed using the software described in Speranza and Canteri [34].

2.4.3. Spectrophotometric Analysis

All solutions produced during the functionalization of microbeads as well as the stock solutions added to the beads were collected and measured at the spectrophotometer (Jasco V-550) to check the amount of aptamer/antibody present. Spectra from 190 to 340 nm were acquired and the absorbance at 260 (aptamers) or 280 nm (antibodies) was used to quantify the amount of aptamers (considering A_{260} ssDNA = 33 µg/mL) and antibodies (considering an approximate ε_{280}IgG = 210,000).

2.5. Binding/Elution Test

The binding of antibiotics to their specific targets, which are bound to microbeads, was studied both in 1.5 mL vials and using fluidic microdevices. The test in vials started from the functionalized microbeads, which are incubated with the antibiotics at fixed concentrations for 1 h at room temperature using a tilting and rotation mixer. At the end of incubation, the vials were mounted on a magnet, the supernatant was collected (unbound antibiotic), and the microbeads were washed for at least three times with the suitable binding buffer. Then, different conditions of ionic strength, pH, and temperature were tested for the elution of the antibiotics. A final washing step was then performed. All fractions were collected for quantification in HPLC.

The binding test performed on the microfluidic chip was adapted from Lunelli et al. [20] and is schematized in Figure 1. Briefly, a microfluidic PMMA chip was mounted on the ChipGenie ®edition P device (microfluidic ChipShop, Jena, Germany; in Figure 1b) and fluids were injected in the chip chamber inlets with two disposable syringes actuated by two syringe pumps (Legato 185, KD Scientific, Holliston, MA, USA; in Figure 1a). The chip chambers were filled with functionalized magnetic beads at the beginning of the experiment (typically 30 µL diluted in 350 µL of suitable buffer), and the solutions containing known amounts of antibiotic were loaded in the microdevice (step in Figure 1c). The antibiotic is captured on the beads' surfaces, while all the unwanted materials present in the raw matrix are washed away with the buffer. The antibiotic is then released with the proper elution condition (high ionic strength solutions, extreme pH, and temperature). Finally, a washing step with buffer is performed. All fractions were collected from the outlet for the quantification via HPLC (step in Figure 1d). Concerning the milk experiment, no preliminary treatments or dilution in the binding buffer was performed before the spiking of TC. Honey instead required a preliminary dilution of 5 times in the binding buffer.

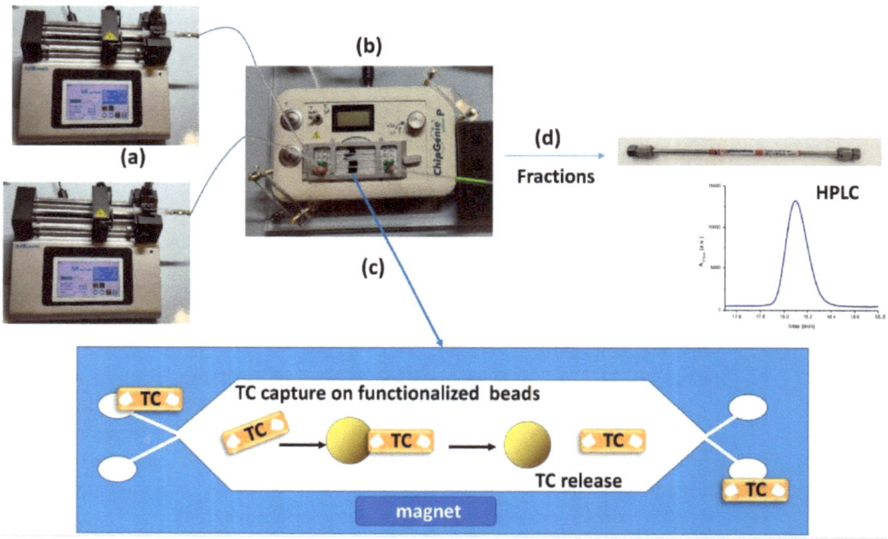

Figure 1. Scheme of the microfluidic purification of antibiotics. Two syringe pumps (**a**) inject the fluids into a microfluidic chip positioned on a device (**b**) equipped with a magnet. The chip, previously filled with the functionalized magnetic microbeads, is fluxed with a solution containing the antibiotics (TC; step (**c**)), which are captured by the beads. When a suitable elution solution is fluxed, the antibiotics are released from the beads and collected for analysis with HPLC (**d**).

2.6. Quantification of Antibiotic Solutions

The antibiotic present in the fractions collected during the binding/elution test (both in 1.5 mL vial and microdevice) was quantified by means of a Shimadzu (Kyoto, Japan) HPLC instrument. The instrument was equipped with the UV/Vis Detector SPD-20A, the Shimadzu prominence Communications Bus Module CBM-20A, the binary pump system Liquid Chromatograph LC- 20AB and with a degasser (Waters^{TM} In-Line Degasser) to improve HPLC performance. Fractions, after centrifugation at 21,000× g for 10 min, were manually injected through a Rheodyne® model 7725i injector connected with a 50 µL loop and passed through a precolumn Security Guard™ by Phenomenex® before separation by means of an inverse phase column (Luna® C18 by Phenomenex®, 5 µm, 10 nm, 250 mm × 3 mm), acquiring data with the software provided with the instrument (LCSolution). The mobile phase for tetracycline was oxalic acid 0.01 M, pH 2.7 (solvent A), and the gradient was obtained with methanol and acetonitrile in a 1.5:1 ratio (solvent B). Flux was set to 0.35 mL/min. The separation method was the following: 10% solvent B for 5 min, and gradient up to 58% of solvent B in 20 min. For the quantification of sulfonamides, 0.05 M sodium acetate pH 4.8 was used as a mobile phase (solvent A) and pure acetonitrile as solvent B. The separation method was gradient from 20% of solvent B to 50% B in 23 min, and 1 min up to 70% B. For the quantification of chloramphenicol, 0.1% formic acid in water was used as solvent A and 0.1% formic acid in acetonitrile as solvent B, while the separation method was 20% of solvent B for 5 min, gradient up to 95% B in 15 min. A calibration curve was obtained for each antibiotic by injecting known amounts of antibiotic in HPLC; the fit of this curve allowed the precise quantification of the antibiotic present in each fraction.

3. Results and Discussion

Different strategies implementable on microfluidic devices were tested, and their performance in terms of antibiotics capture and release were compared. All strategies are based on the use of microbeads exposing different functional groups, which were exploited for the specific functionalization either with aptamers (Figure 2, strategy a), or antibodies (Figure 2, strategy b), or small molecules such as sulfanilic acid (Figure 2, strategy c). The morphology of all the tested microbeads was checked by field-emission scanning electron microscopy before functionalization (Figure S1 of Supplementary Materials). The functionalization of beads was carefully monitored as well as the interaction of aptamers/antibodies with their specific targets before using the functionalized beads for the on-chip purification of antibiotics.

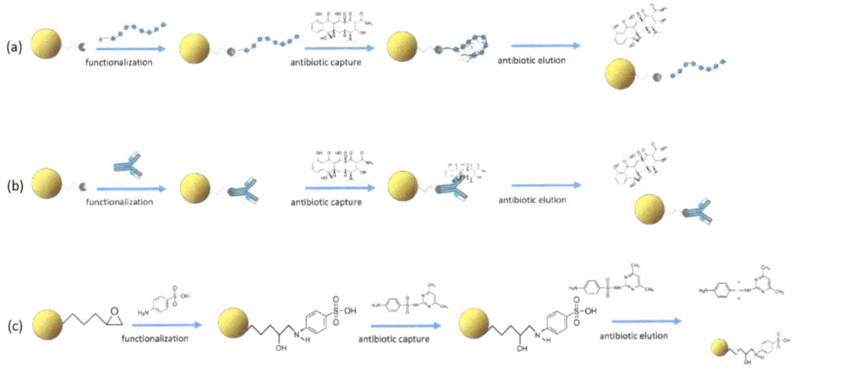

Figure 2. Strategies for the purification of antibiotics. Beads exposing specific functional groups are covalently coupled to the recognition element (functionalization), which binds its antibiotic target (antibiotic capture). Molecules present in the matrix are washed away, while the antibiotic is retained. By changing the environmental conditions (pH, ionic strength, temperature), the antibiotic is released (antibiotic elution) and collected for further analyses. (**a**) Strategy based on beads functionalized with aptamers as recognition element; (**b**) strategy based on beads functionalized with antibodies as recognition element; (**c**) strategy based on beads functionalized with sulfone groups for the purification of sulfonamides.

3.1. Strategy Based on Beads Functionalized with Aptamers

The first strategy explored for the microfluidic purification of antibiotics possibly present as contaminants in food, is based on the specific interaction of aptamers with their target antibiotics. Many aptamers are reported in the literature as specific for antibiotics, in particular, for TC. Here, four aptamers recognizing TC were selected and tested (see Table 1): a-TC76, a DNA-aptamer 76 nucleotides long and two shorter DNA aptamers, a-TC40 and a-TC8, partially sharing the same sequence as a-TC76 [35–37]. A fourth aptamer based on a RNA sequence was also tested [17]. In addition, two aptamers specific for sulfonamides were analyzed, i.e., a-SDM [38] and a-SMZ [18,39].

3.1.1. Aptamers–Antibiotics-Free Interaction in Solution

(a) Fluorescence test

The binding affinity of anti-TC aptamers was tested in solution, exploiting the intrinsic fluorescence of TC [17], which may change when aptamers interact with the antibiotic. Starting from a fixed nanomolar amount of TC, increasing amounts of aptamers were added, and the fluorescence spectra were acquired. When the aptamer binds to TC, the fluorescence signal changes from $F_f \cdot TC$ and tends to $F_b \cdot TC$ at high aptamer concentration. This leads to an increase in fluorescence, if $F_b > F_f$. The experimental binding plots are reported in Figure 3a–c. As can be observed, only the a-TCmu aptamer leads to a measurable increase in fluorescence, obtaining in this case a value of $\simeq$ 60 nM for the equilibrium constant k, and an increase of $\simeq$ 12 times for the tetracycline fluorescence yield, when bound. When a-TC8 (data not shown), a-TC40 and a-TC76 are employed, instead, no measurable changes in fluorescence intensity are detected. Correspondingly, in these cases, $F_b \simeq F_f$ is obtained from the fit and no k values can be obtained for these aptamers with this method.

(b) Equilibrium filtration test

The anti-TC aptamers were then tested with the equilibrium filtration method (Figure 3d–f). Because of its very low MW (2830 dalton), well below the nominal filter cutoff of 10 kD and 3 kD, a-TC8 was not tested. The parameters obtained from the fitting procedure are summarized in Table 3. The value of k obtained for the a-TCmu aptamer compares very well with the value obtained with the fluorescence method. For the a-TC40 aptamer, a k of the same order of magnitude is obtained, while for the a-TC76 aptamer, a much lower binding is measured. Note that when the a-TC40 aptamer is employed, a noticeable filter leakage is detected, as expected considering the MW of this aptamer (13 kD), not far from the filter cutoff of 10 kD.

Table 3. Parameters obtained from data fitting of the equilibrium filtration test for the interaction of aptamers with TC.

Aptamer	k (nM)	λ	$TC_{aspecific}$ (nM)
a-TC40	100	0.53	12.4
a-TC76	1.4×10^4	0	2.9
a-TCmu	64	$\simeq 0$	3.5

The same test was performed on SMZ and SDM sulfonamides, using a-SMZ and a-SDM aptamers, respectively. Unfortunately, as shown by the plots reported in Figure 4, no effect of binding was detected. In fact, the k value of a-SMZ results far higher than the range of explored aptamer concentration, i.e., larger than 2.5 µM. Therefore, in the tested conditions, no apparent interaction of SDM neither of SMZ with their specific aptamers was observed. To be noted, the filtration test with the a-SDM aptamer was performed with filters of 3 kD nominal cutoff since the MW of the a-SDM aptamer is approximately 7.4 kD, i.e., lower than the cutoff of the standard filters used for all other aptamers.

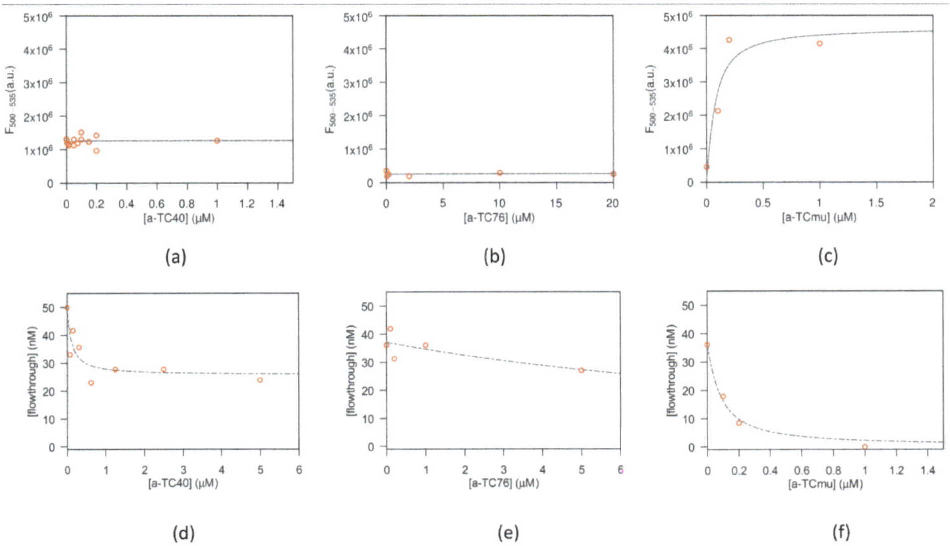

Figure 3. Binding tests of tetracycline and aptamers: circles (experimental data), dashed lines (fit). Fluorescence emission data for TC incubated with aptamer a-TC40 (a), a-TC76 (b) and for a-TCmu (c). Data obtained with a-TC8 are similar to (b). Data obtained with the filtration test for tetracycline and a-TC40 aptamer (d), a-TC76 aptamer (e) and a-TCmu aptamer (f).

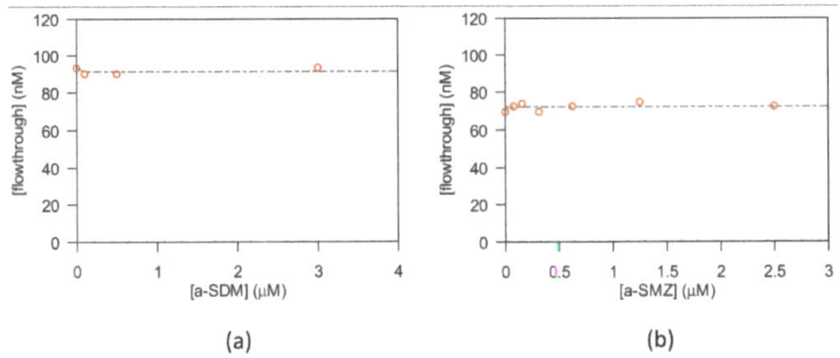

Figure 4. Filtration test of sulfonamides. (a) Data obtained with sulfadimethoxine and a-SDM aptamer. (b) Data obtained with sulfamethazine and a-SMZ aptamer. Y axes show the concentration of antibiotic present in the filter flow through.

3.1.2. Functionalization of Microbeads with Aptamers

The ideal conditions for the interaction aptamer target could be very different when this couple is present in a solution with respect to having the aptamer bound to the microbeads surface. For this reason, aptamers were coupled to beads different for the dimensions, material and exposed chemical group (for reference, see [40]). Amino-terminated aptamers were covalently conjugated to beads exposing epoxy or NHS groups, while thiolated aptamers were covalently conjugated to maleimide beads and biotinilated aptamers to beads coated with streptavidin or neutravidin. The binding was verified with a spectrophotometer, when possible, and by confocal analysis, using labeled aptamers. Figure 5 shows some examples of beads carrying different chemistry (epoxy–amino in (a), streptavidin–biotin in (b) and maleimide–thiol conjugation in (c)), after their functionalization with aptamers. For the confocal analysis, all aptamers employed were synthesized with a suitable reactive group at 5′ end and a fluorophore at the 3′ end. All types of chemistry worked well concerning the binding of the aptamers, independently of the beads dimensions (in the range from few µm to tens of µm) and materials (polymers vs. agarose).

After aptamer binding, all beads were fluorescent (by comparison with pictures of the same area, acquired using transmitted light, data not shown), indicating a good yield of the aptamer conjugation.

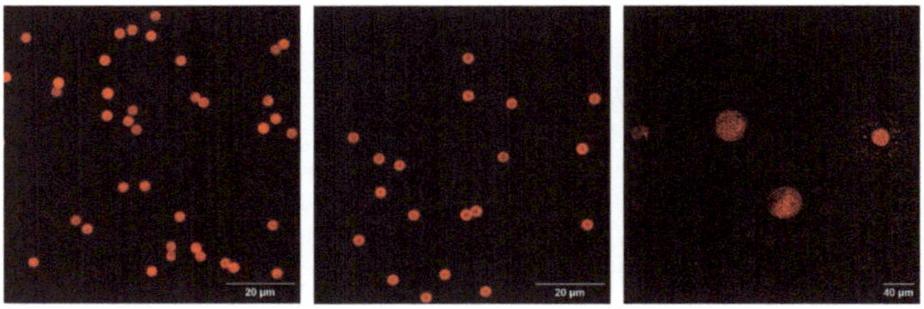

(a) (b) (c)

Figure 5. Confocal analysis of fluorescent aptamers bound to beads: DynaEpoxy beads conjugated with a-TC8 labeled with TAMRA (**a**), DynaSA conjugated with a-TC40 TAMRA (**b**), and PureCube Maleimide with a-TC76 TAMRA. Scale bars represent 20 μm in panels (**a**,**b**), while in (**c**) is referred to 40 μm.

3.1.3. Capture and Release of Antibiotics from Functionalized Microbeads

Upon assessing the conjugation of aptamers to the microbeads, a binding test was setup, firstly in vial and then on-chip. The different microbeads functionalized with anti-antibiotic aptamers were incubated with the corresponding antibiotic and solutions were collected after each step of the test (unbound, washes, elution, and final wash). All the collected fractions were quantified via HPLC in order to estimate the amount of antibiotic captured and released from aptamer-functionalized beads. All the a-TC aptamers bound to DynaEpoxy beads captured a good amount of the target, significantly higher than the nonsense (NS) aptamer (Figure 6, panel a), but no elution was observed. However, when a different elution step, i.e., glycine 100 mM pH 2.3, was implemented for a-TC8 and a-TC40 bound to DynaEpoxy, the percentage of eluted TC was between 9 and 11%, and similar results were obtained with these two aptamers conjugated to PureCubeNHS beads.

Moving on to a-TC aptamers conjugated via thiol chemistry to PureCube Maleimide beads, they seem to capture a quite good amount of TC and also the eluted TC seems in good percentage (Figure 6, panel b). Unfortunately, the NS aptamers gave in this case similar performance results. Finally, among biotinylated aptamers bound to DynaSA beads a-TCmu stands out both for the capture and elution of TC (Figure 6, panel c). This RNA aptamer resulted then as the most promising aptamer for the on-chip purification of TC.

The anti-sulfonamides aptamers were also tested in similar conditions, obtaining a small capture of SMZ (around 2%) by the a-SMZ-b aptamer bound to DynaSA, while no capture of SDM could be measured with the same chemistry, i.e., a-SMZ-b and DynaSA (data not shown). Analogously, as expected, no elution was observed for both the tested sulfonamides. To be noted, both aptamers are conjugated to the beads, as verified with the spectrophotometric analysis. Nevertheless, these results are in good agreement with the data of filtration test (Figure 4).

The binding test performed in vials allowed to select the most promising aptamer for the implementation of the on-chip purification of antibiotics. The a-TCmu aptamer was selected for the purification of tetracycline, binding its biotinylated form to MagarNA or DynaSA beads. The functionalized beads were inserted in the chip chamber and TC was fluxed at the lowest TC concentration allowed in milk (maximum residue limit or MRL of 100 μg/kg or 0.1 ng/μL [8,9]), and then eluted with a high ionic strength solution or with high temperature in order to promote a structural change in the aptamer sequence, releasing TC. Fractions of the flowing solutions were collected during the experiment and quantified (Figure 7). Different amounts of functionalized MagarNA beads were tested (Figure 7, panel a), finding a higher TC recovery when a double amount of beads was used, while 3 times more beads did not further improve the purification performance results. Moreover, two different conditions were tested for TC elution, i.e., 2 M NaCl (high ionic strength) and high temperature (75 °C in water), as shown in Figure 7, panel b. In all the tested conditions, TC not only could be purified from the starting solution but was also concentrated up to two times in the eluted fractions. Interestingly, the elution in water is particularly promising since TC could be further concentrated by evaporating water without any contaminants. This result opens the way for a sensitive detection and analysis of pre-purified antibiotics with modern technologies, such as label-free methods. Moreover, the possibility to concentrate antibiotics with respect to the initial sample could be crucial for detecting minute traces of antibiotic residues.

Given the good results obtained starting from TC spiked in buffer (Figure 7), the on-chip purification was extended to TC in milk or honey. Unfortunately, 0.1 ng/µL of TC spiked in raw milk or in milk diluted with binding buffer up to 10 times gave no positive results, possibly because of the matrix effects on the capture of TC by a-TCmu. A similar protocol was tested for TC spiked in honey diluted 5 times in binding buffer, at 1 ng/µL concentration. In this case, a little amount of TC could be recovered (TC/TC0 for e2 fraction was 0.05; elution was performed at high temperature) but far from the results obtained for the on-chip purification of TC spiked in buffer. A possible explanation of this result is that honey contains many different compounds able to interfere with the aptamer–antibiotic binding [41,42]. Good binding results are indeed reported only when honey is treated to a great extent before analysis [43]. Similarly, analogous data were obtained when TC spiked in raw milk was processed on PureCubeNHS beads functionalized with a-TC8 aptamer (TC/TC0 for the eluted fraction e2). Additionally, milk contains several compounds that could interfere with the aptamer–antibiotic binding [44,45], while this adverse effect is less present when the interaction of antibiotics is mediated by small molecules or even ions [20].

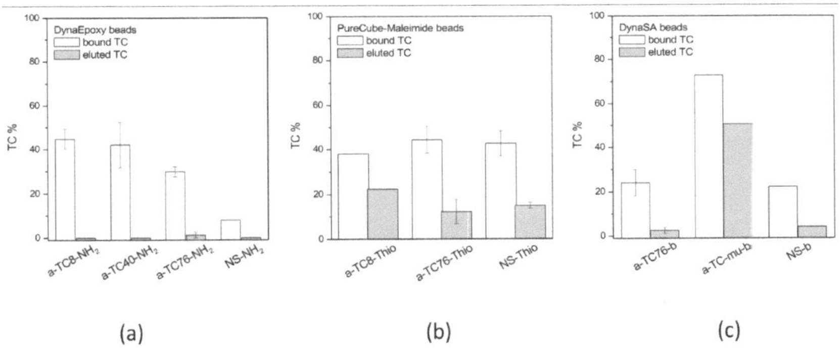

Figure 6. Binding of tetracycline to beads functionalized with aptamers. The percentage of TC is referred to the initial antibiotic added to the beads and set as 100% TC. (**a**) Beads exposing epoxy groups functionalized with amino-terminated aptamers; (**b**) beads exposing maleimide groups functionalized with thiol-terminated aptamers; (**c**) beads exposing streptavidin functionalized with biotin at 5′. The elution step was performed with 2 M NaCl and heating at 80 °C. All reactions were carried out in 1.5 mL vial, fractions of each step were collected and quantified with HPLC. NS: nonsense aptamers. Means and standard errors are reported.

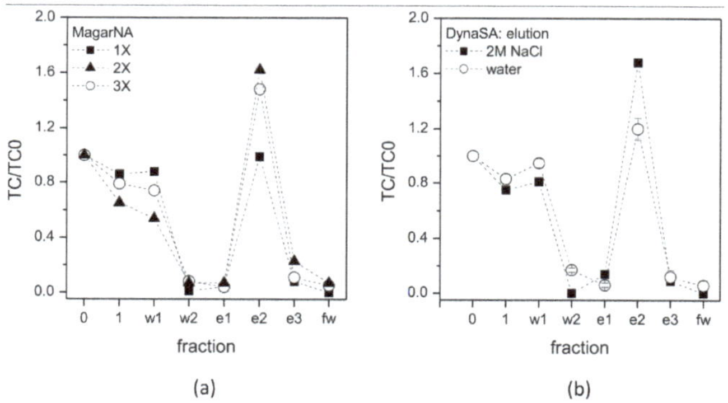

Figure 7. Microfluidic test. MagarNA beads (**a**) or DynaSA beads (**b**) were functionalized with a-TCmu aptamer and inserted in a 100 µL chip chamber before fluxing TC at 0.1 ng/µL concentration (fraction named 0). Consecutive fractions were collected and quantified with HPLC: fraction 1 represents unbound TC, fractions w1 and w2 are two successive washes, fractions e1-e3 represent eluted TC and fraction fw is the final wash. In (**a**) different amount of functionalized beads are tested, while in (**b**) different elution methods are shown, i.e., high ionic strength (2M NaCl) and high temperature (water at 75 °C).

3.2. *Strategy Based on Beads Functionalized with Antibodies*

Besides aptamers, antibodies recognizing antibiotics have been developed despite the difficulty given by the small size of the antibiotic molecules [46,47]. Antibodies or protein G were covalently conjugated to epoxy or NHS beads. When protein G is bound, the antibodies are specifically bound to protein G with their Fc region, obtaining an oriented layer of antibodies (see for details [40]). Three antibodies recognizing antibiotics belonging to three different classes were selected to implement the on-chip purification based on microbeads functionalized with a-TC Ab, a-SMZ Ab and a-CAF Ab, respectively (strategy b in Figure 2).

3.2.1. Antibodies-Antibiotics Free Interaction in Solution

As reported for aptamers in Section 3.1.1, also the antibody–antibiotic interaction was evaluated in solution using the equilibrium filtration method. Figure 8 reports the data obtained for the antibiotics TC, CAF and SMZ. The corresponding values of k are reported in Table 4. While a detectable interaction was obtained for TC and CAF, again, as obtained with the a-SMZ aptamer, no measurable interaction was found for SMZ.

Since the interaction in solution or on the microbeads surface could give different results, the binding test was performed for all the three antibodies conjugated to the respective microbeads.

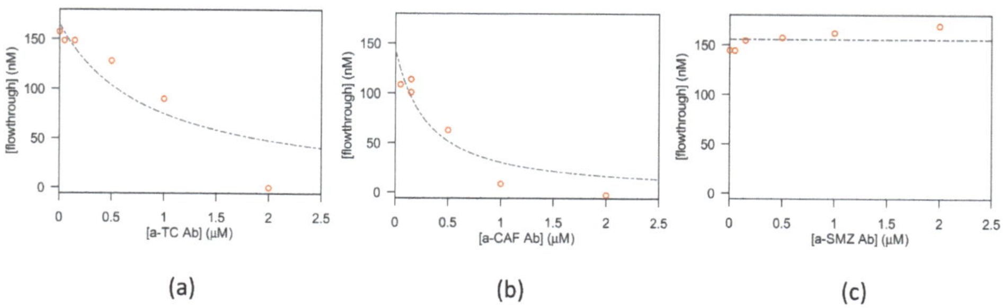

Figure 8. Characterization of binding between TC/CAF/SMZ and their Abs in solution (filtration test).

Table 4. Parameters obtained from data fitting of the equilibrium filtration test for the interaction of antibodies with their target antibiotics. ND: not determined.

Antibody	k (µM)	λ	$L_{aspecific}$ (nM)
a-TC Ab	1.6	0	33
a-CAF Ab	0.6	0	25
a-SMZ Ab	ND	-	-

3.2.2. Characterization of the Antibody Binding to the Microbeads

Before testing the binding of antibiotics to the functionalized beads, the successful conjugation of antibodies to the surface of both DynaEpoxy and PureCubeNHS was verified with the confocal analysis. A fluorescent antibody (Ab-A568) was conjugated following the same protocol developed for DynaEpoxy beads (i.e., binding of ptG to epoxy groups through amino residues, followed by passivation with ethanolamine and incubation with the proper antibody; see Section 2.3) or for PureCubeNHS and then was measured with the confocal microscope. All beads were found to be coated with the fluorescent antibody, as clearly visible in Figure 9), and as resulted from the comparison of same areas acquired using transmitted light (data not shown).

For DynaEpoxy beads, a quantification of the amount of antibodies bound to the beads was possible via spectrophotometric analysis, while unfortunately this was not possible for PureCubeNHS, which released the NHS group during the conjugation process. The NHS group absorbs near the typical protein absorption band (i.e., 260–280 nm), giving as a result a broad peak comprising both phenomena. a-TC Ab was selected for the characterization of the binding to the DynaEpoxy beads, finding a percentage of bound antibody of (33 ± 15) % and (1.2 ± 0.3) $\times 10^{12}$ molecules of antibody/cm^2 of beads surface, starting from the 0.3 mg/mL of antibody added during beads-Ab incubation.

The successful functionalization with antibody was checked for DynaEpoxy beads and a-TC Ab also with XPS measurements, acquiring the N 1s signal. DynaEpoxy indeed presents a small amount of nitrogen on its surface before functionalization (sample named "DynaEpoxy" in table 5), but this quantity increases when protein G is conjugated to the surface and even more when a-TC Ab is bound (Table 5), attesting to the correct binding of molecules after each step of conjugation. Taken together, these results indicate that DynaEpoxy was correctly functionalized with a-TC Ab. Unfortunately, the XPS analysis was not informative for PureCubeNHS since the signal related to the nitrogen content of antibody could not be distinguished from the much stronger nitrogen signal of NHS group, already present on the surface of the untreated beads.

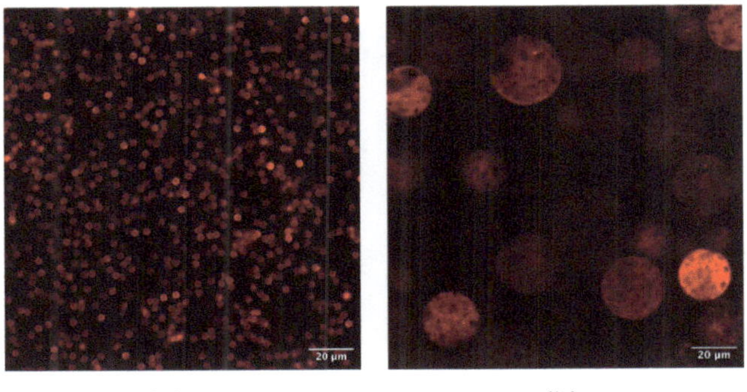

(a) (b)

Figure 9. Characterization of the binding of a fluorescent antibody to beads (confocal analysis). Ab-A568 was conjugated either to DynaEpoxy beads coated with protein G (**a**), or to PureCubeNHS beads (**b**). Scale bars represent 20 µm.

Table 5. Elemental composition (%) determined by XPS analysis. The error does not exceed the 1–2% of the reported value.

Sample	O (%)	N (%)	C (%)	Si (%)
DynaEpoxy	32.0	3.9	61.3	2.8
DynaEpoxy + pt G	28.3	4.1	65.4	2.2
DynaEpoxy + pt G + EA + a-TC Ab	28.9	5.2	64.4	1.6

3.2.3. Binding/Elution Test Using Antibodies

The microbeads functionalized with antibodies were tested for the specific capture and release of their respective antibiotics (Figure 10). Starting from DynaEpoxy beads conjugated with a-TC Ab, the conditions for TC elution were studied in terms of temperature, ionic strength, extreme pH and ion chelators (Figure 10a). All these conditions are expected to perturb the interaction between a-TC Ab and TC, releasing the antibiotic in solution for further analyses. Among the different conditions tested, the elution with citrate, i.e., 100 mM citrate pH 3, gave better TC recovery, and therefore, citrate was selected as the preferred elution buffer for further tests. a-TC Ab was also used to functionalize PureCubeNHS beads to compare the performance of the two types of microbeads in similar conditions (1 mg/mL a-TC Ab, 1 ng/µL TC, elution with citrate; Figure 10a,b). The capture and especially the release of TC was higher when PureCubeNHS was used, resulting then in being more promising for evaluating the performance of the other antibodies. PureCubeNHS was therefore functionalized with the three different antibodies at the same initial concentration, i.e., 0.3 mg/mL (Figure 10b). The same amount of the three antibiotics (i.e., 1 ng/µL) was incubated with the matching antibody-functionalized beads, finding both good binding and good recovery for a-TC Ab and a-CAF Ab, while for a-SMZ Ab binding, and especially elution, lower performance results were observed. These data are indeed in good agreement with the results obtained with the filtration test (Figure 8 and Table 4).

Beside tests in vials, a-TC Ab was selected as case study for setting up the on-chip purification of TC through both types of microbeads (Figure 10c). A solution of 0.1 ng/µL of TC spiked in binding buffer, i.e., the lower amount allowed by regulations, was flowed in the chip filled with the functionalized microbeads, and the collected fractions were quantified by HPLC. Again, the PureCubeNHS beads are able to release slightly more TC than DynaEpoxy. However, these findings clearly indicate a much worse performance, when compared with those obtained with aptamers in similar conditions (Figure 7), suggesting not to proceed with testing the on-chip purification of antibiotics spiked in raw food.

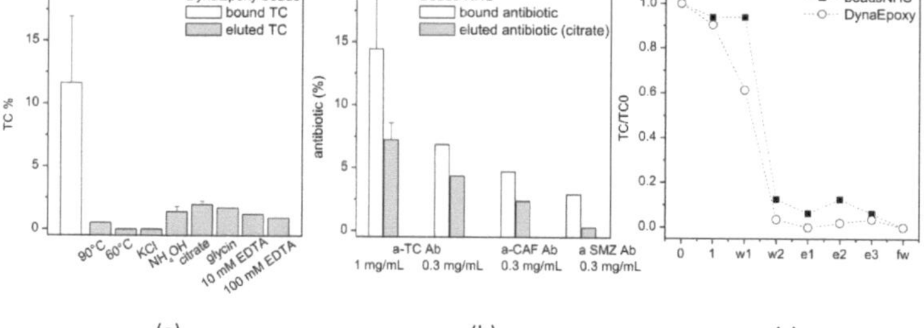

Figure 10. Binding and release of antibiotics from beads functionalized with antibodies. (**a**) Binding to DynaEpoxy beads functionalized with a-TC-Ab through protein G and elution of TC with different conditions. The percentage of TC is referred to the initial amount of antibiotic added to beads, set as 100%. (**b**) Binding of different antibiotics to PureCubeNHS beads functionalized with the respective antibodies and elution with citrate. (**c**) On-chip purification of TC captured by a-TC Ab bound to PureCubeNHS (solid squares) or DynaEpoxy (open circles).

3.3. Strategy Based on Small Molecules (Sulfanilic Acid)

The third strategy considered for capturing antibiotics on the functionalized microbeads relies on the binding of small molecules to the beads surface in order to expose suitable chemical groups (Figure 2, strategy c). This strategy was demonstrated as very promising for TC purification by means of microbeads exposing copper ions [20], with the only limit to be valid solely for tetracyclines.

3.3.1. Characterization of Sulfanilic Acid Binding

Here, sulfanilic acid was reacted with DynaEpoxy or PureCubeNHS beads in order to introduce a strong negatively charged group (SO_3^-), able to capture sulfonamides in acid environment. The correct functionalization was verified for DynaEpoxy treated with sulfanilic acid via XPS analysis (Table 6). Untreated DynaEpoxy does not contain any sulfur, and therefore the presence of S 2p is only due to the reaction of sulfanilic acid. On treated DynaEpoxy beads, indeed, a small but detectable amount of sulfur is found (Table 6).

XPS analysis was performed also on PureCubeNHS beads before and after the treatment with sulfanilic acid, but in this case, no differences were observed, indicating that sulfur, if present, is below the sensitivity of this technique.

Table 6. Elemental composition (%) determined by XPS analysis. The error does not exceed the 1–2% of the reported value.

Sample	O (%)	N (%)	C (%)	S (%)
DynaEpoxy	28.7	4.1	67.2	0.0
DynaEpoxy + sulfanilic acid	27.5	4.4	68.0	0.1

3.3.2. Binding/Elution Test

DynaEpoxy beads functionalized with sulfanilic acid were tested for the binding and elution of two sulfonamides, i.e., SMZ and SDM (Figure 11). The best performance reuslts in terms of both the capture and release of antibiotics were obtained for SDM, while SMZ was bound in a little amount (8%) to the beads and eluted in an even minor amount (2%). The binding test was performed also with PureCubeNHS beads and SMZ, finding an even smaller binding of the antibiotic (4 ± 2%) and no detectable elution. The quite good result obtained using DynaEpoxy beads and SDM, however, requires harsh binding conditions (HCl 0.1 M) as well as quite harsh elution conditions (NaOH 28%/methanol in 35/65 v/v ratio), making this strategy quite difficult for raw food processing. Moreover, these harsh conditions hinder the use of the eluted antibiotic for further analyses without any treatment, making, in this way, the whole idea behind the purification process ineffective.

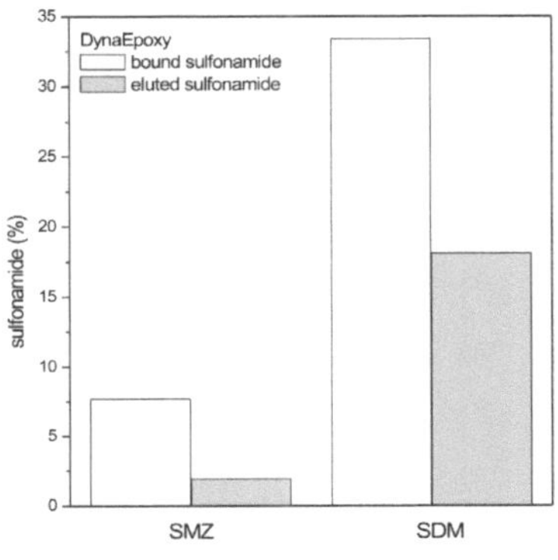

Figure 11. Binding and elution test of sulfonamides from DynaEpoxy beads functionalized with sulfanilic acid. The percentage of sulfonamides is referred to the initial amount of antibiotic present in the reaction.

4. Conclusions

The microfluidic purification of antibiotics could be a powerful instrument to detect traces of residual antibiotics in food before they enter the food chain, becoming a great issue for human health and for the environment. Microbeads inserted into a microchamber lead to an increased surface-to-volume ratio of the microdevice. Therefore, if properly functionalized, they have a huge potential in purifying and concentrating also diluted antibiotics for easier quantification. Combining the two technologies, i.e., microfluidic and functionalized microbeads, could offer an improved approach toward an efficient, rapid, automated and easy-to-use analysis. Here, three strategies were evaluated for the functionalization of microbeads and used for binding and eluting antibiotics. Every strategy has its own pros and cons, but taking together all the presented results, the most promising strategy for the on-chip purification of antibiotics seems to be the strategy based on aptamers. Aptamers gave the highest yield in antibiotic capture and, at least one aptamer (i.e., a-TCmu), was able to concentrate almost two times the initial amount of antibiotics. Unfortunately, all the strategies evaluated in this paper showed some weak points when applied to the purification of antibiotics directly from raw food, leaving room for improvement in the interactions between the target and its ligand in complex food matrices, such as milk or honey.

Supplementary Materials: The following supporting information [29–32] can be downloaded at: https://www.mdpi.com/article/10.3390/bios13030325/s1. Methods: Fit of the antibiotic-macromolecule titration; Methods: Functionalization of microbeads. Figure S1: FE-SEM analysis of the morphology of all types of microbeads used in this study.

Author Contributions: Conceptualization, L.L. and C.P.; methodology, L.L. and C.P.; software, L.L.; validation, all authors; investigation, all authors; writing—original draft preparation, C.P.; writing—review and editing, all authors; visualization, L.L. and C.P.; project administration, L.L. and C.P. All authors have read and agreed to the published version of the manuscript.

Funding: This work was partially supported by the Food-Drug-Free project in the framework of the "Piattaforma Tecnologica: Bioeconomia" funded by Regione Piemonte (POR-FESR 2014–2020): "Sviluppo di una piattaforma tecnologica miniaturizzata portatile per l'identificazione di farmaci nei prodotti alimentari".

Institutional Review Board Statement: Not applicable.

Informed Consent Statement: Not applicable.

Data Availability Statement: The data presented in this study are available from the corresponding author upon request.

Acknowledgments: Authors thank M. Fedrizzi for his valuable help with electron microscopy measurements.

Conflicts of Interest: The authors declare no conflict of interest.

Abbreviations

The following abbreviations are used in this manuscript:

TC	Tetracycline
CAF	Chloramphenicol
SMZ	Sulfamethazine
SDM	Sulfadimethoxine
EA	Ethanolamine
ptG	Protein G

References

1. Ana, K.M.S.; Madriaga, J.; Espino, M.P. β -Lactam antibiotics and antibiotic resistance in Asian lakes and rivers: An overview of contamination, sources and detection methods. *Environ. Pollut.* **2021**, *275*, 116624. [CrossRef] [PubMed]
2. Thakali, A.; MacRae, J.D. A review of chemical and microbial contamination in food: What are the threats to a circular food system? *Environ. Res.* **2021**, *194*, 110635. [CrossRef] [PubMed]
3. Wang, H.; Ren, L.; Yu, X.; Hu, J.; Chen, Y.; He, G.; Jiang, Q. Antibiotic residues in meat, milk and aquatic products in Shanghai and human exposure assessment. *Food Control* **2017**, *80*, 217–225. [CrossRef]
4. Baazize-Ammi, D.; Dechicha, A.S.; Tassist, A.; Gharbi, I.; Hezil, N.; Kebbal, S.; Morsli, W.; Beldjoudi, S.; Saadaoui, M.R.; Guetarni, D. Screening and quantification of antibiotic residues in broiler chicken meat and milk in the central region of Algeria. *Rev. Sci. Tech. (Int. Off. Epizoot.)* **2019**, *38*, 863–877. [CrossRef] [PubMed]
5. Sachi, S.; Ferdous, J.; Sikder, M.; Hussani, S. Antibiotic residues in milk: Past, present, and future. *J. Adv. Vet. Anim. Res.* **2019**, *6*, 315. [CrossRef] [PubMed]
6. Chen, J.; Ying, G.G.; Deng, W.J. Antibiotic residues in food: Extraction, analysis, and human health concerns. *J. Agric. Food Chem.* **2019**, *67*, 7569–7586. [CrossRef] [PubMed]
7. Nnadozie, C.F.; Odume, O.N. Freshwater environments as reservoirs of antibiotic resistant bacteria and their role in the dissemination of antibiotic resistance genes. *Environ. Pollut.* **2019**, *254*, 113067. [CrossRef] [PubMed]
8. European Commission. Commission Regulation (EU) No 37/2010 of 22 December 2009 on pharmacologically active substances and their classification regarding maximum residue limits in foodstuffs of animal origin. *Off. J. Eur. Union* **2010**, *15*, 1.
9. Authority, E.F.S. *Report for 2017 on the Results from the Monitoring of Veterinary Medicinal Product Residues and Other Substances in Live Animals and Animal Products*; EFSA Technical Report; EFSA: Parma, Italy, 2019.
10. Companyó, R.; Granados, M.; Guiteras, J.; Prat, M.D. Antibiotics in food: Legislation and validation of analytical methodologies. *Anal. Bioanal. Chem.* **2009**, *395*, 877–891; Erratum in *Anal. Bioanal. Chem.* **2010**, *396*, 949–949. [CrossRef] [PubMed]
11. Jin, Y.; Dou, M.; Zhuo, S.; Li, Q.; Wang, F.; Li, J. Advances in microfluidic analysis of residual antibiotics in food. *Food Control* **2022**, *136*, 108885. [CrossRef]
12. Majdinasab, M.; Mishra, R.K.; Tang, X.; Marty, J.L. Detection of antibiotics in food: New achievements in the development of biosensors. *Trends Anal. Chem.* **2020**, *127*, 115883. [CrossRef]
13. Lin, Z.; He, L. Recent advance in SERS techniques for food safety and quality analysis: A brief review. *Curr. Opin. Food Sci.* **2019**, *28*, 82–87. [CrossRef]
14. Huang, C.; Wang, H.; Ma, S.; Bo, C.; Ou, J.; Gong, B. Recent application of molecular imprinting technique in food safety. *J. Chromatogr. A* **2021**, *1657*, 462579. [CrossRef] [PubMed]
15. Zhao, M.; Li, X.; Zhang, Y.; Wang, Y.; Wang, B.; Zheng, L.; Zhang, D.; Zhuang, S. Rapid quantitative detection of chloramphenicol in milk by microfluidic immunoassay. *Food Chem.* **2021**, *339*, 127857. [CrossRef]
16. Stevenson, H.S.; Shetty, S.S.; Thomas, N.J.; Dhamu, V.N.; Bhide, A.; Prasad, S. Ultrasensitive and Rapid-Response Sensor for the Electrochemical Detection of Antibiotic Residues within Meat Samples. *ACS Omega* **2019**, *4*, 6324–6330. [CrossRef] [PubMed]
17. Müller, M.; Weigand, J.E.; Weichenrieder, O.; Suess, B. Thermodynamic characterization of an engineered tetracycline-binding riboswitch. *Nucleic Acids Res.* **2006**, *34*, 2607–2617. [CrossRef] [PubMed]
18. He, Y.; Zhang, B.; Fan, Z. Aptamer based fluorometric sulfamethazine assay based on the use of graphene oxide quantum dots. *Microchim. Acta* **2018**, *185*, 163. [CrossRef]
19. Evtugyn, G.; Porfireva, A.; Tsekenis, G.; Oravczova, V.; Hianik, T. Electrochemical Aptasensors for Antibiotics Detection: Recent Achievements and Applications for Monitoring Food Safety. *Sensors* **2022**, *22*, 3684. [CrossRef] [PubMed]
20. Lunelli, L.; Germanis, M.; Vanzetti, L.; Tatti, R.; Potrich, C.; Pederzolli, C. On-Chip Purification of Tetracyclines Based on Copper Ions Interaction. *Sensors* **2021**, *21*, 7236. [CrossRef] [PubMed]
21. Chen, T.; Cheng, G.; Ahmed, S.; Wang, Y.; Wang, X.; Hao, H.; Yuan, Z. New methodologies in screening of antibiotic residues in animal-derived foods: Biosensors. *Talanta* **2017**, *175*, 435–442. [CrossRef]
22. Luan, Y.; Wang, N.; Li, C.; Guo, X.; Lu, A. Advances in the Application of Aptamer Biosensors to the Detection of Aminoglycoside Antibiotics. *Antibiotics* **2020**, *9*, 787. [CrossRef] [PubMed]
23. Lan, L.; Yao, Y.; Ping, J.; Ying, Y. Recent advances in nanomaterial-based biosensors for antibiotics detection. *Biosens. Bioelectron.* **2017**, *91*, 504–514. [CrossRef] [PubMed]
24. Ding, R.; Chen, Y.; Wang, Q.; Wu, Z.; Zhang, X.; Li, B.; Lin, L. Recent advances in quantum dots-based biosensors for antibiotics detection. *J. Pharm. Anal.* **2022**, *12*, 355–364. [CrossRef] [PubMed]
25. Massaglia, G.; Frascella, F.; Chiadò, A.; Sacco, A.; Marasso, S.L.; Cocuzza, M.; Pirri, C.F.; Quaglio, M. Electrospun Nanofibers: from Food to Energy by Engineered Electrodes in Microbial Fuel Cells. *Nanomaterials* **2020**, *10*, 523. [CrossRef] [PubMed]
26. Wang, X.; Xie, Y.; Lin, L. Recent development of microfluidic biosensors for the analysis of antibiotic residues. *Trends Anal. Chem.* **2022**, *157*, 116797. [CrossRef]
27. Jenison, R.D.; Gill, S.C.; Pardi, A.; Polisky, B. High-resolution molecular discrimination by RNA. *Science* **1994**, *263*, 1425–1429. [CrossRef] [PubMed]
28. Zimmermann, G.R.; Wick, C.L.; Shields, T.P.; Jenison, R.D.; Pardi, A. Molecular interactions and metal binding in the theophylline-binding core of an RNA aptamer. *RNA* **2000**, *6*, 659–667. [CrossRef] [PubMed]

29. Cantor, C.R.; Schimmel, P.R. *Biophysical Chemistry: Part III: The Behavior of Biological Macromolecules*; Macmillan: New York, NY, USA, 1980.
30. Eaton, J.W.; Bateman, D.; Hauberg, S.; Wehbring, R. *GNU Octave Version 6.1.0 Manual: A High-Level Interactive Language for Numerical Computations*; Octave: San Francisco, CA, USA, 2020.
31. van der Hoeven, J. Towards semantic mathematical editing. *J. Symb. Comput.* **2015**, *71*, 1–46. [CrossRef]
32. Hirsch, D.B.; Baieli, M.F.; Urtasun, N.; Lázaro-Martínez, J.M.; Glisoni, R.J.; Miranda, M.V.; Cascone, O.; Wolman, F.J. Sulfanilic acid-modified chitosan mini-spheres and their application for lysozyme purification from egg white. *Biotechnol. Prog.* **2018**, *34*, 387–396. [CrossRef] [PubMed]
33. Vanzetti, L.; Pasquardini, L.; Potrich, C.; Vaghi, V.; Battista, E.; Causa, F.; Pederzolli, C. XPS analysis of genomic DNA adsorbed on PEI-modified surfaces. *Surf. Interface Anal.* **2016**, *48*, 611–615. [CrossRef]
34. Speranza, G.; Canteri, R. RxpsG a new open project for Photoelectron and Electron Spectroscopy data processing. *SoftwareX* **2019**, *10*, 100282. [CrossRef]
35. Niazi, J.H.; Lee, S.J.; Gu, M.B. Single-stranded DNA aptamers specific for antibiotics tetracyclines. *Bioorganic Med. Chem.* **2008**, *16*, 7245–7253. [CrossRef] [PubMed]
36. Jeong, S.; Paeng, I.R. Sensitivity and Selectivity on Aptamer-Based Assay: The Determination of Tetracycline Residue in Bovine Milk. *Sci. World J.* **2012**, *2012*, 1–10. [CrossRef] [PubMed]
37. Kwon, Y.S.; Raston, N.H.A.; Gu, M.B. An ultra-sensitive colorimetric detection of tetracyclines using the shortest aptamer with highly enhanced affinity. *Chem. Commun.* **2014**, *50*, 40–42. [CrossRef] [PubMed]
38. Song, K.M.; Jeong, E.; Jeon, W.; Jo, H.; Ban, C. A coordination polymer nanobelt (CPNB)-based aptasensor for sulfadimethoxine. *Biosens. Bioelectron.* **2012**, *33*, 113–119. [CrossRef] [PubMed]
39. Yang, L.; Ni, H.; Li, C.; Zhang, X.; Wen, K.; Ke, Y.; Yang, H.; Shi, W.; Zhang, S.; Shen, J.; et al. Development of a highly specific chemiluminescence aptasensor for sulfamethazine detection in milk based on in vitro selected aptamers. *Sensors Actuators B Chem.* **2019**, *281*, 801–811. [CrossRef]
40. Hermanson, G.T. *Bioconjugate Techniques*; Academic Press: San Diego, CA, USA, 1996.
41. Miguel, M.G.; Antunes, M.D.; Faleiro, M.L. Honey as a Complementary Medicine. *Integr. Med. Insights* **2017**, *12*, 117863371770286. [CrossRef]
42. da Silva, P.M.; Gauche, C.; Gonzaga, L.V.; Costa, A.C.O.; Fett, R. Honey: Chemical composition, stability and authenticity. *Food Chem.* **2016**, *196*, 309–323. [CrossRef]
43. Huang, Y.; Yan, X.; Zhao, L.; Qi, X.; Wang, S.; Liang, X. An aptamer cocktail-based electrochemical aptasensor for direct capture and rapid detection of tetracycline in honey. *Microchem. J.* **2019**, *150*, 104179. [CrossRef]
44. Craig, A.L.; Gordon, A.W.; Hamill, G.; Ferris, C.P. Milk Composition and Production Efficiency within Feed-To-Yield Systems on Commercial Dairy Farms in Northern Ireland. *Animals* **2022**, *12*, 1771. [CrossRef]
45. Schwendel, B.H.; Wester, T.J.; Morel, P.C.H.; Tavendale, M.H.; Deadman, C.; Shadbolt, N.M.; Otter, D.E. Invited review: organic and conventionally produced milk-an evaluation of factors influencing milk composition. *J. Dairy Sci.* **2015**, *98*, 721–746. [CrossRef] [PubMed]
46. Zhang, Y.; Lu, S.; Liu, W.; Zhao, C.; Xi, R. Preparation of Anti-Tetracycline Antibodies and Development of an Indirect Heterologous Competitive Enzyme-Linked Immunosorbent Assay to Detect Residues of Tetracycline in Milk. *J. Agric. Food Chem.* **2007**, *55*, 211–218. [CrossRef] [PubMed]
47. Bienenmann-Ploum, M.; Korpimäki, T.; Haasnoot, W.; Kohen, F. Comparison of multi-sulfonamide biosensor immunoassays. *Anal. Chim. Acta* **2005**, *529*, 115–122. [CrossRef]

biosensors

MDPI

Article

Establishment and Validation of an Integrated Microfluidic Step Emulsification Chip Supporting Droplet Digital Nucleic Acid Analysis

Gangyin Luo [1,2], Ying Zhang [3], Shun Wang [1], Xinbei Lv [4], Tianhang Yang [1,*] and Jinxian Wang [1,2,*]

1 Suzhou Institute of Biomedical Engineering and Technology, Chinese Academy of Sciences, Suzhou 215163, China; luogy@sibet.ac.cn (G.L.); wangs@sibet.ac.cn (S.W.)
2 School of Biomedical Engineering (Suzhou), Division of Life Sciences and Medicine, University of Science and Technology of China, Hefei 230026, China
3 JiHua Laboratory, Foshan 528251, China; zhangyin@jihualab.ac.cn
4 Qingdao Innovation and Development Base, Harbin Engineering University, Qingdao 266000, China; lvxinbei@hrbeu.edu.cn
* Correspondence: yangth@sibet.ac.cn (T.Y.); wangjx@sibet.ac.cn (J.W.); Tel.: +86-132-8451-1292 (T.Y.); +86-176-0594-0611 (J.W.)

Abstract: Uniform and stable droplet generation is critical for accurate and efficient digital nucleic acid analysis (dNAA). In this study, an integrated microfluidic step emulsification device with wide-range droplet generation capability, small device dimensions, convenient fabrication strategy, low contamination and high robustness was developed. A tree-shaped droplet generation nozzle distribution design was proposed to increase the uniformity of droplet generation by equating flow rates, and the flow field in the design was numerically simulated. Theoretical analysis and comparative experiments on droplet size were performed regarding the influences of nozzle dimensions and surface properties. With incubation and hydrophobic reagent treatment, droplets as small as 73.1 μm were generated with multiplex nozzles of 18 μm (h) × 80 μm (w). The droplets were then collected into a standard PCR tube and an on-chip monolayer droplet collection chamber, without manual transfer and sample contamination. The oil-to-sample volume ratio in the PCR tube was recorded during collection. In the end, the droplets generated and collected using the microfluidic device proved to be stable and uniform for nucleic acid amplification and detection. This study provides reliable characteristic information for the design and fabrication of a micro-droplet generation device, and represents a promising approach for the realization of a three-in-one dNAA device under a step emulsification method.

Keywords: step emulsification; micro-droplet; integrated microfluidic chip; nucleic acid detection

Citation: Luo, G.; Zhang, Y.; Wang, S.; Lv, X.; Yang, T.; Wang, J. Establishment and Validation of an Integrated Microfluidic Step Emulsification Chip Supporting Droplet Digital Nucleic Acid Analysis. *Biosensors* **2023**, *13*, 888. https://doi.org/10.3390/bios13090888

Received: 19 June 2023
Revised: 31 August 2023
Accepted: 11 September 2023
Published: 18 September 2023

1. Introduction

Nucleic acid amplification techniques can increase target nucleic acids into millions of copies in dozens of amplification cycles. It has been widely used in scenarios such as food safety evaluation, forensic identification, clinical diagnosis, and particularly during epidemiological investigation under the coronavirus disease 2019 (COVID-19) pandemic [1,2].

DNAA is a method used for the absolute quantification of nucleic acid molecules. Samples are divided into enormous amounts of micro units with homogeneous volumes, and then every unit undergoes an amplification process. The unit signals are read out based on fluorescence detection, and the ratio of positive droplets to total droplets can be obtained to calculate target nucleic acid concentrations [3].

Based on the insolubility between oil and water, droplet dNAA (ddNAA) is one of the dNAA methods that disperses sample solutions into micro-droplets in an oil phase fluid, using microfluidic approaches such as flow focusing, coaxial flow, step emulsification, ink-jetting, ultrasound excitation, oscillating, etc. [4–9]. For a certain volume of sample solution, generating droplets with smaller sizes means larger numbers of droplets would be obtained, which leads to a higher detection sensibility. For a 20 μL sample liquid, typically 20,000–50,000 droplets are generated in the ddNAA process [10]. Therefore, high-throughput droplet generation is significantly essential for investigation applications that require fast test results.

Commercial microfluidic chips using the flow focusing method can disperse 20 μL of sample mixture into 90 μm–120 μm diameter droplets in 1.5 min–5 min, mainly under shearing forces induced by the oil phase's hydraulic force, but this method usually requires a channel width of less than 80 μm [11]. The fabrication difficulty and cost at such scale are relatively high, regardless of soft lithography or precision machining, especially for lab research and product prototyping applications. For microfluidic chips with larger microchannel scales, small droplets can be obtained by raising the oil-to-sample volume ratio. However, this results in wasted oil phase and poor amplification performance. Moreover, the droplet size is extremely sensitive to the flow rates of the two phases in the flow focusing method and T-junction method, which requires a precise flow rate control group [12,13]. Step emulsification takes advantage of the density difference between the two phases to enhance the Plateau–Rayleigh instability [14], and it is easier for the disperse phase to form droplets under surface tension. Furthermore, the generated droplet diameter is not sensitive to small flow rates until it encounters a sudden rise at a flow rate threshold where the droplet generation principle transitions. Compared to generating droplets with the same diameter by the flow focusing method, larger channel dimensions and easier flow rate control in a step emulsification system can satisfy the same requirement.

Research involving multiple forms of devices has been conducted to help realize ddNAA with a step emulsification method in recent years. In 2017, Ofner fabricated a high throughput chip via glass etching [15]. The chip contains 364 microfluidic channels (step from 20 μm to 120 μm) in a comb-shaped distribution, and the total generation flow rate can reach 25 mL/h. Droplets of 81 μm diameter were generated. The glass chip is reusable for its compatibility with a high-temperature autoclave, but hydrophilic glass is not suitable for nucleic acid solution droplet generation. In 2021, Shi proposed a microfluidic emulsification device that relied on centrifugal force [16]. A microfluidic nozzle (step from 40 μm to 180 μm) fabricated with two-step soft lithography was connected with a sample chamber, and a 2 cm × 2 cm droplet chamber was preloaded with the oil phase. With a centrifuge rotor, the solution for SARS-CoV-2 N gene detection via droplet digital loop-mediated isothermal amplification (ddLAMP) can be dispersed into 130–175 μm diameter droplets. In 2020, Li generated droplets based on an asymmetrical beveled capillary [17]. The droplet diameter was proven to have a linear relationship with the capillary inner diameter. Droplets of deoxyribonucleic acid (DNA) samples were collected and successfully amplified. Furthermore, in 2020, Schulz designed a cartridge with eight emulsification nozzles in a brush-shaped distribution, and the cartridge can fit into a 2 mL standard polymerase chain reaction (PCR) tube [18]. As many as 6×10^5 droplets of a 66 μm diameter were generated by centrifuging the tube in less than 10 min. The droplets were then amplified directly in the 2 mL tube with droplet dPCR (ddPCR) and ddLAMP reagents. Later in 2021, Schlenker, from the same group, proposed a four-plex ddPCR based on a LabDisk [19]. The LabDisk contains 12 ddPCR units, and for each ddPCR unit, 6 μL of emulsification oil and 8 μL of PCR oil were used. Droplets of 82.7 μm can be generated from the single-step emulsification nozzle in the unit, and then amplified and read out within a pyramidal PCR collecting chamber. The droplets were generated under centrifugal force, using a centrifugal device instead of syringe pumps, which is inconvenient for miniaturization and portability. Peng, in 2021, proposed a compact massive microfluidic device that can be integrated into a PCR tube [20]. The PDMS devices, in three shapes, contained 18–24 nozzles in a circumferential distribution, and were fabricated based on two-step lithography. Droplets can be generated continuously without extra oil supply after premium addition in Peng's design, but the droplets need to be manually transferred to a PCR tube for amplification. Among all of the above research examples, using in-tube cartridges involves contamination processes, such as cartridge removal or droplets pipetting before/after amplification, whereas on-chip droplets storage usually leads to limited droplet amounts or bulk device volumes.

Moreover, most collecting tubes and monolayer droplet collecting chambers are customized, and are incompatible with commercial amplification devices. In developing a step emulsification microfluidic device integrated with droplet generation, droplet amplification and droplet detection (three-in-one device) facilities would be convenient for the realization of low-contamination ddNAA nucleic acid quantification.

To improve the performance of step emulsification devices for nucleic acid analysis applications, this study proposed a novel microfluidic chip that can be manufactured with a convenient and cost-efficient method, and operates with fewer manual transfer processes. The chip mold fabrication requires only one-step lithography, domestic printing and a domestic UV exposure device. The pressure and flow rate imbalances at the nozzles caused by nozzle distribution were studied and optimized, since these imbalances lead to irregular droplet sizes and/or only a few nozzles that can emulsify the sample [21]. Via simulation, the optimized 16-channel tree-shaped nozzle distribution was proven to have a better flow rate distribution. The entire chip design is compact, and the microstructure is compatible with an on-chip collecting chamber and PCR tube. The influences of microfluidic channel dimensions and surface treatments on the sample droplet size were recorded and analyzed. The collected droplets have a wide size range, a good sample-to-oil ratio, and remain stable after amplification. The experimental results prove that the homemade PDMS chip can be used as a highly efficient, reliable and integrated device for digital droplet nucleic acid analysis applications.

2. Materials and Methods

2.1. Characterization of Droplet Generation in Microchannel

During step emulsification, the water phase samples were separated into microscale monodisperse droplets in the oil phase, under the combined effect of Laplace pressure difference, gravitational forces induced by the density difference of the two phases, inertial forces, surface tension forces and viscous forces, among which the Laplace pressure difference has less influence than the others in our study. From previous research and our preliminary experiments, micro-droplets generation would transform into extra-large droplets generation or jet flow in specific parameter ranges. Dimensionless number Ca (capillary number), which represents the influence of viscous forces over surface tension forces; Bo (Bond number), which represents the influence of gravitational forces over surface tension forces; and We (Weber number), which represents the influence of inertial forces over surface tension forces, are introduced in our study to characterize the droplet generation processes [22–25]. The dimensionless numbers are defined as follows:

$$Ca = \mu_{\text{dis}} v / \gamma \tag{1}$$

$$Bo = (\rho_{\text{con}} - \rho_{\text{dis}}) G D^2 \tag{2}$$

$$We = \rho_{\text{dis}} D v^2 / \gamma \tag{3}$$

where μ_{dis}, v, γ, ρ_{con}, ρ_{dis}, G and D are the viscosity of the continuous phase, the velocity of the dispersed phase, the interfacial tension between the continuous and dispersed phases, the density of the continuous phase, the density of the dispersed phase, the acceleration of gravity and the droplet diameter, respectively.

The critical Ca, Bo and We values are seen as the constraints on the dripping mode to the jetting mode transition for multi-phase flow. It is commonly admitted in previous research that the critical Ca and We are $\sim 6 \times 10^{-3}$ and $\sim 10^{-2}$–10^{-1}, respectively [26]. Droplet generation conditions with Ca and We lower than their critical values are considered to be stable, as the surface tension governs the flow. However, the critical value of Bo number varies in different studies. Since buoyancy is involved in this research, we assumed ∼10^{-3}–1 to be the critical Bo number [20,27].

2.2. Numerical Simulation for Flow Rate in Microchannel

Variations in the flow rates in different nozzles can lead to unevenness in the droplet sizes. It is difficult to measure the flow rate on a microscale level. Therefore, we conducted a laminar flow simulation (COMSOL Multiphysics 5.6) to simulate the flow conditions inside the microfluidic chip. The flow of sample fluid in microfluidic channels is governed by the Navier–Stokes equation and continuity equation:

$$\rho(\mu\cdot\nabla)\, u = \nabla\cdot(-p + \mu\nabla\, u) \tag{4}$$

$$\rho\nabla\cdot u = 0 \tag{5}$$

where ρ, u, p and μ are the density, fluid velocity, pressure and viscosity, respectively. The mass flow at the inlet was set as 0.2 μL/s, and the pressure at the nozzle outlets was set as atmospheric pressure.

2.3. Nucleic Acid Sample and Reagents

Droplet generation oil for probes (BIO-RAD, Hercules, CA, USA) was chosen as the continuous phase in this study. The samples for PCR amplification were the dispersed phase. In a standard PCR sample with a volume of 15 μL, there were 7.5 μL ddPCR supermix for probes (no dUTP, BIO-RAD, Hercules, California, CA, USA), 0.75 μL λ DNA (SD0011, Thermo Fisher Scientific™, Waltham, MA, USA) as the PCR template, 2.25 μL of customized probe and primer (Invitrogen Trading (Shanghai) Co., Ltd., Shanghai, China) and 4.5 μL of water (Sangon, Shanghai, China) (Table S1). The details of the material properties used for the study are listed in Table 1.

Table 1. Material properties used for the study.

μ_{dis}	ρ_{con}	ρ_{dis}	γ
1.005 mPa·s	1614 kg/m^3	998.2 kg/m^3	8.1 mN/m

2.4. Fabrication Process of the Microdevice

The microfluidic chip designed for droplet generation consists three parts: a glass slide spin-coated with PDMS (polydimethylsiloxane, Sylgard™ 184, Dow, Midland, MI, USA) thin film as a substrate; a 3 mm thick PDMS layer with tree-shaped microchannels and an ascending reservoir for step emulsification; and a 1 mm thick PDMS layer as a top layer for sealing and droplet collection. In this study, the nozzles of the microfluidic channels were designed with several different widths for pattern study. The heights of the microchannels are 18 μm, 28 μm, and 38 μm. The width-to-height ratios (w/h) are larger than 3.5 for all the nozzles, in order to achieve better monodispersive generation [28–30]. Molds for the microfluidic channel with heights of 18 μm (nozzle widths = 80, 90, 100, 110, 120, 130, 140, 150 μm) and 28 μm (nozzle width = 100, 110, 120, 130, 140, 150, 160, 170 μm) were made with classic one-step lithography with SU8. The mold with a height of 38 μm (nozzle width = 100, 110, 120, 130, 140, 150, 160, 170, 180 μm) was home-made, with photoresistive dry film (Dupont, Wilmington, DE, USA), a high-precision ink-jet film printer and a UV light source, which is convenient for prototyping. The inlet of the tree-shaped microchannel is connected with a syringe pump containing sample solution. The 16 nozzles of the tree-shaped microchannel are connected with a large reservoir. The sample fluid is divided uniformly by four-stage binary divisions of the tree-shaped microchannel. The microfluidic structures were fabricated with soft lithography, and the large reservoir was manually carved out after the PDMS was cured. For on-chip monolayer droplets collection, a 15 mm × 11 mm square collection chamber with a height of 120 μm was added on the top PDMS layer. The height of the chamber should be smaller than 1.5 times the droplet diameter to prevent the droplets from overlapping and inaccurate quantification. Cylinder pillars were added in the chamber to support the chamber ceiling and help fluid spread evenly via surface tension. Four branches linked to the chip outlet were also attached at the end edge of the chamber to help the droplets spread adequately and uniformly in the chamber. The detailed design and dimensions of the soft lithography masks are provided in Figure S1.

The three layers were irreversibly bonded together using O_2 plasma treatment (60 s, 80 W). Small assembled chip dimensions (three-layer integrated chip: 22 mm (l) × 18 mm (w) × 5 mm (h)) and one-step lithography are significant advantages of this study's design, compared to other emulsification droplet generation devices. Since PDMS becomes hydrophilic after O_2 plasma treatment, and hydrophily is unfavorable for sample droplet generation, the internal surfaces of the chips were treated with three different methods after bonding which were the following: incubating the chip 48 h at 120 °C; perfusing the channels with hydrophobic reagent (1H, 1H, 2H, 2H -perfluorodecyltriethoxysilane (PFDTES, P122385, Aladdin, Shanghai, China): engineered fluid (3M™ Novec™ 7500, 3M, Saint Paul, MI, USA) (*v*/*v*) = 2%) and maintaining the chip at 120 °C for 5 h; and perfusing the channels with hydrophobic reagent and keeping the chip at 120 °C for 48 h [31]. The complete chip fabrication and treatment process is shown in Figure 1.

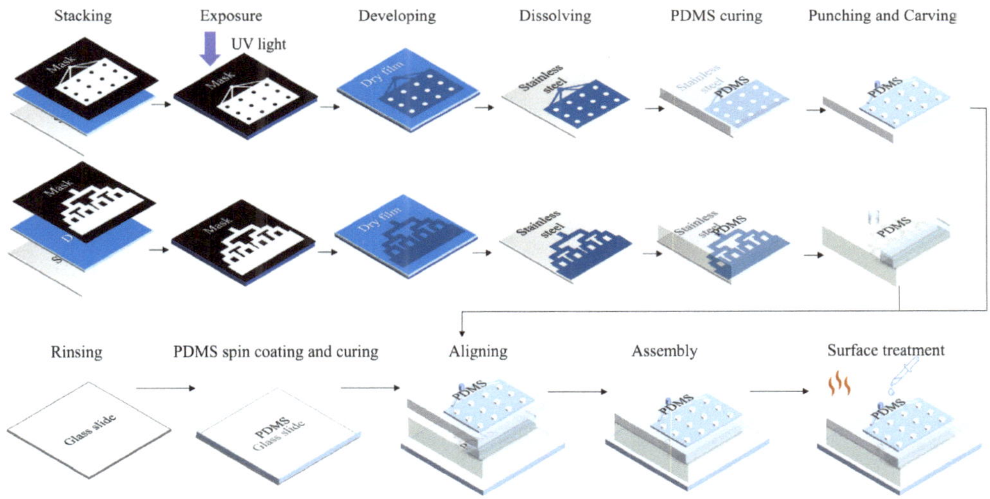

Figure 1. Fabrication of the integrated microfluidic chip.

2.5. Experimental Setup

For all of the characterization experiments in this study, sample solution was pumped into the microfluidic chip with a syringe pump (500 µL ball-end syringe, Tecan, San Jose, CA, USA) at 12 µL/min. PCR oil was preloaded in the reservoir for droplet generation. The sample solutions were pumped through the tree structure microfluidic channel and dispersed from the nozzles into the oil reservoir. Droplets generated from the emulsification were floated into the collection unit, since the amplification reaction solution has a lighter density than the oil phase (Figure 2A). Droplets can be collected either into a standard PCR tube through tubing (Figure 2B,C), or into an on-chip monolayer droplets collecting chamber (Figure 2D). The droplets collected in the standard PCR tube were then amplified in a commercial ddPCR thermal cycling device, and detected in a fluorescence flow cytometry device. The droplet collecting chamber was overlapped on the tree-shaped microchannels so that the 3D microstructure allowed a higher degree of device miniaturization, as illustrated in Figure 2D. An optical upright microscope and a camera (99 fps) were used to record the generation process. Then, the droplets were transferred into a flat monolayer observation chip, and microscopic photos were taken for droplet size statistics. The observation chamber has a height of 120 μm, and droplets with diameters larger than 120 μm were converted as a drum-shaped model to a spherical model for equivalent diameter calculation. Droplets collected in the chamber can be in situ amplified in the future if temperature control units are integrated on the chip substrate and/or materials/reagents are modified. Extra treatments (Figure S2) were also applied on the chip, and the chip containing the liquid was placed in a 60 °C incubator for 60 min to demonstrate the feasibility of the chip in amplification with low-temperature reaction reagents. The positive and negative droplets array in the on-chip chamber proved to be distinguishable with fluorescence microscopy (Axio Observer A1, Carl Zeiss AG, Oberkochen, Baden-Wuerttemberg, Germany). However, due to the restrictions of PDMS, only droplet generation and droplet imaging were performed in this study.

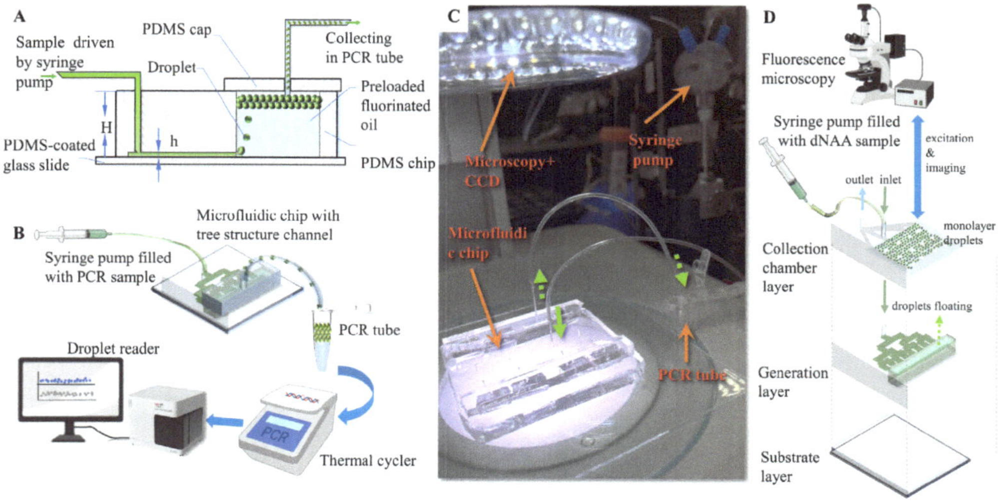

Figure 2. Experimental setup for droplet generation. (**A**) Cross-section illustration of the microfluidic chip. (**B**) Schematic diagram of ddPCR process with microfluidic step emulsification chip. (**C**) Experimental setup. (**D**) Illustration of ddNAA process with integrated three-in-one step emulsification chip.

2.6. Performance Verification of the Collected Droplets

In order to verify the stability and biocompatibility of the microfluidic chip after the fabrication, surface treatment and emulsification processes, droplets were collected and amplified. Droplet amplifications were performed using a commercial PCR thermal cycler (two-step amplification, 40 cycles, Genesy 96E, TIANLONG, Xi'an, Shaanxi, China). FAM channel fluorescence droplet readouts were performed using a droplet reader (DS100 Digital PCR Reader, ZK-Medical, Suzhou, Jiangsu, China) with flow cytometry principles. The target nucleic acid concentration of the sample can be obtained via the Poisson distribution law, with numbers of fluorescence positive droplets and negative droplets [32]. Control group droplets were generated with the same sample solution using a commercial droplet generation device (DS100 Digital PCR Generator, ZK-Medical, Suzhou, Jiangsu, China).

3. Results and Discussion

3.1. Simulation Results of Nozzle Pressure Distribution

In this study, finite element models were built of the tree-shaped and three classic nozzle distribution designs; each design had 16 nozzles. Their flow rate distributions were simulated and compared. The mass flow at the inlet was set as 0.2 μL/s, and the pressure at the nozzle outlets was set as atmospheric pressure. The cross-sections of all of the nozzles were rectangular (28 μm (h) × 150 μm (w)).

The results of the simulations are shown in Figure 3. The mass flow rates are obtained by integrating the sample velocity on the cross-section of the nozzle outlets. Among the tree-shaped distribution, the comb-shaped distribution, the brush-shaped distribution and the circumferential distribution, it could be seen that the circumferential distribution has the largest non-uniformity of flow rates at the nozzles. The tree-shaped distribution has a slightly more uniform distribution than the comb-shaped distribution and the brush-shaped distribution. Therefore, droplets generated with the tree-shape distribution theoretically have better uniformity than the other designs.

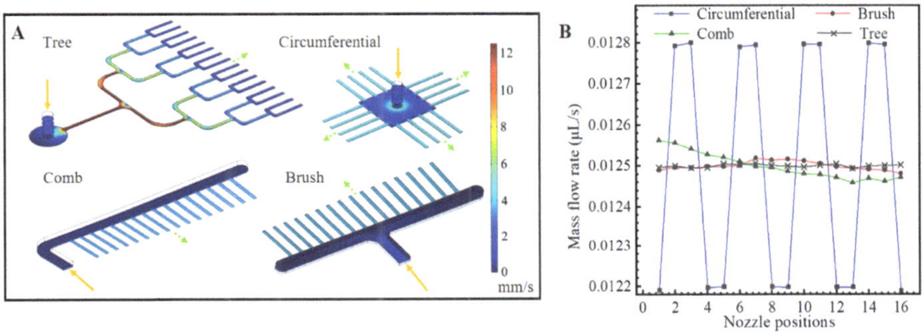

Figure 3. Flow simulation of different chip designs. (**A**) Flow rate distributions in microfluidic structures with four microchannel arrangements. (**B**) Average flow rates at droplet generation nozzles.

3.2. Droplet Generation

Multiple droplet generation tests were performed in our microfluidic chips. A dyed sample fluid phase with a slow flow rate (6 μL/min) was used to better observe the droplet emulsification process at the freshly processed hydrophobic nozzles. As shown in Figure 4, the sample liquid was gradually pushed out of the nozzles under hydraulic pressure with increasing upward Laplace force. In a very short time, the sample liquid shrunk at the nozzles, broke at its "neck" under Rayleigh–Plateau instability, formed into a droplet and floated upward due to buoyancy. Droplets with an average diameter of 101.2 μm were generated at two droplets/s at each nozzle (width = 170 μm). Although large flow rates at the nozzles would lead to jetting mode and result in sudden increases in droplet diameter or droplet generation failure, the droplet generation rate can be compensated by increasing the number of channels.

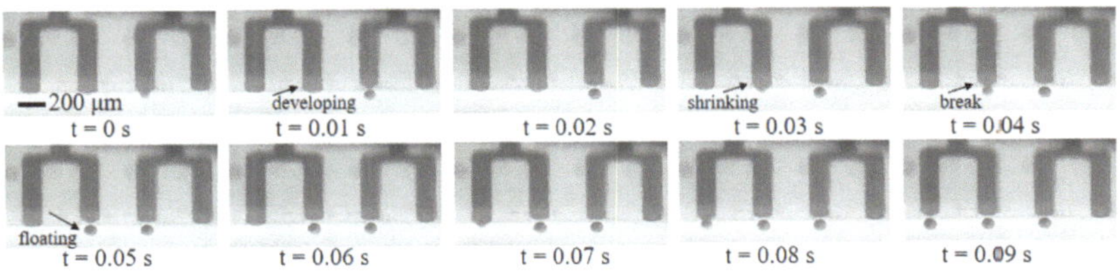

Figure 4. Droplets generation process at four adjacent nozzles.

Figure 5 shows the characterization results of step emulsification with various nozzle heights, nozzle widths and surface treatment methods. In all of the tested conditions, the Ca is between 2.27×10^{-4} and 1.08×10^{-3}, which is smaller than the critical Ca ($\sim 6 \times 10^{-3}$) in previous studies. The Bo is between 3.97×10^{-3} and 1.8×10^{-2}. The We is between 6.4×10^{-5} and 6.8×10^{-4}. The Bo and We are both smaller than their critical values. Therefore, the droplet generation processes in the tests should be stable in the dripping mode. As shown in Figure 5A–C, the droplet diameter increases linearly with increasing channel width. Figure 5D shows that the droplet diameter also increases with increasing channel height. With the combination of incubation and hydrophobic reagent, the droplet diameter decreases to 77–89% compared to only the hydrophobic reagent treatment, and 84–96% compared to only the incubation treatment, which indicates improvement in the dispersing performance. The decreasing degree is higher with lower channel heights. Figure 5E compares droplets collected in eight typical generation conditions, with an average droplet diameter that ranges from 73.1 μm to 155.9 μm; these results can meet the droplet diameter requirements of most ddNAA applications. The coefficients of variation (CV) are all smaller than 5%, and this proves that the droplet generation processes from the 16 channels are all stable and uniform.

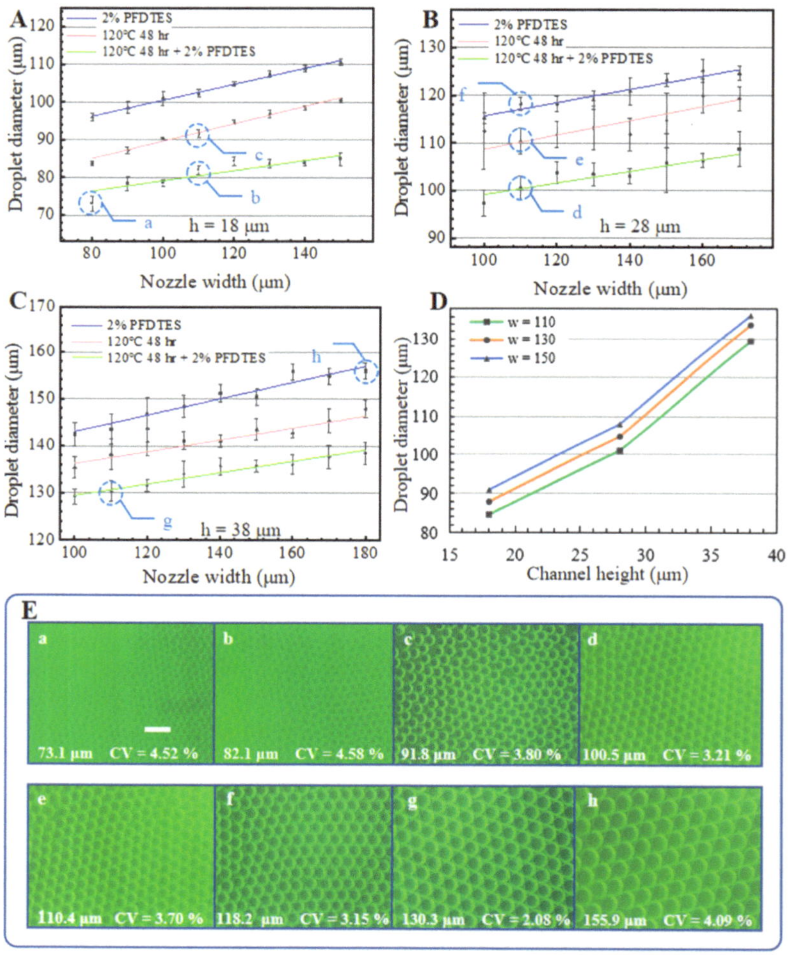

Figure 5. Characterization of droplets generation with the integrated microfluidic chip. Droplets diameter changes with different surface treatments, nozzle widths and nozzle heights (h = 18 μm (**A**), 28 μm (**B**), 38 μm (**C**), respectively). (**D**) shows the increasing droplet diameter with increasing channel height under the same flow rate and same surface treatment method (120 °C 48 h incubation with 2% PFDTES). Error bars represent standard deviations of the droplet diameter. (**E**) The corresponding droplet images of the generation conditions are marked in (**A–C**). The scale bar measures 200 μm in (**E**).

3.3. Droplet Collection and Detection

Figure 6 shows images of the droplet collection tests of two different amplification and detection approaches. In Figure 6A, after being generated at 12 µL/min, the droplets were pumped out of the chip at 24 µL/min and collected in a standard PCR tube through tubing connected with the outlet of the chip. The sample droplets were amplified in the tube, and then detected using fluorescence flow cytometry. It is obvious that the oil-to-sample volume ratio is small (less than one) at the generation stage (1–40 s), compared to a ratio of about two during most flow-focusing ddPCR applications, which indicates less oil phase consumption. Figure 6B shows a picture of the three-layer PDMS chip with on-chip droplet collection, and 6C shows the generation and collection process with dyed droplets filling the on-chip flat collection chamber above the tree-shaped generation structures. The sub-figure in Figure 6C shows details of the droplets distribution as being in a good monolayer array fashion. The on-chip generation and collection processes are shown in Video S1. The 11 mm (l) × 15 mm (w) × 120 µm (h) chamber can contain up to 23,500 droplets of a 90 µm diameter with a total sample volume of 8.9 µL, and 13,200 droplets of a 120 µm diameter with total sample volume of 11.9 µL. While remaining in the same flat chamber, droplets with target nucleic acid can be in situ amplified, then be detected using fluorescence photography and be counted later using image processing. Comparing to previous studies, this integrated collection approach realized in situ amplification and detection in a compact structure, and eliminated manual transfer tasks and possible contamination.

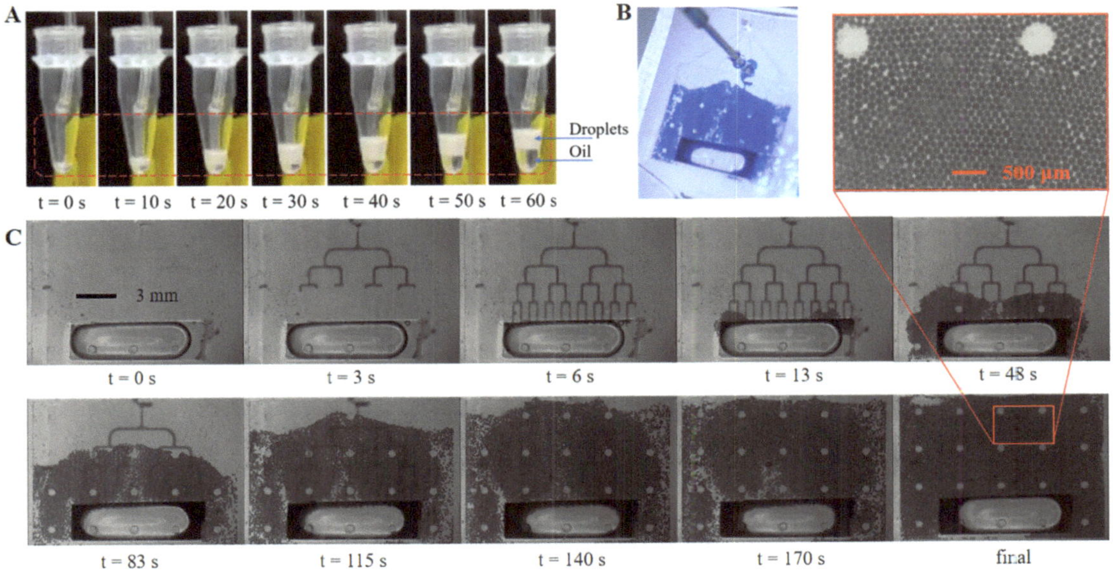

Figure 6. Droplets collection. (**A**) Droplets collection with a standard PCR tube; (**B**) droplets spreading in the on-chip collection chamber; (**C**) droplets generation and collection process with on-chip collection chamber under microscopy.

Droplets (diameter = 131 µm) containing λDNA templates were generated by the proposed step emulsification device, collected in a standard PCR tube and amplified with the ddPCR method. The control group were droplets (diameter = 128 µm) that were generated by the commercial DS100 Digital PCR Generator. DdPCR amplification and fluorescence readouts for both droplets were completed with the devices mentioned in Section 2.6. Figure 7 shows the droplet fluorescence distributions, where positive droplets (containing amplified target gene) have obviously higher fluorescence intensities than negative droplets (without target gene). Our readout results and the control readout results were both statistically analyzed under the Poisson distribution law to acquire the quantitative nucleic acid concentrations. The concentration results show 614.63 copies/µL with droplets generated by DS100 Digital PCR Generator and 624.22 copies/µL with our device-generated droplets. The results of the two methods are consistent, demonstrating that our integrated chip can realize stable droplet generation for accurate nucleic acid quantitative measurement.

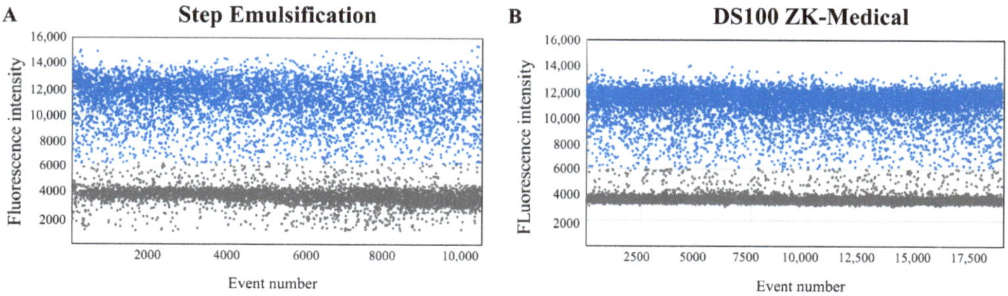

Figure 7. Fluorescence distribution readouts using droplets generated by proposed microfluidic device (**A**), compared with readouts using droplets generated by commercial DS100 (**B**).

The droplets array in the on-chip chamber can be counted using fluorescence imaging and image stitching. Since the liquid in the chamber will vaporize and leak out during PCR denaturation because PDMS is a porous material, on-chip monolayer amplification was not performed in this study. However, the evaporation can be prevented for 60 min at 60 °C in a chip with extra treatments, which could satisfy the time and temperature requirements of ddRPA and ddLAMP (Figure S2). Multiplex fluorescence of negative and positive droplets can also be detected with pre-amplified droplets injected into the collection chamber (Figure S3), which indicates that the multilayer PDMS structure is compatible with multiplex droplets in situ detection. These results could support the idea that with thermal stable plastic chip and/or low-temperature digital amplification reagents, on-chip in-situ amplification and detection are highly promising with our device.

4. Conclusions

Absolute nucleic acid quantification can be realized by dividing nucleic acid samples into individual micro reaction units and analyzing the nucleic acid signal of each unit. Dividing the liquid sample into appropriate volumes with commonly used microfluidic flow-focusing droplet generation methods requires high precision in fabrication and flow control, whereas step emulsification of droplets of similar volumes requires less precision in both. Therefore, this study presented an integrated step emulsification ddNAA microfluidic chip that avoids second-time lithography, regulates the unevenness in the multiplex flow rate, collects droplets with two convenient options and operates with a high degree of convenience, compared to previous step emulsification devices.

In this study, a step emulsification structure with 16 nozzles in a tree-shaped distribution was fabricated simply with one-step lithography. Through flow field simulations, the tree-shaped distribution proved to have a better flow rate distribution compared to the other three distributions. With the microfluidic chip, we then conducted characterization tests and data analysis on the generated droplet diameters. Surface treatments with PFDTES and incubation are helpful for droplet generation by increasing the surface hydrophobicity, which means smaller droplets can be obtained with the same microchannel dimensions. The influences of surface properties, channel height and channel width on droplet sizes were compared in further experiments. Droplets with diameters of 71.3 µm–155.9 µm were generated using microchannels with height of 18, 28 and 38 µm, and widths ranging from 73 µm to180 µm, which could be a reference while choosing an appropriate droplet size for droplet-based nucleic acid detection applications. Two droplet collection methods were proposed in this study. The droplets that were automatically collected in a PCR tube were amplified and successfully read. PCR amplification with on-chip-collected droplets in a monolayer array was restricted because PDMS is gas-permeable, but the three-in-one device was demonstrated to be applicable with low-temperature amplification approaches in the future. Therefore, the research on this interesting step emulsification chip with tree-shaped nozzles and its characterization results are referential for ddNAA applications.

Supplementary Materials: The following supporting information can be downloaded at: https://www.mdpi.com/article/10.3390/bios13090888/s1, Table S1: The sequences of primers and probe; Figure S1: Mask film designs and dimensions of the microfluidic layer and the collection chamber; Figure S2: Verification tests of on-chip amplification; Figure S3: Microscopic images of droplets in PDMS–glass chip at fixed position; Video S1: On-chip progress of droplets collection. References [33,34] are cited in the Supplementary Materials.

Author Contributions: G.L.: formal analysis, writing—original draft, and conceptualization. Y.Z.: validation and data curation. S.W.: investigation and data curation. X.L.: software and resources. T.Y.: writing—review and editing and funding acquisition. J.W.: investigation and Writing—review and editing. All authors have read and agreed to the published version of the manuscript.

Funding: This research was funded by the China Postdoctoral Science Foundation, grant number 2022M722336, and the Suzhou Pilot Project of Basic Research, grant number SJC2022015.

Institutional Review Board Statement: Not applicable.

Informed Consent Statement: Not applicable.

Data Availability Statement: Not applicable.

Conflicts of Interest: The authors declare no conflict of interest.

References

1. Ding, X.; Mu, Y. Research and application progress of digital nucleic acid amplification detection techniques. *Chin. J. Anal. Chem.* **2016**, *44*, 512–521. [CrossRef]
2. Li, Q.; Zhou, X.; Wang, Q.; Liu, W.; Chen, C. Microfluidics for COVID-19: From current work to future perspective. *Biosensors* **2023**, *13*, 163. [CrossRef] [PubMed]
3. Xiang, X.; Shang, Y.; Zhang, J.; Ding, Y.; Wu, Q. Advances in improvement strategies of digital nucleic acid amplification for pathogen detection. *TrAC Trends Anal. Chem.* **2022**, *149*, 116568. [CrossRef]
4. Zeng, W.; Yang, S.; Liu, Y.; Yang, T.; Tong, Z.; Shan, X.; Fu, H. Precise monodisperse droplet generation by pressure-driven microfluidic flows. *Chem. Eng. Sci.* **2022**, *248*, 117206. [CrossRef]
5. Ma, C.; Sun, Y.; Huang, Y.; Gao, Z.; Huang, Y.; Pandey, I.; Jia, C.; Feng, S.; Zhao, J. On-chip nucleic acid purification followed by ddPCR for SARS-CoV-2 detection. *Biosensors* **2023**, *13*, 517. [CrossRef]
6. Zhang, W.; Li, N.; Koga, D.; Zhang, Y.; Zeng, H.; Nakajima, H.; Lin, J.; Uchiyama, K. Inkjet printing based droplet generation for integrated Online Digital Polymerase Chain Reaction. *Anal. Chem.* **2018**, *90*, 5329–5334. [CrossRef]
7. Liu, L.; Xiang, N.; Ni, Z.; Huang, X.; Zheng, J.; Wang, Y.; Zhang, X. Step emulsification: High-throughput production of monodisperse droplets. *Biotechniques* **2020**, *68*, 114–116. [CrossRef]
8. Ye, S.; Li, C.; Zheng, X.; Huang, W.; Tao, Y.; Yu, Y.; Yang, L.; Lan, Y.; Ma, L.; Bian, S.; et al. OsciDrop: A versatile deterministic droplet generator. *Anal. Chem.* **2022**, *94*, 2918–2925. [CrossRef]
9. Zhang, X.; Wang, J.; Lyu, X.; Yang, T.; Liu, W.; Wang, B.; Luo, G. Hundred-micron droplet ejection by focused ultrasound for genomic applications. *Sens. Actuators A Phys.* **2023**, *354*, 114271. [CrossRef]
10. Wei, C.; Yu, C.; Li, S.; Meng, J.; Li, T.; Cheng, J.; Li, J. A droplet-based multivolume microfluidic device for digital polymerase chain reaction. *Sens. Actuators B Chem.* **2022**, *371*, 132473. [CrossRef]
11. Zhan, A.; Liu, Z.; Jiang, S.; Zhu, C.; Ma, Y.; Fu, T. Comparison of formation of bubbles and droplets in step-emulsification microfluidic devices. *J. Ind. Eng. Chem.* **2022**, *106*, 469–481. [CrossRef]
12. Zeng, W.; Fu, H. Quantitative measurements of the somatic cell count in fat-free milk based on droplet microfluidics. *J. Mater. Chem.* **2020**, *8*, 13770–13776. [CrossRef]
13. Cubaud, T.; Mason, T.G. Capillary threads and viscous droplets in square microchannels. *Phys. Fluids* **2008**, *20*, 053302. [CrossRef]
14. Eggersdorfer, M.L.; Seybold, H.; Ofner, A.; Weitz, D.A.; Studart, A.R. Wetting controls of droplet formation in step emulsification. *Proc. Natl. Acad. Sci. USA* **2018**, *115*, 9479–9484. [CrossRef] [PubMed]

15. Ofner, A.; Moore, D.G.; Ruhs, P.A.; Schwendimann, P.; Eggersdorfer, M.L.; Amstad, E.; Weitz, D.A.; Studart, A.R. High-throughput step emulsification for the production of functional materials using a glass microfluidic device. *Macromol. Chem. Phys.* **2017**, *218*, 1600472. [CrossRef]
16. Shi, Z.; Dong, N.; Lai, X.; Yu, H.; Li, D. Centrifugal step emulsification microfluidics supporting droplet digital loop-mediated isothermal amplification (LAMP) of SARS-CoV-2 N gene. In Proceedings of the 2021 21st International Conference on Solid-State Sensors Actuators and Microsystems (Transducers), Orlando, FL, USA, 20–24 June 2021; pp. 976–979.
17. Li, T.; Wang, H.F.; Wang, Y.; Pan, J.Z.; Fang, Q. A minimalist approach for generating picoliter to nanoliter droplets based on an asymmetrical beveled capillary and its application in digital PCR assay. *Talanta* **2020**, *217*, 120997. [CrossRef]
18. Schulz, M.; Probst, S.; Calabrese, S.; Homann, A.R.; Borst, N.; Weiss, M.; von Stetten, F.; Zengerle, R.; Paust, N. Versatile tool for droplet generation in standard reaction tubes by centrifugal step emulsification. *Molecules* **2020**, *25*, 1914. [CrossRef]
19. Schlenker, F.; Kipf, E.; Borst, N.; Paust, N.; Zengerle, R.; von Stetten, F.; Juelg, P.; Hutzenlaub, T. Centrifugal microfluidic integration of 4-plex ddPCR demonstrated by the quantification of cancer-associated point mutations. *Processes* **2021**, *9*, 97. [CrossRef]
20. Zhao, S.H.; Zhang, Z.M.; Hu, F.; Wu, J.J.; Peng, N.C. Massive droplet generation for digital PCR via a smart step emulsification chip integrated in a reaction tube. *Analyst* **2021**, *146*, 1559–1568. [CrossRef]
21. Nie, M.; Zheng, M.; Li, C.; Shen, F.; Liu, M.; Luo, H.; Song, X.; Lan, Y.; Pan, J.; Du, W. Assembled step emulsification device for multiplex droplet digital polymerase chain reaction. *Anal. Chem.* **2019**, *91*, 1779–1784. [CrossRef]
22. Barkley, S.; Weeks, E.R.; Dalnoki-Veress, K. Snap-off production of monodisperse droplets. *Eur. Phys. J. E* **2015**, *38*, 138. [CrossRef] [PubMed]
23. Barkley, S.; Scarfe, S.J.; Weeks, E.R.; Dalnoki-Veress, K. Predicting the size of droplets produced through Laplace pressure induced snap-off. *Soft Matter* **2016**, *12*, 7398–7404. [CrossRef] [PubMed]
24. Erb, R.M.; Obrist, D.; Chen, P.W.; Studer, J.; Studart, A.R. Predicting sizes of droplets made by microfluidic flow-induced dripping. *Soft Matter* **2011**, *7*, 8757–8761. [CrossRef]
25. Stolovicki, E.; Ziblat, R.; Weitz, D.A. Throughput enhancement of parallel step emulsifier devices by shear-free and efficient nozzle clearance. *Lab Chip* **2018**, *18*, 132–138. [CrossRef]
26. Vladisavljević, G.T.; Kobayashi, I.; Nakajima, M. Effect of dispersed phase viscosity on maximum droplet generation frequency in microchannel emulsification using asymmetric straight-through channels. *Microfluid. Nanofluid.* **2011**, *10*, 1199–1209. [CrossRef]
27. Schulz, M.; von Stetten, F.; Zengerle, R.; Paust, N. Centrifugal step emulsification: How buoyancy enables high generation rates of monodisperse droplets. *Langmuir* **2019**, *35*, 9809. [CrossRef]
28. Xu, X.; Yuan, H.; Song, R.; Yu, M.; Chung, H.; Hou, Y.; Shang, Y.; Zhou, H.; Yao, S. High aspect ratio induced spontaneous generation of monodisperse picolitre droplets for digital PCR. *Biomicrofluidics* **2018**, *12*, 014103. [CrossRef]
29. Vladisavljevic, G.T.; Ekanem, E.E.; Zhang, Z.L.; Khalid, N.; Kobayashi, I.; Nakajima, M. Long-term stability of droplet production by microchannel (step)emulsification in microfluidic silicon chips with large number of terraced microchannels. *Chem. Eng. J.* **2018**, *333*, 380–391. [CrossRef]
30. Opalski, A.S.; Makuch, K.; Lai, Y.K.; Derzsi, L.; Garstecki, P. Grooved step emulsification systems optimize the throughput of passive generation of monodisperse emulsions. *Lab Chip* **2019**, *19*, 1183–1192. [CrossRef]
31. de Givenchy, E.P.T.; Amigoni, S.; Martin, C.; Andrada, G.; Caillier, L.; Geribaldi, S.; Guittard, F. Fabrication of superhydrophobic PDMS surfaces by combining acidic treatment and perfluorinated monolayers. *Langmuir* **2009**, *25*, 6448–6453. [CrossRef]
32. Pinheiro, L.B.; Coleman, V.A.; Hindson, C.M.; Herrmann, J.; Hindson, B.J.; Bhat, S.; Emslie, K.R. Evaluation of a droplet digital polymerase chain reaction format for DNA copy number quantification. *Anal. Chem.* **2012**, *84*, 1003–1011. [CrossRef] [PubMed]
33. Bhat, S.; Curach, N.; Mostyn, T.; Bains, G.S.; Griffiths, K.R.; Emslie, K.R. Comparison of methods for accurate quantification of DNA mass concentration with traceability to the international system of units. *Anal. Chem.* **2010**, *82*, 7185–7192. [CrossRef] [PubMed]
34. Zhu, Q.; Qiu, L.; Yu, B.; Xu, Y.; Gao, Y.; Pan, T.; Tian, Q.; Song, Q.; Jin, W.; Jin, Q.; et al. Digital PCR on an integrated self-priming compartmentalization chip. *Lab Chip* **2014**, *14*, 1176–1185. [CrossRef] [PubMed]

biosensors

MDPI

Review

Microfluidic Mechanoporation: Current Progress and Applications in Stem Cells

Rubing Wang [1], Ziqi Wang [2], Lingling Tong [2], Ruoming Wang [3], Shuo Yao [3], Di Chen [2,4,5,6,*] and Huan Hu [1,*]

1 Zhejiang University-University of Illinois Urbana-Champaign Institute (ZJU-UIUC Institute), International Campus, Haining 314400, China; rubing.21@intl.zju.edu.cn
2 Center for Regeneration and Cell Therapy of Zhejiang University-University of Edinburgh Institute (ZJU-UoE Institute), Zhejiang University School of Medicine, Zhejiang University, Hangzhou 310003, China; ziqiw.20@intl.zju.edu.cn (Z.W.); lingling.21@intl.zju.edu.cn (L.T.)
3 Zhejiang University-University of Edinburgh Institute (ZJU-UoE Institute), International Campus, Zhejiang University, Haining 314400, China; ruoming.21@intl.zju.edu.cn (R.W.); shuo.22@intl.zju.edu.cn (S.Y.)
4 Center for Reproductive Medicine, The Second Affiliated Hospital, School of Medicine, Zhejiang University, Hangzhou 310003, China
5 Dr. Li Dak Sum & Yip Yio Chin Center for Stem Cell and Regenerative Medicine, Zhejiang University, Hangzhou 310003, China
6 National Key Laboratory of Biobased Transportation Fuel Technology, Haining 314400, China
* Correspondence: dichen@intl.zju.edu.cn (D.C.); huanhu@intl.zju.edu.cn (H.H.)

Abstract: Intracellular delivery, the process of transporting substances into cells, is crucial for various applications, such as drug delivery, gene therapy, cell imaging, and regenerative medicine. Among the different approaches of intracellular delivery, mechanoporation stands out by utilizing mechanical forces to create temporary pores on cell membranes, enabling the entry of substances into cells. This method is promising due to its minimal contamination and is especially vital for stem cells intended for clinical therapy. In this review, we explore various mechanoporation technologies, including microinjection, micro–nano needle arrays, cell squeezing through physical confinement, and cell squeezing using hydrodynamic forces. Additionally, we highlight recent research efforts utilizing mechanoporation for stem cell studies. Furthermore, we discuss the integration of mechanoporation techniques into microfluidic platforms for high-throughput intracellular delivery with enhanced transfection efficiency. This advancement holds potential in addressing the challenge of low transfection efficiency, benefiting both basic research and clinical applications of stem cells. Ultimately, the combination of microfluidics and mechanoporation presents new opportunities for creating comprehensive systems for stem cell processing.

Keywords: microfluidic; intracellular delivery; mechanoporation; stem cells

Citation: Wang, R.; Wang, Z.; Tong, L.; Wang, R.; Yao, S.; Chen, D.; Hu, H. Microfluidic Mechanoporation: Current Progress and Applications in Stem Cells. *Biosensors* **2024**, *14*, 256. https://doi.org/10.3390/bios14050256

Received: 10 April 2024
Revised: 8 May 2024
Accepted: 12 May 2024
Published: 17 May 2024

1. Introduction

Intracellular delivery is an important technique in molecular and cell biology research, which introduces biomaterials into cells for investigating the regulation of gene expression, functions of genes of interest, protein–protein interactions, the sub-cellular localization of proteins, and for genome editing and gene therapy [1]. In recent decades, many effective methods have been established to achieve intracellular transport with higher efficiency.

Cell transfections with DNA can be divided into transient transfections and stable transfections according to the expression duration of the exogenous biomaterials in cells. In transient transfections, exogenous DNA does not integrate into the host chromosome and only lasts for several days because of the dilution upon cell divisions [2]. On the contrary, the stable transfection of exogenous DNA could be integrated into host genomes and then express target genes or proteins constitutively in the cells depending on the regulatory sequences applied to drive the expression [3]. Currently, traditional transection methods can be divided into three categories [4]: (1) biological transfection methods mediated by viral vectors, including lentivirus, adenovirus, and adeno-associated viruses [5–9]; (2) chemical transfection methods using different transfection media, such as calcium phosphate [10], liposome [11], and cationic polymers [12,13]; and (3) mechanical methods to achieve transfections through the disturbance or destruction of cell membranes, such as electroporation [14], microinjection [15,16], the gene gun method [17], and acoustic hole effect-mediated transfections [18,19].

The transfection of biomaterials into cells has greatly improved our understanding of gene function and regulation. However, the above methods have their own shortcomings that limit the applications of transfections into different types of cells, especially for stem cells. For a viral vector-mediated transfection, although it exhibits high transfection efficiency with the continuous and stable expression of exogenous genes [20], there are safety concerns, because the insertion site of the viral vectors into the host genome is uncertain. This uncertain gene integration may cause the activation of proto-oncogenes, the inactivation of oncogenes, RNA splicing, and gene fusion, thus posing a risk of carcinogenesis [21–24]. For chemical transfection methods, although the transfection efficiency has been improved after liposome modification [25], it is still expensive, and the transfection efficiency is still low for stem cells. Mechanical transfection methods have been successfully applied to different cell types with high efficiency, including stem cells [15,18,26,27]. However, they require specific equipment and complex operation processes, which significantly increase cell mortality [16,18,26,28]. Therefore, we expect an intracellular transfection method that is suitable for many cell types, with high transfection efficiency, cell biosafety, economy, ease of operation, and so on. Emerging microfluidic technology [29,30] is promising due to its low solvent consumption, low counter dose, small cell-like volume, and a relatively high transfection efficiency and cell survival rate. It can be applied to a wide range of applications. In addition, the microfluidic environment is close to the diameter of cells, which is conducive to single-cell research and even in situ visual observations and real-time monitoring [31–36].

Although there exist several excellent reviews on mechanoporation [37–39], there is no review particularly targeting stem cells. We particularly selected mechanoporation approaches that are integrated with microfluidic chips for intracellular delivery to stem cells with high throughput and a low dead rate. In this review, we introduce different methods based on microfluidic transfections, including microinjection [31,32,40–43], micro/nanoneedle arrays [44–48], cell squeezing based on mechanical confinement [33,34,49], and cell squeezing based on hydrodynamic manipulation [35,36]. Furthermore, we briefly introduce the current progress for applying microfluidic methods in stem cell research, highlighting the advantages and limitations.

2. Microfluidic-Based Mechanoporation

As a critical step in microfluidic cell transfections, membrane disruption-based intracellular delivery methods drew a lot of attention from researchers [50,51] and can be classified into electroporation [52–57], optoporation [58–62], magnetoporation [63–65], acoustoporation [66–69], and mechanoporation [34,48,70–73] based on pore creation mechanisms. While each technique above possesses its own set of advantages, it is important to note that all except for mechanoporation rely on an external energy field to disrupt the cell membrane. However, this dependence on external energy fields can potentially impact the biological function and viability of the cells being manipulated [74–76]. Therefore, this review focuses on the mechanoporation techniques that are independent of an external energy field. We will discuss four different mechanoporation techniques that employ only mechanical structures without causing severe damage to cell membranes. Based on the different microfluidic device structures, microfluidic-based mechanoporation methods can be classified as microinjection (Figure 1a), micro/nanoneedle arrays (Figure 1b), cell squeezing based on mechanical confinement (Figure 1c), and cell squeezing based on hydrodynamic manipulation (Figure 1d). All these methods exhibit advantages, such as high transfection efficiency, high throughput, ease of handling, and high cell viability [34,48,70–73].

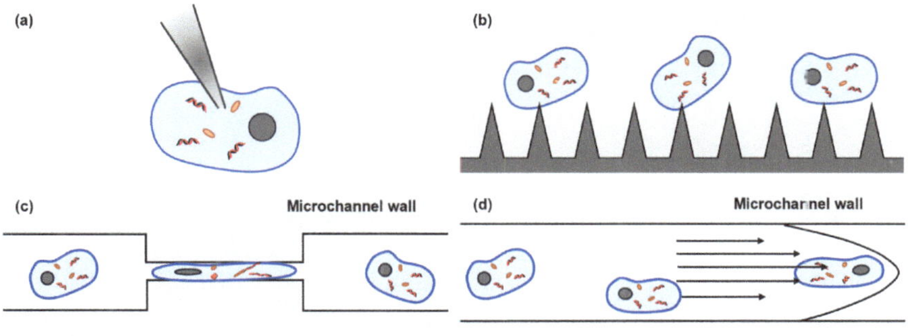

Figure 1. Microfluidic-based mechanoporation methods. (**a**) Microinjection; (**b**) micro/nanoneedle arrays; (**c**) cell squeezing based on mechanical confinement; (**d**) cell squeezing based on hydrodynamic manipulation.

2.1. Microinjection

Microinjection, ever since its inception in the previous century [77], has remained a commonly utilized method for single-cell transfections due to both its straightforward concept and ease of manipulation. By inserting a glass micropipette into specific positions of individual cells [78], almost any cargo can be successfully delivered into the cells via microinjection. This versatile technique finds applications in various areas, including in vitro fertilization and nuclear transfer for cloning [79]. As one of the traditional microinjection schemes, the AFM (atomic force microscopy)-based microinjection is adopted for the precise intracellular delivery to single cells by functionalizing antibodies to the AFM probe [38,80] or through the hollow AFM cantilever [81]. Benefiting from a size of 200–300 nm and a high aspect ratio structure, the AFM tip can penetrate the cell and adhere to the substrate with proper force and cause little or no damage to the cell membrane [39]. However, one major limitation of this approach is its low throughput and limited suitability for suspended types of cells, since it depends on the surface adhesion property of cells. The introduction of microfluidic techniques provides a platform to better manipulate all types of cells for microinjections, improving the intracellular delivery throughput and enabling suspension cell transfections [31,32,40–43].

By integrating microinjection and microfluidic techniques, Adamo and Jensen proposed a microneedle-immobilized microfluidic microinjection device [31]. As shown in Figure 2(ai), in this device, single cells were driven by fluid streams from channel A to channel B and transfected by immobilized microneedles while valve 1 was opened and valve 2 was closed. After cell transfection, the cells were driven by fluid streams from channel B to channel C by closing valve 1 and opening valve 2. The experimental results showcase an approximate throughput of 1 cell in 5 min, conducted with HeLa cells [31]. To enhance the throughput of the microinjection system, Liu and Sun presented a vacuum-based cell-holding device for single-cell immobilization and applied this device to a mouse zygote microinjection [40]. In this study, mouse zygotes were immobilized into arrays of 5×5 through-holes (Figure 2b), making cell capture and immobilization easier and allowing for the transfection of a total amount of 200 s at a speed of 9 cells/min, substantially improving the throughput of traditional microinjection methods. The experimental results demonstrate the progression of zygotes into the blastocyst stage after microinjection, providing evidence for the claim that the microneedle-immobilized microfluidic microinjection device would not affect embryo survival and development [40].

To improve injection automation for effective transfections [32,41–43], an automated quantitative microinjection platform was developed by Chow et al., showcasing the ability to deliver precise quantities of materials into cells [32]. By immobilizing cells in a microfluidic chip and injecting a certain amount of substances through an injection pressure- and time-controlled micropipette to cells one by one (Figure 2c), this microinjection platform achieved a precise single-cell microinjection. This microinjection platform, which was applied to human foreskin fibroblast cells, achieved about 80% transfection efficiency and 82.1% cell viability. However, this microinjection still suffers from low throughput, limiting its application to larger amounts of cells.

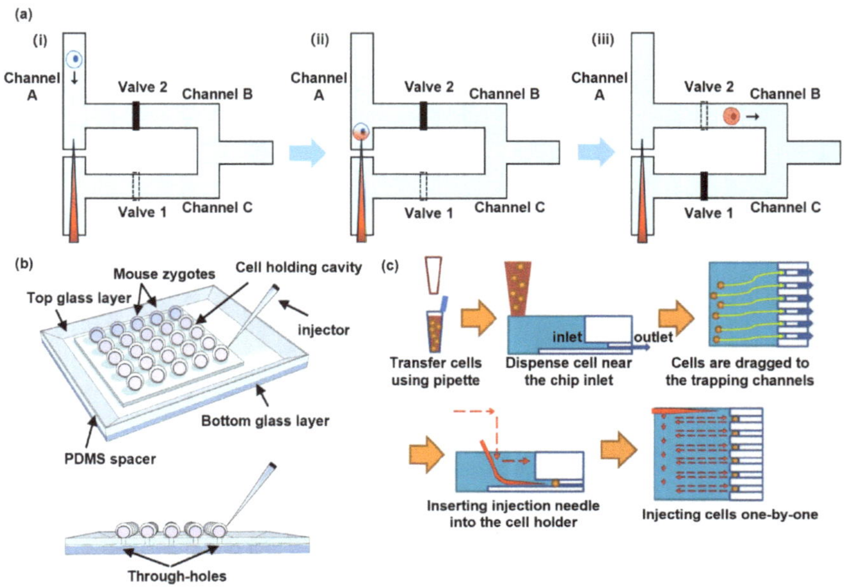

Figure 2. Microinjection. (**a**) A schematic illustration of the microfluidic-based single-cell microinjection system. (**i**) Cells are driven from Channel A to Channel B by fluid stream. (**ii**) Cells are transfected by immobilized microneedle. (**iii**) Cells are driven from Channel B to Channel C by fluid stream. Reprinted and modified from Ref. [31]. (**b**) The vacuum-based cell-holding device for single-cell immobilization. Reprinted and modified from Ref. [40]. (**c**) A workflow illustration of the automated quantitative microinjection platform. Reprinted and modified with permission from Ref. [32].

2.2. Micro/Nanoneedle Arrays

Compared with microinjections with a single pipette, the integration of micro/nanoneedles into microfluidic-based devices could achieve high-throughput and efficient single-cell transfections. Microfabrication techniques can fabricate different micro/nanoneedle array structures in a straightforward and convenient fashion [44,45].

Zhang et al. proposed a microfluidic microneedle device with massively parallel microinjector arrays, enabling a superhigh throughput microinjection [46]. As shown in Figure 3a, this device operates by attracting the cell onto the hollow penetrator during aspiration-based captures. Subsequently, exogenous cargos are injected into the cell through the resulting membrane pore before the cells are released by a positive aspiration flow. Each microinjector in the microneedle array incorporates a hollow penetrator with a sub-micron tip with a base of approximately 1–2 µm in diameter. In this device, the negative and positive aspiration flows ensure the minimal force required for cell capture and penetration, since they allow for the minimal stress of the sub-micro tip to penetrate the cell membrane. Moreover, the massively parallel microinjector array, which refers to 100 × 100 capture sites, realizes an ultrahigh throughput microinjection. Further experimental results exhibit a transfection efficiency of approximately 93% at a flow rate of 40 µL/min using an immortalized human T lymphocyte cell line-applied propidium iodide dye [47]. In the case of delivering a green fluorescent protein plasmid, efficiency rates of 82% in the primary human T cells were achieved, with over 87% cell viability. Overall, this microfluidic microneedle device demonstrates high efficiency and throughput capabilities for microinjections, showcasing its potential in various cellular transfection applications [47].

Furthermore, Huang et al. devised a microfluidic nanoneedle device including a silicon nanoneedle array along with the staggered herringbone channel design [48]. In this design, as depicted in Figure 3b, a PDMS structure featured a channel on its top surface, which was composed of periodically staggered herringbone grooves. By incorporating two asymmetrically shifted groups of staggered herringbone grooves, this configuration facilitated the chaotic mixing of the substances introduced through the inlet port. Unlike microinjections which directly deliver exogenous cargos into the cells, exogenous cargos are diffused into the cell after the cells collide with the nanoneedle array and then form pores on the cell membrane. The experimental results, achieved with human embryonic kidney cells, demonstrate transfection efficiency of over 20% and cell viability exceeding 95% while transfected with GFP-expressing plasmids [48].

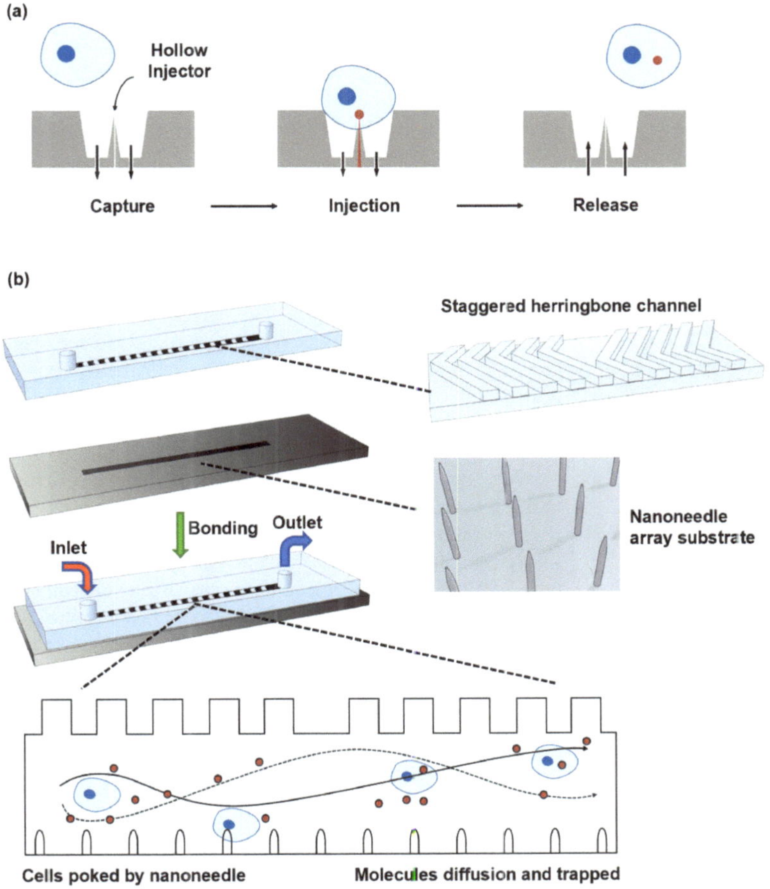

Figure 3. Micro/nanoneedle arrays. (**a**) A schematic of the microneedle arrays for the single-cell capture site. The arrows denote the flow direction and magnitude. Reprinted and modified from Ref. [46]. (**b**) A schematic illustration of the PDMS-based nanoneedle arrays. Reprinted and modified from Ref. [48].

2.3. *Cell Squeezing Based on Mechanical Confinement*

In response to the drawbacks of microinjections and micro/nanoneedle arrays potentially causing irreversible damage to cell membranes, researchers developed mechanical confinement-based cell-squeezing strategies. In these methods, the cell membrane undergoes rapid mechanical deformation when passing through a microfluidic constriction smaller than its size, leading to the formation of transient holes. These holes are recoverable, meaning that the damage caused to the cell membrane is almost negligible.

Sharei et al. [33] demonstrated cell squeezing based on the mechanical confinement method for cell delivery, in which multiple cells undergo mechanical squeezing simultaneously when passing through parallel micro-constriction channels. Figure 4a clearly demonstrates that when cells were subjected to a constriction channel narrower than their size, a temporary disruption of the cell membrane was observed. Transient pores were generated, which promoted intracellular delivery based on the diffusion of biomaterials into the cell. This method achieves an average throughput rate of 20,000 cells per second, which is significantly higher than that of the microfluidic device that employs the aforementioned microinjections and micro/nanoneedle arrays, exhibiting about 75% delivery efficiency and a maximum of 95% cell viability while transferring blue-labeled 3 kDa dextran molecules into HeLa cells [33]. By introducing key transcription factors (Oct4, Sox2, c-Myc, and Klf-4) required for stem cell pluripotency into human fibroblast cells [82], Sharei et al. implemented cell reprogramming. The identification of transformed colonies expressing embryonic stem cell markers reveals the morphological transformation in human fibroblast cells and the effect on gene expression, providing more possibilities for cell therapy and regenerative medicine.

To further enhance cell delivery efficiency, Modaresi et al. introduced a microfluidic platform to perform double cell deformation [34]. Figure 4b illustrates two microfluidic device designs, one allowing for single deformation (Figure 4(bi)) and the other allowing for double deformation (Figure 4(bii)). In the case of the first design, cells were subjected to continuous paralleled constrictions, which were 20 μm in length and 8 μm in width, permitting single deformation. Conversely, the second device translated one side of the narrow channel in the first design to create staggered squeeze constrictions with an 8 μm gap, enabling double transformation. The experiments showed that the double-deformation approach resulted in the higher delivery efficiency of biomaterials into cells compared to the single deformation method while applying human adipose-derived stem cells that were transfected with Dex-FITC. This device, which allows for cell double deformation, is superior for delivering small-sized exogenous materials, achieving an 85% delivery efficiency and improving cell viability to 95%, while maintaining a higher throughput. Furthermore, it did not induce the cell apoptosis associated with the single-deformation method.

Joo et al. proposed a microfluidic device for droplet mechanoporation, where cells encapsulated with biomolecules in one droplet are transported through multiple constrictions to prevent cell damage and increase cell viability [49]. This device, as illustrated in Figure 4(ci), comprises two parts: the droplet generator and the cell-squeezing sections. By injecting oil through separated inlet channels and utilizing a droplet generation technique [83], cell-biomolecule-encapsulated droplets are formed, leading to an increasing localized concentration of biomolecules that enhances cell delivery efficiency. As cells traverse through the constrictions within droplets, they experience a synergistic effect of convection and diffusion-mediated transport. This dynamic combination enables the efficient delivery of biomolecules through the cell membrane (Figure 4(ciii)). This method maximizes transfection efficiency, with a remarkable 98% achieved in a high throughput of 1 million cells per minute, and provides a minimum cell survival rate of 80%. Moreover, since each droplet carries the required cargo and most of the microchannel is occupied by carrier oil, significantly less cargo is utilized, minimizing the risk of clogging issues.

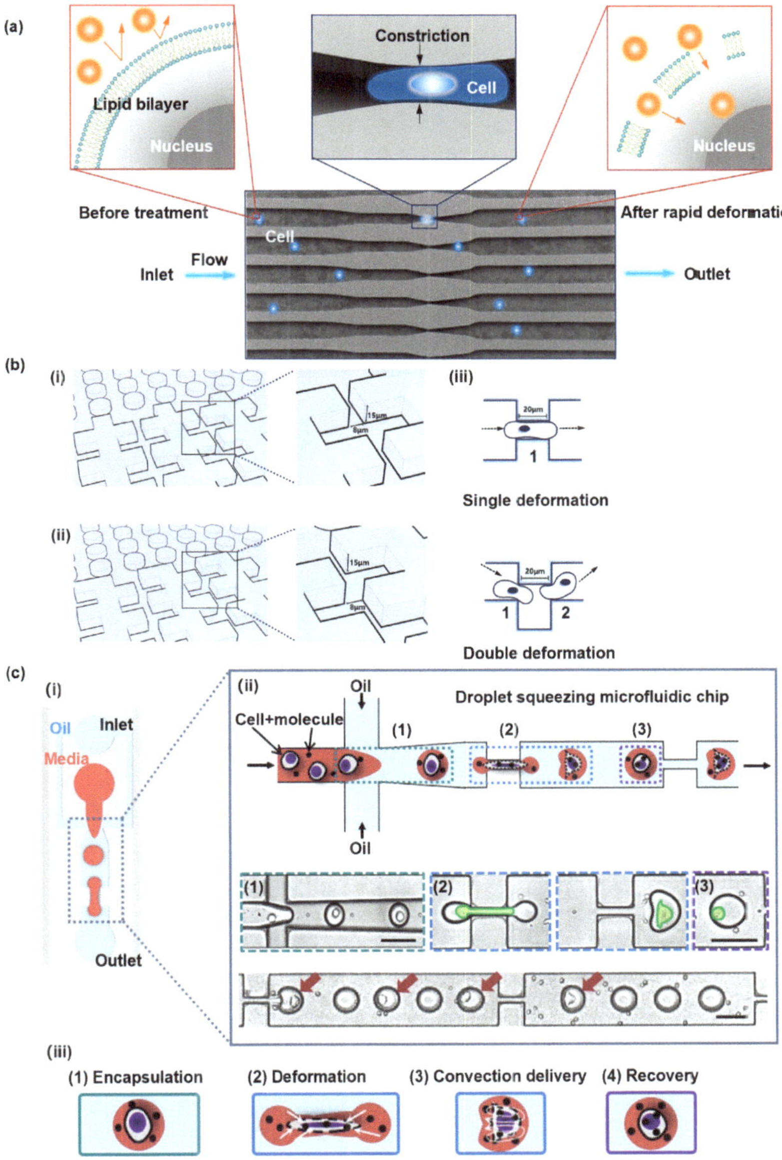

Figure 4. Cell squeezing based on mechanical confinement. (**a**) An illustration of the delivery hypothesis, whereby the rapid deformation of a cell, as it passes through a microfluidic constriction, generates transient membrane holes. Reprinted and modified with permission from Ref. [33]. (**b**) Two designs of a microfluidic cell deformation device. (**i**,**ii**) Schemata of the two designs. (**iii**) An illustration of the cell-squeezing process in two different cell deformation devices. Reprinted and modified from Ref. [34]. (**c**) The droplet squeezing platform design. (**i**) A schematic of a droplet squeezing microfluidic device. (**ii**) An illustration of the working flow of the platform and high-speed microscope images that show the three stages of the cell in the platform; (1) encapsulation, (2) deformation, and (3) restoration. (**iii**) A schematic diagram illustrating the delivery mechanism of droplet squeezing, owing to a convection-based cargo transport. Reprinted and modified with permission from Ref. [49].

2.4. Cell Squeezing Based on Hydrodynamic Manipulation

As explained in Section 2.3, although cell-squeezing-type microfluidic devices based on channel confinement can achieve high-throughput cell deformation, they often suffer from cell membrane damage caused by the narrow channels as well as device failure due to clogging. To overcome these challenges, researchers explored hydrodynamic forces to control single cells stretching or squeezing in a microchannel. During this process, transient pores are generated on the cell membrane, facilitating the delivery of exogenous material through a blend of fluid convection and diffusion. The risk of microchannel clogging and cell lysis is significantly reduced, since the cells are not squeezed using constriction channels. Hydrodynamic techniques for creating transient nanopores offer several advantages, such as a simple design, inexpensive equipment requirements, and the capability to achieve the high-throughput intracellular delivery of diverse biomaterials into a broad spectrum of cells.

Kizer et al. reported a hydrodynamic manipulation-based cell-stretching approach [35] that effectively eliminated the possibility of the device clogging observed in earlier designs (Figure 5(ai)). In this proposed system, transient pores were formed on the cell membrane through the rapid hydrodynamic shearing of the cells, and the stagnation point, at which the transient fluid velocity is zero, is generated by two fluids with the same velocity but in opposite directions. As the cells approached the cross-section, they experienced hydrodynamic stretching and reached the maximum degree of deformation at the stagnation point, leading to the generation of transient membrane nanopores (Figure 5(aii)). Due to the rapid exchange of cytosol and external fluids across the cell membrane, this method facilitated convection-based intracellular delivery during the cell-stretching process (Figure 5(aii)), showing that transfection efficiency increased as the flow rates (i.e., Reynolds number) increased, while the cell viability decreased as the Reynolds number increased. Therefore, a suitable Reynolds number is the key to balancing transfection efficiency and cell viability. As a result, the experimental results demonstrate the successful delivery of DNA into various cell types, such as K-562, MDA-MB-231, HeLa cells, and so on, with a transfection efficiency of over 90%, an approximately 80% cell viability, and a remarkable throughput of over 1,600,000 cells per minute when the Reynolds number equals to 189. This approach showcased its effectiveness in achieving efficient delivery while maintaining cell viability by carefully controlling the Reynolds number to optimize the performance.

To simplify operations and improve the efficiency of material transportation, J. Hur et al. introduced a hydrodynamic manipulation-based cell-stretching intracellular delivery platform [36]. The device contained a T-shaped microchannel equipped with a small cavity, which provided the intrinsic inertial flow to deform the cell passing by. In the T-shaped microchannel, cells are exposed to elongational flows, enabling their lateral migration toward the center of the channel through intrinsic inertial flow. This mechanism allows for the uniform stretching of cells. As illustrated in Figure 5b, each cell hit into the cavity due to the force of elongational flows, leading to a collision-induced deformation. Subsequently, the cells were released from the cavity and reached maximum deformation at the stagnation point and then underwent slight cell stretching downstream while moving to the outlet. This study applied the cell delivery mechanism, which involves a mixture of a convection and diffusion-based solution exchange across the cell membrane during the cell-stretching and -recovering processes. J. Hur et al. achieved a knockdown of the ITGA1 gene by delivering siRNA into Hela cells using this cell-stretching device. Cells subjected to this microfluidic cell-stretching device exhibited a near-complete suppression of the ITGA1 gene expression, with a knockdown efficiency of 97% [36], indicating the tremendous potential of this technique in genetic editing. Overall, this intracellular delivery platform offers several advantages, including a high delivery efficiency of up to 98%, a high throughput of up to 1 million cells per minute, simplicity in operation, low material costs, and the ability to deliver various cell types and biomaterials.

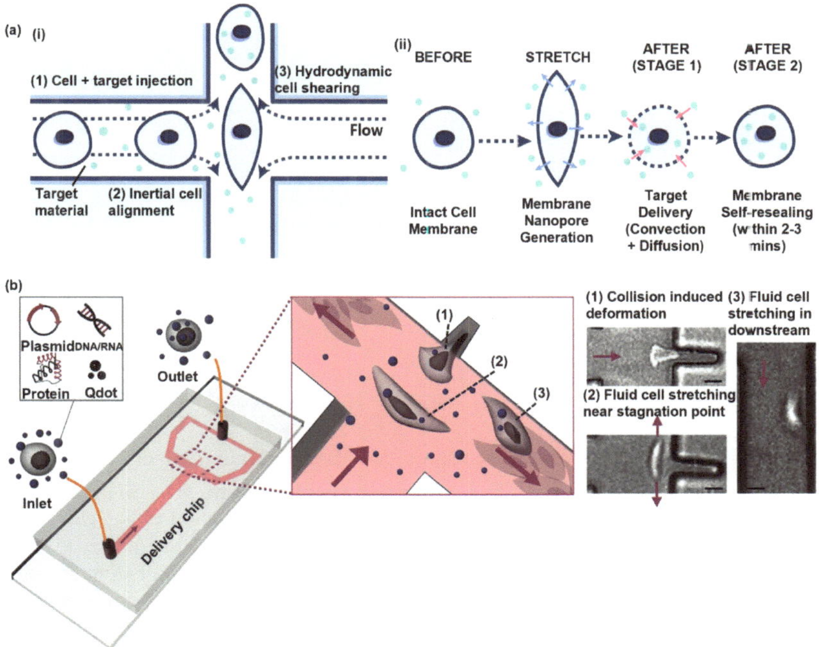

Figure 5. Cell squeezing based on hydrodynamic manipulation. (**a**) Hydroporator. (**i**) A schematic illustration of the design and operating principles of the vector-free intracellular delivery system. (**ii**) An illustration of the delivery mechanism. Reprinted and modified from Ref. [35]. (**b**) A schematic of the fluidic cell-stretching platform. High-speed microscope images showing the three stages of cell deformation (1)~(3). All arrows indicate the main flow direction (scale bars: 15 μm). Reprinted and modified with permission from [36]. Copyright (2020), American Chemical Society.

2.5. Summary

In this section, four different microfluidic mechanoporation methods were discussed, and each of them has its advantages and disadvantages (Table 1). Due to the accurate cargo delivery by inserting the micropipette into cells, microinjections ensure uniform transfections and achieve high transfection efficiency. Nevertheless, this method has limitations, such as low throughput rates and a high cost, which arises from the need for specialized expertise and expensive devices. Compared with microinjections, micro/nanoneedle arrays offer advantages, such as higher throughput and ease of use, as they allow for the simultaneous perforation of multiple cells. However, the manufacturing process of micro/nanoneedle arrays is usually complex and costly. Also, the effectiveness of this approach is dependent on the cell type utilized, as optimal results can typically be achieved with adherent cells. As for cell squeezing based on mechanical confinement, this has the advantage of high throughput and the ability to achieve intracellular delivery for a wide range of cells on the one hand and has the disadvantages of device clogging and non-uniform transfections. Instead of transforming cells by narrow channels in the microfluidic device, cell squeezing based on hydrodynamic manipulation avoids the issue of device clogging and maintains high throughput at the same time. Meanwhile, it also has the drawback of non-uniform transfections.

Table 1. Comparison of four microfluidic mechanoporation methods.

Microfluidic Mechanoporation Method	Advantages	Disadvantages	Throughput (Cells/min)	Cell Viability
Microinjection	Uniform transfection High transfection efficiency	High cost Low throughput rates	<100	82.5%
Micro/nanoneedle arrays	Higher throughput than microinjection Ease of use	Complex and costly manufacturing Dependent on the cell type	>10,000	95%
Cell squeezing based on mechanical confinement	High throughput Suitable for a wide range of cells	Device clogging Non-uniform transfections	>1,000,000	95%
Cell squeezing based on hydrodynamic manipulation	High throughput No device clogging	Non-uniform transfections	>1,600,000	80%

3. Application of Mechanoporation in Stem Cells

Stem cells are critical for the homeostasis of tissues and organs. Upon each cell division, the daughter cells can maintain as stem cells (self-renewal) or initiate a differentiation program for functional cells to replace the old, dead, or damaged cells. Understanding how the self-renewal and differentiation of stem cells is balanced is critical for clinical applications. Exogeneous gene expression and genome editing are both crucial for not only understanding how stem cells are regulated but also for the application of gene-edited cells for clinical purposes, highlighting the importance of delivering biomaterials into stem cells.

They possess self-renewal and differentiation capabilities, thus holding broad prospects for basic research and clinical applications. Human embryonic stem cells (hESCs) were first isolated in 1998, and since then, several adult stem cells, induced pluripotent stem cells (iPSCs), have been isolated as important models for basic research [84–86]. Due to the unique self-renewal and differentiation potential of stem cells, stem cell therapy has the potential to treat diseases, such as heart disease and type I diabetes [87,88].

Intracellular transfection technology is a crucial step for applications of stem cells, since it can introduce exogenous genes or small molecule drugs into stem cells, thereby changing the transcriptome state and physiological functions of stem cells for different purposes. For example, transfecting CRISPR-related components into cells can be used for the gene editing of stem cells [89],, and transfecting small molecules into stem cells can be used to mark them, which can be applied to stem cell therapy [90,91].

3.1. Comparison of Different Transfection Methods in Stem Cells

In order to conduct stem cell research and application, many technical problems must be solved, one of which is how to efficiently deliver external genes or drugs into stem cells through intracellular transfection. To improve delivery efficiency, scientists have established different means to optimize the delivery scheme (Figure 6). Common stem cell-transfection techniques include chemical transfections, electroporation, and viral vector transfections [92–95]. Although these methods can successfully introduce exogenous genes or small molecules into stem cells, there are many limitations (Table 2).

Chemical transfections suffer from the cytotoxicity of the transfection material and the low transformation efficiency in primary cells and stem cells [96]. Electroporation applies an electric field across the cells to perforate the cell membrane, achieving higher transfection efficiency, although specialized equipment and manual handling are required. In addition, electrical cell perforation causes a high cell death rate and low stability [56,97,98]. Viral vector-based transfections can achieve high-efficiency transfections, but there are biological safety issues, and the random insertion of viral vectors into the genome may lead to a disruption of local genes, resulting in unexpected risks, such as cell death [99]. It is essential to apply the transfection method with high transfection efficiency, a high survival rate, easy operation, low cost, and the large-scale operation of cells for different types of stem cells, enabling the application of stem cells and stem cell-derived functional cells for regenerative medicine.

As mentioned above, mechanoporation is a new type of transfection method, which promotes cell deformation through mechanical force, resulting in increased membrane permeability. This strategy improves the incorporation of therapeutic substances, such as DNA, RNA, and drugs, into the cells. Mechanoporation increases the transfection efficiency and improves the survival rate of cells, thereby facilitating the research of stem cells [36,90,91,100]. Therefore, mechanoporation may represent the best strategy for various fields of stem cell research.

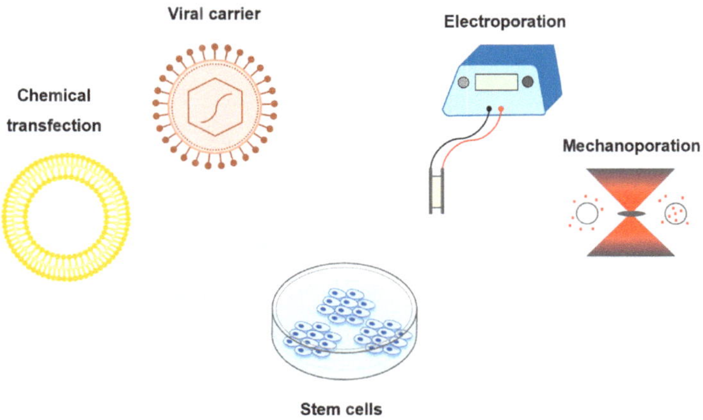

Figure 6. Schematic diagram of different stem cell-transfection methods.

Table 2. Comparison of different stem cell-transfection methods.

Transfection Method	Advantages	Disadvantages
Chemical Transfection	Simple operation	Low transfection efficiency Toxicity of the biomaterial
Viral Carrier	High transfection efficiency	Low biosecurity Instability due to random insertion
Electroporation	High transfection efficiency High mortality rate	Low stability High cost due to difficult operation
Mechanoporation	High transfection efficiency Capable of mass operation High cell survival rate High stability	

3.2. Application of Mechanoporation in Stem Cells

Transfection efficiency and cell viability are critical for applying stem cells for clinical applications. Due to the characteristics of less cell damage, a high cell survival rate, and high transfection efficiency, mechanoporation has shown great prospects in the clinical treatment of regenerative medicine.

Adipose tissue-derived stem cells (ADSCs) are one of the well-studied stem cells for clinical applications. As shown in Figure 7, Jung et al. employed mechanically perforated ADSCs for rapid labeling for PET/MRI imaging [90]. The patient's own stem cells can be used to repair or regenerate damaged joint tissue. In this context, these transplanted cells need to be labeled with in vivo molecular imaging tools to distinguish them from the host cells. In the follow-up treatment, it is necessary to monitor and observe the implantation, survival, migration, and differentiation activities to achieve the purpose of predicting the therapeutic effects. Jung et al. applied mechanical compression to transport iron oxide nanoparticles and 18F-FDG into ADSCs for subsequent identification by PET/MRI. The labeling process can be completed in a very short time during the operation, and the labeling efficiency is similar to that achieved by passive incubation for 30 min. The detection of labeled cells found that, compared with unlabeled cells, the survival rate of labeled cells reached 94%, and there was no increase in long-term toxicity, and even DNA damage was lower than that of passive incubation methods.

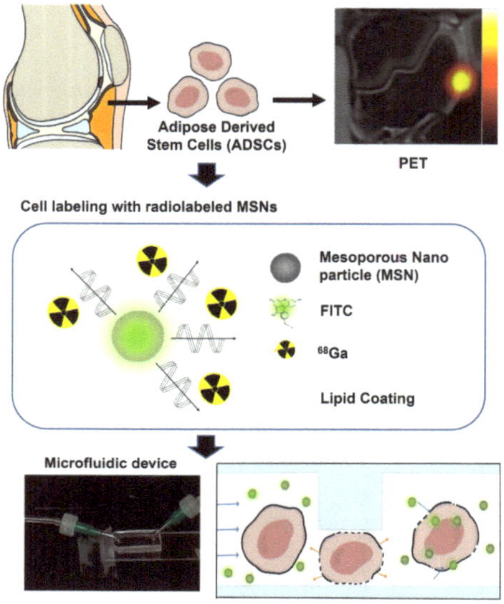

Figure 7. Experimental scheme of mechanoporation, which enables the rapid and efficient radiolabeling of adipose tissue-derived stem cells (ADSCs) for PET imaging. Reprinted with permission from Ref. [90].

On the other hand, stem cells cultured in vitro are an important research system for applications in regenerative medicine. Intracellular transfection technology can be used to introduce specific regulatory proteins or modified enzymes to explore the regulatory mechanisms governing the self-renewal and differentiation of stem cells. Intracellular transfection combined with gene editing technology has also been widely used in stem cell research. Therefore, the difficult-to-transfect feature has become a major challenge in stem cell research.

Chung et al. used fluidic cell mechanoporation to successfully transfect primary human stem cells with plasmid DNA. Cell viability, after mechanoporation, was significantly higher than electroporation [36]. Garcia et al. demonstrated a novel microfluidic device for the successful transfection of mRNA into human primary T cells, natural killer (NK) cells, and CD34+ hematopoietic stem and progenitor cells (HSPCs) via volume exchange for convective transfections (VECTs) [100]. In addition to the role of intracellular transport, studies have found that mechanical stretching can promote the reprogramming efficiency of fibroblasts to functional cells, even skipping the process of reprogramming into stem cells. When suspension cells pass through a narrow microfluidic channel, the nucleus undergoes a rapid extrusion. This reversible nuclear deformation process can significantly reduce the methylation level of histone H3K9 and DNA, thereby improving chromatin accessibility. Finally, it promotes the reprogramming efficiency of fibroblasts to neurons [101].

Through these stem cell studies, scientists can better understand the properties and functions of stem cells and apply them in various fields of regenerative medicine and disease treatment. Mechanoporation technology is an important tool for stem cell research, which helps scientists better utilize the potential of stem cells to achieve more effective therapeutic effects (Figure 8). Although this technology may still have some limitations and challenges at present, we believe that with continuous development and updating, mechanical perforation technology will exert more potential and make greater contributions in applying stem cells in regenerative medicine.

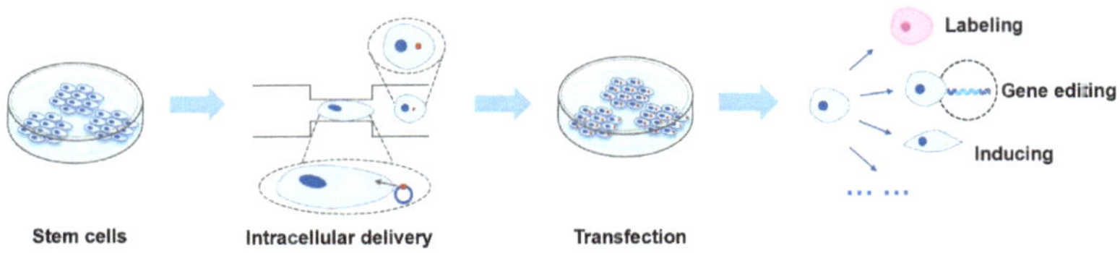

Figure 8. Transfection of the stem cell and its main application.

4. Outlook and Conclusions

Microfluidic mechanoporation, as an emerging intracellular delivery approach, offers huge potential in stem cell applications due to the following advantages.

First, mechanoporation offers the precious feature of no contamination or cytotoxicity, which is critical for stem cell study or therapy. For stem cell therapy, there must be product control, meaning the gene-edited stem cells should be exactly edited as we expect. But other intracellular delivery methods can introduce potential complications, such viral vectors, chemicals, and electrical stimulations, which may bring gene mutation or contamination. Mechanoporation only uses mechanical forces to disrupt cell membranes introducing no chemicals or viruses. In addition, a microfluidic chip can achieve the whole function of stem cell therapy by integrating sample processing, such as filtering and purification, gene editing using mechanoporation, quality checks, etc., in one personalized chip, holding promising applicational potential. Even single-cell manipulation, gene editing, and quality checks can all be realized in one microfluidic chip.

Second, microfluidic mechanoporation has great versatility, simplicity, and low fabrication costs. It allows for different types of cargos, such as nucleic acids, small molecules, and proteins to be delivered to various types of cells. It does not require any external energy sources, such as an electrical field or an acoustic field, and is straightforward to use. The fabrication of a microfluidic mechanoporation device is quite mature using conventional soft lithography or standard silicon microfabrication.

Third, microfluidic mechanoporation can offer high delivery efficiency and high cell viability, which is critical for gene editing and stem cell therapy. Low transfection efficiency in stem cells is one of the main bottlenecks for both basic research and clinic applications. Human pluripotent stem cells, including embryonic stem cells and induced pluripotent stem cells, possess the capacity to differentiate into different kinds of functional cells for cell therapy. However, the transfection efficiency is extremely low compared with cancerous cell lines. Furthermore, the transfection of large-size plasmid DNA or proteins is also inefficient. Cell state, extracellular components, and other cell membrane characteristics may "protect" stem cells from taking exogenous materials, making the traditional methods inefficient. Mechanoporation may substantially increase the transfection efficiency, because it stretches cells by mechanical forces. High transfection efficiency and the potential for delivering large cargoes will trigger both the mechanistic research and translational applications of stem cells.

Microfluidic mechanoporation technology still has a few challenges to overcome, including a lack of precise pore size control, a relatively low transfection efficiency for some cell types, and potential cell damages, which could be solved with a further understanding of the mechanoporation mechanisms and optimized device and operation parameters.

With a fast-growing interest in gene-editing-relevant applications, we can foresee more research development and commercialization in microfluidic mechanoporation technology toward more ideal technology that features high throughput, low costs, high cell viability, excellent reliability, and a more straightforward utilization.

Author Contributions: D.C. and H.H. planned and led the whole project; R.W. (Rubing Wang), Z.W. and L.T. were principally responsible for the writing of the manuscript; R.W. (Ruoming Wang) and S.Y. were responsible for the reprinting of the figures; D.C. and H.H. commented on and revised the manuscript. All authors have read and agreed to the published version of the manuscript.

Funding: This research was funded by the Human Space X program at the international campus of Zhejiang University, Center of Pathogen Detection in the Dynamic Research Enterprise for Multidisciplinary Engineering Sciences (DREMES), ZJU-UIUC Institute startup funding, as well as ZJU-UoE Institute startup funding.

Conflicts of Interest: The authors declare no conflicts of interest.

References

1. Wurm, F.M. Production of recombinant protein therapeutics in cultivated mammalian cells. *Nat. Biotechnol.* **2004**, *22*, 1393–1398. [CrossRef] [PubMed]
2. Recillas-Targa, F. Multiple strategies for gene transfer, expression, knockdown, and chromatin influence in mammalian cell lines and transgenic animals. *Mol. Biotechnol.* **2006**, *34*, 337–354. [CrossRef]
3. Glover, D.J.; Lipps, H.J.; Jans, D.A. Towards safe, non-viral therapeutic gene expression in humans. *Nat. Rev. Genet.* **2005**, *6*, 299–310. [CrossRef] [PubMed]
4. Karra, D.; Dahm, R. Transfection techniques for neuronal cells. *J. Neurosci.* **2010**, *30*, 6171–6177. [CrossRef] [PubMed]
5. Bakota, L.; Brandt, R.; Heinisch, J.J. Triple mammalian/yeast/bacterial shuttle vectors for single and combined Lentivirus- and Sindbis virus-mediated infections of neurons. *Mol. Genet. Genom.* **2012**, *287*, 313–324. [CrossRef] [PubMed]
6. Guo, Z.; Hong, S.; Jin, X.; Luo, Q.; Wang, Z.; Wang, Y. Study on the multidrug resistance 1 gene transfection efficiency using adenovirus vector enhanced by ultrasonic microbubbles in vitro. *Mol. Biotechnol.* **2011**, *48*, 138–146. [CrossRef] [PubMed]
7. Hiltunen, M.O.; Ruuskanen, M.; Huuskonen, J.; Mahonen, A.J.; Ahonen, M.; Rutanen, J.; Kosma, V.M.; Mahonen, A.; Kroger, H.; Yla-Herttuala, S. Adenovirus-mediated VEGF-A gene transfer induces bone formation in vivo. *FASEB J.* **2003**, *17*, 1147–1149. [CrossRef] [PubMed]
8. Royo, N.C.; Vandenberghe, L.H.; Ma, J.Y.; Hauspurg, A.; Yu, L.; Maronski, M.; Johnston, J.; Dichter, M.A.; Wilson, J.M.; Watson, D.J. Specific AAV serotypes stably transduce primary hippocampal and cortical cultures with high efficiency and low toxicity. *Brain Res.* **2008**, *1190*, 15–22. [CrossRef]
9. Washbourne, P.; McAllister, A.K. Techniques for gene transfer into neurons. *Curr. Opin. Neurobiol.* **2002**, *12*, 566–573. [CrossRef]
10. Jordan, M.; Wurm, F. Transfection of adherent and suspended cells by calcium phosphate. *Methods* **2004**, *33*, 136–143. [CrossRef]
11. Dubey, S.K.; Tripathi, A.K.; Upadhyay, S.N. Exploration of soil bacterial communities for their potential as bioresource. *Bioresour. Technol.* **2006**, *97*, 2217–2224. [CrossRef] [PubMed]
12. Zhang, Z.; Cao, X.; Zhao, X.; Withers, S.B.; Holt, C.M.; Lewis, A.L.; Lu, J.R. Controlled delivery of antisense oligodeoxynucleotide from cationically modified phosphorylcholine polymer films. *Biomacromolecules* **2006**, *7*, 784–791. [CrossRef]
13. Zhao, X.; Zhang, Z.; Pan, F.; Ma, Y.; Armes, S.P.; Lewis, A.L.; Lu, J.R. DNA immobilization using biocompatible diblock phosphorylcholine copolymers. *Surf. Interface Anal.* **2006**, *38*, 548–551. [CrossRef]
14. Takeuchi, J.K.; Koshiba-Takeuchi, K.; Matsumoto, K.; Vogel-Hopker, A.; Naitoh-Matsuo, M.; Ogura, K.; Takahashi, N.; Yasuda, K.; Ogura, T. Tbx5 and Tbx4 genes determine the wing/leg identity of limb buds. *Nature* **1999**, *398*, 810–814. [CrossRef] [PubMed]
15. Ikawa, M.; Kominami, K.; Yoshimura, Y.; Tanaka, K.; Nishimune, Y.; Okabe, M. A rapid and non-invasive selection of transgenic embryos before implantation using green fluorescent protein (GFP). *FEBS Lett.* **1995**, *375*, 125–128. [CrossRef] [PubMed]
16. Gueroussov, S.; Tarnawsky, S.P.; Cui, X.A.; Mahadevan, K.; Palazzo, A.F. Analysis of mRNA nuclear export kinetics in mammalian cells by microinjection. *J. Vis. Exp.* **2010**, *46*, 2387. [CrossRef]
17. Klein, R.M.; Wolf, E.D.; Wu, R.; Sanford, J.C. High-velocity microprojectiles for delivering nucleic acids into living cells. 1987. *Biotechnology* **1992**, *24*, 384–386. [PubMed]
18. Rodamporn, S.; Harris, N.R.; Beeby, S.P.; Boltryk, R.J.; Sanchez-Elsner, T. HeLa cell transfection using a novel sonoporation system. *IEEE Trans. Biomed. Eng.* **2011**, *58*, 927–934. [CrossRef]
19. Qiu, Y.; Zhang, C.; Tu, J.; Zhang, D. Microbubble-induced sonoporation involved in ultrasound-mediated DNA transfection in vitro at low acoustic pressures. *J. Biomech.* **2012**, *45*, 1339–1345. [CrossRef]
20. Cockrell, A.S.; Kafri, T. Gene delivery by lentivirus vectors. *Mol. Biotechnol.* **2007**, *36*, 184–204. [CrossRef]
21. Hacein-Bey-Abina, S.; Le Deist, F.; Carlier, F.; Bouneaud, C.; Hue, C.; De Villartay, J.P.; Thrasher, A.J.; Wulffraat, N.; Sorensen, R.; Dupuis-Girod, S.; et al. Sustained correction of X-linked severe combined immunodeficiency by ex vivo gene therapy. *N. Engl. J. Med.* **2002**, *346*, 1185–1193. [CrossRef] [PubMed]
22. Pfeifer, A.; Verma, I.M. Gene therapy: Promises and problems. *Annu. Rev. Genom. Hum. Genet.* **2001**, *2*, 177–211. [CrossRef] [PubMed]
23. Roesler, J.; Brenner, S.; Bukovsky, A.A.; Whiting-Theobald, N.; Dull, T.; Kelly, M.; Civin, C.I.; Malech, H.L. Third-generation, self-inactivating gp91(phox) lentivector corrects the oxidase defect in NOD/SCID mouse-repopulating peripheral blood-mobilized CD34+ cells from patients with X-linked chronic granulomatous disease. *Blood* **2002**, *100*, 4381–4390. [CrossRef] [PubMed]

24. Woods, N.B.; Muessig, A.; Schmidt, M.; Flygare, J.; Olsson, K.; Salmon, P.; Trono, D.; von Kalle, C.; Karlsson, S. Lentiviral vector transduction of NOD/SCID repopulating cells results in multiple vector integrations per transduced cell: Risk of insertional mutagenesis. *Blood* **2003**, *101*, 1284–1289. [CrossRef] [PubMed]
25. Zhou, C.; Yu, B.; Yang, X.; Huo, T.; Lee, L.J.; Barth, R.F.; Lee, R.J. Lipid-coated nano-calcium-phosphate (LNCP) for gene delivery. *Int. J. Pharm.* **2010**, *392*, 201–208. [CrossRef] [PubMed]
26. Arabsolghar, R.; Rasti, M. Optimal Electroporation Condition for Small Interfering RNA Transfection into MDA-MB-468 Cell Line. *Iran. J. Med. Sci.* **2012**, *37*, 187–193. [PubMed]
27. Uchida, M.; Li, X.W.; Mertens, P.; Alpar, H.O. Transfection by particle bombardment: Delivery of plasmid DNA into mammalian cells using gene gun. *Biochim. Biophys. Acta* **2009**, *1790*, 754–764. [CrossRef]
28. Valero, A.; Post, J.N.; van Nieuwkasteele, J.W.; Ter Braak, P.M.; Kruijer, W.; van den Berg, A. Gene transfer and protein dynamics in stem cells using single cell electroporation in a microfluidic device. *Lab A Chip* **2008**, *8*, 62–67. [CrossRef]
29. Groisman, A.; Enzelberger, M.; Quake, S.R. Microfluidic memory and control devices. *Science* **2003**, *300*, 955–958. [CrossRef]
30. Quake, S.R.; Scherer, A. From micro- to nanofabrication with soft materials. *Science* **2000**, *290*, 1536–1540. [CrossRef]
31. Adamo, A.; Jensen, K.F. Microfluidic based single cell microinjection. *Lab A Chip* **2008**, *8*, 1258–1261. [CrossRef] [PubMed]
32. Chow, Y.T.; Chen, S.; Wang, R.; Liu, C.; Kong, C.-w.; Li, R.A.; Cheng, S.H.; Sun, D. Single Cell Transfection through Precise Microinjection with Quantitatively Controlled Injection Volumes. *Sci. Rep.* **2016**, *6*, 24127. [CrossRef] [PubMed]
33. Sharei, A.; Zoldan, J.; Adamo, A.; Sim, W.Y.; Cho, N.; Jackson, E.; Mao, S.; Schneider, S.; Han, M.J.; Lytton-Jean, A.; et al. A vector-free microfluidic platform for intracellular delivery. *Proc. Natl. Acad. Sci. USA* **2013**, *110*, 2082–2087. [CrossRef]
34. Modaresi, S.; Pacelli, S.; Subham, S.; Dathathreya, K.; Paul, A. Intracellular Delivery of Exogenous Macromolecules into Human Mesenchymal Stem Cells by Double Deformation of the Plasma Membrane. *Adv. Ther.* **2020**, *3*, 1900130. [CrossRef]
35. Kizer, M.E.; Deng, Y.; Kang, G.; Mikael, P.E.; Wang, X.; Chung, A.J. Hydroporator: A hydrodynamic cell membrane perforator for high-throughput vector-free nanomaterial intracellular delivery and DNA origami biostability evaluation. *Lab A Chip* **2019**, *19*, 1747–1754. [CrossRef] [PubMed]
36. Hur, J.; Park, I.; Lim, K.M.; Doh, J.; Cho, S.G.; Chung, A.J. Microfluidic Cell Stretching for Highly Effective Gene Delivery into Hard-to-Transfect Primary Cells. *ACS Nano* **2020**, *14*, 15094–15106. [CrossRef]
37. Chakrabarty, P.; Gupta, P.; Illath, K.; Kar, S.; Nagai, M.; Tseng, F.G.; Santra, T.S. Microfluidic mechanoporation for cellular delivery and analysis. *Mater. Today Bio* **2022**, *13*, 100193. [CrossRef] [PubMed]
38. Kaladharan, K.; Kumar, A.; Gupta, P.; Illath, K.; Santra, T.S.; Tseng, F.G. Microfluidic Based Physical Approaches towards Single-Cell Intracellular Delivery and Analysis. *Micromachines* **2021**, *12*, 631. [CrossRef] [PubMed]
39. Kumar, A.; Mohan, L.; Shinde, P.; Chang, H.-Y.; Nagai, M.; Santra, T.S. Mechanoporation: Toward Single Cell Approaches. In *Handbook of Single-Cell Technologies*; Santra, T.S., Tseng, F.-G., Eds.; Springer: Singapore, 2022; pp. 31–59.
40. Liu, X.; Sun, Y. Microfabricated glass devices for rapid single cell immobilization in mouse zygote microinjection. *Biomed. Microdevices* **2009**, *11*, 1169–1174. [CrossRef]
41. Delubac, D.; Highley, C.B.; Witzberger-Krajcovic, M.; Ayoob, J.C.; Furbee, E.C.; Minden, J.S.; Zappe, S. Microfluidic system with integrated microinjector for automated Drosophila embryo injection. *Lab A Chip* **2012**, *12*, 4911–4919. [CrossRef]
42. Adamo, A.; Roushdy, O.; Dokov, R.; Sharei, A.; Jensen, K.F. Microfluidic jet injection for delivering macromolecules into cells. *J. Micromech. Microeng.* **2013**, *23*, 035026. [CrossRef] [PubMed]
43. Chow, Y.T.; Chen, S.; Liu, C.; Liu, C.; Li, L.; Kong, C.W.M.; Cheng, S.H.; Li, R.A.; Sun, D. A High-Throughput Automated Microinjection System for Human Cells with Small Size. *IEEE/ASME Trans. Mechatron.* **2016**, *21*, 838–850. [CrossRef]
44. Kim, Y.C.; Park, J.H.; Prausnitz, M.R. Microneedles for drug and vaccine delivery. *Adv. Drug Deliv. Rev.* **2012**, *64*, 1547–1568. [CrossRef] [PubMed]
45. Van der Maaden, K.; Luttge, R.; Vos, P.J.; Bouwstra, J.; Kersten, G.; Ploemen, I. Microneedle-based drug and vaccine delivery via nanoporous microneedle arrays. *Drug Deliv. Transl. Res.* **2015**, *5*, 397–406. [CrossRef] [PubMed]
46. Zhang, Y.; Ballas, C.B.; Rao, M.P. Towards ultrahigh throughput microinjection: MEMS-based massively-parallelized mechanoporation. *Annu. Int. Conf. IEEE Eng. Med. Biol. Soc.* **2012**, *2012*, 594–597. [CrossRef] [PubMed]
47. Dixit, H.G.; Starr, R.; Dundon, M.L.; Pairs, P.I.; Yang, X.; Zhang, Y.; Nampe, D.; Ballas, C.B.; Tsutsui, H.; Forman, S.J.; et al. Massively-Parallelized, Deterministic Mechanoporation for Intracellular Delivery. *Nano Lett.* **2020**, *20*, 860–867. [CrossRef] [PubMed]
48. Huang, D.; Zhao, D.; Li, J.; Wu, Y.; Du, L.; Xia, X.H.; Li, X.; Deng, Y.; Li, Z.; Huang, Y. Continuous Vector-free Gene Transfer with a Novel Microfluidic Chip and Nanoneedle Array. *Curr. Drug Deliv.* **2019**, *16*, 164–170. [CrossRef] [PubMed]
49. Joo, B.; Hur, J.; Kim, G.B.; Yun, S.G.; Chung, A.J. Highly Efficient Transfection of Human Primary T Lymphocytes Using Droplet-Enabled Mechanoporation. *ACS Nano* **2021**, *15*, 12888–12898. [CrossRef]
50. Du, X.; Wang, J.; Zhou, Q.; Zhang, L.; Wang, S.; Zhang, Z.; Yao, C. Advanced physical techniques for gene delivery based on membrane perforation. *Drug Deliv.* **2018**, *25*, 1516–1525. [CrossRef]
51. Hamann, A.; Nguyen, A.; Pannier, A.K. Nucleic acid delivery to mesenchymal stem cells: A review of nonviral methods and applications. *J. Biol. Eng.* **2019**, *13*, 7. [CrossRef]
52. Canoy, R.J.; Andre, F.; Shmakova, A.; Wiels, J.; Lipinski, M.; Vassetzky, Y.; Germini, D. Easy and robust electrotransfection protocol for efficient ectopic gene expression and genome editing in human B cells. *Gene Ther.* **2023**, *30*, 167–171. [CrossRef] [PubMed]

53. Jordan, E.T.; Collins, M.; Terefe, J.; Ugozzoli, L.; Rubio, T. Optimizing electroporation conditions in primary and other difficult-to-transfect cells. *J. Biomol. Tech.* **2008**, *19*, 328–334. [PubMed]
54. Liew, A.; Andre, F.M.; Lesueur, L.L.; De Menorval, M.A.; O'Brien, T.; Mir, L.M. Robust, efficient, and practical electrogene transfer method for human mesenchymal stem cells using square electric pulses. *Hum. Gene Ther. Methods* **2013**, *24*, 289–297. [CrossRef] [PubMed]
55. Prasanna, G.L.; Panda, T. Electroporation: Basic principles, practical considerations and applications in molecular biology. *Bioprocess. Eng.* **1997**, *16*, 261–264. [CrossRef]
56. Shi, J.; Ma, Y.; Zhu, J.; Chen, Y.; Sun, Y.; Yao, Y.; Yang, Z.; Xie, J. A Review on Electroporation-Based Intracellular Delivery. *Molecules* **2018**, *23*, 3044. [CrossRef]
57. Stroh, T.; Erben, U.; Kuhl, A.A.; Zeitz, M.; Siegmund, B. Combined pulse electroporation—A novel strategy for highly efficient transfection of human and mouse cells. *PLoS ONE* **2010**, *5*, e9488. [CrossRef]
58. Gupta, P.; Kar, S.; Kumar, A.; Tseng, F.G.; Pradhan, S.; Mahapatra, P.S.; Santra, T.S. Pulsed laser assisted high-throughput intracellular delivery in hanging drop based three dimensional cancer spheroids. *Analyst* **2021**, *146*, 4756–4766. [CrossRef] [PubMed]
59. Mohan, L.; Kar, S.; Nagai, M.; Santra, T.S. Electrochemical fabrication of TiO(2) micro-flowers for an efficient intracellular delivery using nanosecond light pulse. *Mater. Chem. Phys.* **2021**, *267*, 124604. [CrossRef] [PubMed]
60. Pylaev, T.; Vanzha, E.; Avdeeva, E.; Khlebtsov, B.; Khlebtsov, N. A novel cell transfection platform based on laser optoporation mediated by Au nanostar layers. *J. Biophotonics* **2019**, *12*, e201800166. [CrossRef]
61. Santra, T.S.; Kar, S.; Chen, T.C.; Chen, C.W.; Borana, J.; Lee, M.C.; Tseng, F.G. Near-infrared nanosecond-pulsed laser-activated highly efficient intracellular delivery mediated by nano-corrugated mushroom-shaped gold-coated polystyrene nanoparticles. *Nanoscale* **2020**, *12*, 12057–12067. [CrossRef]
62. Shinde, P.; Kar, S.; Loganathan, M.; Chang, H.Y.; Tseng, F.G.; Nagai, M.; Santra, T.S. Infrared Pulse Laser-Activated Highly Efficient Intracellular Delivery Using Titanium Microdish Device. *ACS Biomater. Sci. Eng.* **2020**, *6*, 5645–5652. [CrossRef]
63. Arruebo, M.; Fernández-Pacheco, R.; Ibarra, M.R.; Santamaría, J. Magnetic nanoparticles for drug delivery. *Nano Today* **2007**, *2*, 22–32. [CrossRef]
64. Moysidis, S.N.; Alvarez-Delfin, K.; Peschansky, V.J.; Salero, E.; Weisman, A.D.; Bartakova, A.; Raffa, G.A.; Merkhofer, R.M., Jr.; Kador, K.E.; Kunzevitzky, N.J.; et al. Magnetic field-guided cell delivery with nanoparticle-loaded human corneal endothelial cells. *Nanomedicine* **2015**, *11*, 499–509. [CrossRef]
65. Wang, Y.; Shang, S.; Li, C. Comparison of different kinds of nonviral vectors for gene delivery to human periodontal ligament stem cells. *J. Dent. Sci.* **2015**, *10*, 414–422. [CrossRef]
66. Liang, H.D.; Tang, J.; Halliwell, M. Sonoporation, drug delivery, and gene therapy. *Proc. Inst. Mech. Eng. H* **2010**, *224*, 343–361. [CrossRef]
67. Meng, L.; Liu, X.; Wang, Y.; Zhang, W.; Zhou, W.; Cai, F.; Li, F.; Wu, J.; Xu, L.; Niu, L.; et al. Sonoporation of Cells by a Parallel Stable Cavitation Microbubble Array. *Adv. Sci.* **2019**, *6*, 1900557. [CrossRef]
68. Ramesan, S.; Rezk, A.R.; Dekiwadia, C.; Cortez-Jugo, C.; Yeo, L.Y. Acoustically-mediated intracellular delivery. *Nanoscale* **2018**, *10*, 13165–13178. [CrossRef]
69. Zarnitsyn, V.G.; Meacham, J.M.; Varady, M.J.; Hao, C.; Degertekin, F.L.; Fedorov, A.G. Electrosonic ejector microarray for drug and gene delivery. *Biomed. Microdevices* **2008**, *10*, 299–308. [CrossRef] [PubMed]
70. Liu, A.; Islam, M.; Stone, N.; Varadarajan, V.; Jeong, J.; Bowie, S.; Qiu, P.; Waller, E.K.; Alexeev, A.; Sulchek, T. Microfluidic generation of transient cell volume exchange for convectively driven intracellular delivery of large macromolecules. *Mater. Today* **2018**, *21*, 703–712. [CrossRef]
71. Han, X.; Liu, Z.; Jo, M.C.; Zhang, K.; Li, Y.; Zeng, Z.; Li, N.; Zu, Y.; Qin, L. CRISPR-Cas9 delivery to hard-to-transfect cells via membrane deformation. *Sci. Adv.* **2015**, *1*, e1500454. [CrossRef]
72. Xing, X.; Pan, Y.; Yobas, L. A Low-Backpressure Single-Cell Point Constriction for Cytosolic Delivery Based on Rapid Membrane Deformations. *Anal. Chem.* **2018**, *90*, 1836–1844. [CrossRef] [PubMed]
73. Hao, R.; Yu, Z.; Du, J.; Hu, S.; Yuan, C.; Guo, H.; Zhang, Y.; Yang, H. A High-Throughput Nanofluidic Device for Exosome Nanoporation to Develop Cargo Delivery Vehicles. *Small* **2021**, *17*, e2102150. [CrossRef] [PubMed]
74. Pinero, J.; Lopez-Baena, M.; Ortiz, T.; Cortes, F. Apoptotic and necrotic cell death are both induced by electroporation in HL60 human promyeloid leukaemia cells. *Apoptosis* **1997**, *2*, 330–336. [CrossRef] [PubMed]
75. Kim, T.K.; Eberwine, J.H. Mammalian cell transfection: The present and the future. *Anal. Bioanal. Chem.* **2010**, *397*, 3173–3178. [CrossRef] [PubMed]
76. Mali, S. Delivery systems for gene therapy. *Indian J. Hum. Genet.* **2013**, *19*, 3–8. [CrossRef] [PubMed]
77. Barber, M.A. A Technic for the Inoculation of Bacteria and Other Substances Into Living Cells. *J. Infect. Dis.* **1911**, *8*, 348–360. [CrossRef]
78. Ludtke, J.J.; Sebestyen, M.G.; Wolff, J.A. The effect of cell division on the cellular dynamics of microinjected DNA and dextran. *Mol. Ther.* **2002**, *5*, 579–588. [CrossRef] [PubMed]
79. Stewart, M.P.; Langer, R.; Jensen, K.F. Intracellular Delivery by Membrane Disruption: Mechanisms, Strategies, and Concepts. *Chem. Rev.* **2018**, *118*, 7409–7531. [CrossRef] [PubMed]

80. Silberberg, Y.R.; Mieda, S.; Amemiya, Y.; Sato, T.; Kihara, T.; Nakamura, N.; Fukazawa, K.; Ishihara, K.; Miyake, J.; Nakamura, C. Evaluation of the actin cytoskeleton state using an antibody-functionalized nanoneedle and an AFM. *Biosens. Bioelectron.* **2013**, *40*, 3–9. [CrossRef]
81. Meister, A.; Gabi, M.; Behr, P.; Studer, P.; Voros, J.; Niedermann, P.; Bitterli, J.; Polesel-Maris, J.; Liley, M.; Heinzelmann, H.; et al. FluidFM: Combining atomic force microscopy and nanofluidics in a universal liquid delivery system for single cell applications and beyond. *Nano Lett.* **2009**, *9*, 2501–2507. [CrossRef]
82. Liu, L.; Michowski, W.; Kolodziejczyk, A.; Sicinski, P. The cell cycle in stem cell proliferation, pluripotency and differentiation. *Nat. Cell Biol.* **2019**, *21*, 1060–1067. [CrossRef] [PubMed]
83. Teh, S.Y.; Lin, R.; Hung, L.H.; Lee, A.P. Droplet microfluidics. *Lab A Chip* **2008**, *8*, 198–220. [CrossRef] [PubMed]
84. Takahashi, K.; Tanabe, K.; Ohnuki, M.; Narita, M.; Ichisaka, T.; Tomoda, K.; Yamanaka, S. Induction of pluripotent stem cells from adult human fibroblasts by defined factors. *Cell* **2007**, *131*, 861–872. [CrossRef]
85. Thomson, J.A.; Itskovitz-Eldor, J.; Shapiro, S.S.; Waknitz, M.A.; Swiergiel, J.J.; Marshall, V.S.; Jones, J.M. Embryonic stem cell lines derived from human blastocysts. *Science* **1998**, *282*, 1145–1147. [CrossRef] [PubMed]
86. Yu, J.; Vodyanik, M.A.; Smuga-Otto, K.; Antosiewicz-Bourget, J.; Frane, J.L.; Tian, S.; Nie, J.; Jonsdottir, G.A.; Ruotti, V.; Stewart, R.; et al. Induced pluripotent stem cell lines derived from human somatic cells. *Science* **2007**, *318*, 1917–1920. [CrossRef] [PubMed]
87. Chhabra, P.; Brayman, K.L. Stem cell therapy to cure type 1 diabetes: From hype to hope. *Stem Cells Transl. Med.* **2013**, *2*, 328–336. [CrossRef]
88. Garbern, J.C.; Lee, R.T. Cardiac stem cell therapy and the promise of heart regeneration. *Cell Stem Cell* **2013**, *12*, 689–698. [CrossRef] [PubMed]
89. Patmanathan, S.N.; Gnanasegaran, N.; Lim, M.N.; Husaini, R.; Fakiruddin, K.S.; Zakaria, Z. CRISPR/Cas9 in Stem Cell Research: Current Application and Future Perspective. *Curr. Stem Cell Res. Ther.* **2018**, *13*, 632–644. [CrossRef] [PubMed]
90. Jung, K.O.; Theruvath, A.J.; Nejadnik, H.; Liu, A.; Xing, L.; Sulchek, T.; Daldrup-Link, H.E.; Pratx, G. Mechanoporation enables rapid and efficient radiolabeling of stem cells for PET imaging. *Sci. Rep.* **2022**, *12*, 2955. [CrossRef] [PubMed]
91. Kiru, L.; Zlitni, A.; Tousley, A.M.; Dalton, G.N.; Wu, W.; Lafortune, F.; Liu, A.; Cunanan, K.M.; Nejadnik, H.; Sulchek, T.; et al. In vivo imaging of nanoparticle-labeled CAR T cells. *Proc. Natl. Acad. Sci. USA* **2022**, *119*, e2102363119. [CrossRef]
92. Liu, J.; Jones, K.L.; Sumer, H.; Verma, P.J. Stable transgene expression in human embryonic stem cells after simple chemical transfection. *Mol. Reprod. Dev.* **2009**, *76*, 580–586. [CrossRef]
93. Yalvac, M.E.; Ramazanoglu, M.; Gumru, O.Z.; Sahin, F.; Palotas, A.; Rizvanov, A.A. Comparison and optimisation of transfection of human dental follicle cells, a novel source of stem cells, with different chemical methods and electro-poration. *Neurochem. Res.* **2009**, *34*, 1272–1277. [CrossRef] [PubMed]
94. Moore, J.C.; Atze, K.; Yeung, P.L.; Toro-Ramos, A.J.; Camarillo, C.; Thompson, K.; Ricupero, C.L.; Brenneman, M.A.; Cohen, R.I.; Hart, R.P. Efficient, high-throughput transfection of human embryonic stem cells. *Stem Cell Res. Ther.* **2010**, *1*, 23. [CrossRef]
95. Coutant, F.; Frenkiel, M.P.; Despres, P.; Charneau, P. Protective antiviral immunity conferred by a nonintegrative lentiviral vector-based vaccine. *PLoS ONE* **2008**, *3*, e3973. [CrossRef]
96. Matosevic, S. Viral and Nonviral Engineering of Natural Killer Cells as Emerging Adoptive Cancer Immunotherapies. *J. Immunol. Res.* **2018**, *2018*, 4054815. [CrossRef]
97. Kumar, P.; Nagarajan, A.; Uchil, P.D. Electroporation. *Cold Spring Harb. Protoc.* **2019**, *2019*, 519–525. [CrossRef] [PubMed]
98. Batista Napotnik, T.; Polajžer, T.; Miklavčič, D. Cell death due to electroporation—A review. *Bioelectrochemistry* **2021**, *141*, 107871. [CrossRef]
99. Mellott, A.J.; Forrest, M.L.; Detamore, M.S. Physical Non-Viral Gene Delivery Methods for Tissue Engineering. *Ann. Biomed. Eng.* **2013**, *41*, 446–468. [CrossRef] [PubMed]
100. Loo, J.; Sicher, I.; Goff, A.; Kim, O.; Clary, N.; Alexeev, A.; Sulchek, T.; Zamarayeva, A.; Han, S.; Calero-Garcia, M. Microfluidic transfection of mRNA into human primary lymphocytes and hematopoietic stem and progenitor cells using ultra-fast physical deformations. *Sci. Rep.* **2021**, *11*, 21407. [CrossRef] [PubMed]
101. Song, Y.; Soto, J.; Chen, B.; Hoffman, T.; Zhao, W.; Zhu, N.; Peng, Q.; Liu, L.; Ly, C.; Wong, P.K.; et al. Transient nuclear deformation primes epigenetic state and promotes cell reprogramming. *Nat. Mater.* **2022**, *21*, 1191–1199. [CrossRef]

biosensors

MDPI

Review

Sustainable Sensing with Paper Microfluidics: Applications in Health, Environment, and Food Safety

Sanjay Kumar [1,*], Jyoti Bala Kaushal [2] and Heow Pueh Lee [3]

1 Durham School of Architectural Engineering and Construction, University of Nebraska-Lincoln, Scott Campus, Omaha, NE 68182-0816, USA
2 Department of Biochemistry and Molecular Biology, University of Nebraska Medical Center, Omaha, NE 68198, USA
3 Department of Mechanical Engineering, National University of Singapore, 9 Engineering Drive 1, Singapore 117575, Singapore; mpeleehp@nus.edu.sg
* Correspondence: skumar13@unl.edu

Abstract: This manuscript offers a concise overview of paper microfluidics, emphasizing its sustainable sensing applications in healthcare, environmental monitoring, and food safety. Researchers have developed innovative sensing platforms for detecting pathogens, pollutants, and contaminants by leveraging the paper's unique properties, such as biodegradability and affordability. These portable, low-cost sensors facilitate rapid diagnostics and on-site analysis, making them invaluable tools for resource-limited settings. This review discusses the fabrication techniques, principles, and applications of paper microfluidics, showcasing its potential to address pressing challenges and enhance human health and environmental sustainability.

Keywords: paper microfluidics; biosensors; healthcare, wearable sensors; environmental monitoring; food safety

Citation: Kumar, S.; Kaushal, J.B.; Lee, H.P. Sustainable Sensing with Paper Microfluidics: Applications in Health, Environment, and Food Safety. *Biosensors* **2024**, *14*, 300. https://doi.org/10.3390/bios14060300

Received: 13 May 2024
Revised: 4 June 2024
Accepted: 5 June 2024
Published: 7 June 2024

1. Introduction

In the past few decades, the increasing demand for portable, cost-effective, and environmentally friendly sensing technologies has driven the rapid advancement of paper-based microfluidics. Leveraging the inherent properties of paper, such as its low cost, biocompatibility, and ease of fabrication, researchers have developed innovative sensing platforms capable of performing complex analytical tasks with minimal resources.

One of the most compelling aspects of paper-based biosensors is their ability to meet the ASSURED criteria outlined by the World Health Organization (WHO) for point-of-care testing. These criteria emphasize the importance of tests being affordable, sensitive, specific, user-friendly, rapid and robust, equipment-free, and deliverable to those in need [1–3]. The paper microfluidics concepts are prominently utilized in healthcare, where it has revolutionized diagnostic testing, particularly in resource-limited settings. By integrating various biochemical assays and detection methods onto paper substrates, clinicians can now perform rapid and accurate diagnoses of various diseases, ranging from infectious diseases like malaria and HIV to chronic conditions such as diabetes and cancer. Moreover, the simplicity and portability of paper-based diagnostic devices make them well-suited for decentralized healthcare delivery, enabling point-of-care testing in remote or underserved communities.

Beyond healthcare, paper microfluidics plays a crucial role in environmental monitoring by providing cost-effective solutions for detecting pollutants and contaminants in air, water, and soil. By functionalizing paper with specific reagents or sensors, researchers can develop portable devices capable of detecting various environmental pollutants, including heavy metals, pesticides, and toxic gases. These paper-based sensors offer real-time monitoring capabilities and can be deployed in field settings to assess environmental quality and identify potential hazards.

Additionally, paper microfluidics contributes significantly to food safety and security by enabling the rapid and on-site detection of foodborne pathogens, adulterants, and contaminants. With the global food supply chain becoming increasingly complex and vulnerable to contamination, there is a growing need for robust and cost-effective methods to ensure the safety and integrity of food products. Paper-based sensors offer a promising solution by providing rapid, sensitive, and user-friendly tools for detecting harmful substances in food samples, helping prevent foodborne illnesses, and mitigating economic losses.

In summary, sustainable sensing with paper microfluidics holds immense promise for addressing critical challenges in healthcare, environmental monitoring, and food safety. Leveraging the unique attributes of paper, researchers are continuously innovating and introducing new sensing platforms that have wide-ranging implications for enhancing human well-being and preserving the environment. This manuscript provides a thorough overview of the current advancements in paper-based sensing technologies and their diverse health, environmental, and food-contaminant-detection applications. This comprehensive review is a valuable resource for researchers, practitioners, and enthusiasts in microfluidics, biotechnology, and environmental science, offering insights into the current state and future directions of biodegradable paper microsystems for health and environmental applications.

2. Fundamentals of Paper Microfluidics

2.1. Paper Types and Their Characteristics

Paper-based sensors leverage a wide variety of paper substrates, such as filter papers [4–6], nitrocellulose membranes [7–10], office papers (70–180 gsm) [11–14], tissue paper [15], photo (e.g., glossy) papers [16,17], waterproof papers [18], polyester papers [19], flexible polyethylene naphthalate sheets [20,21], and chromatography paper [22,23]. Notably, Whatman brand chromatography papers are among the most extensively utilized choices. This preference stems from the exceptional wicking ability that Whatman papers exhibit. Whatman offers a range of fibrous filter papers, such as Grade 1 to Grade 4, each characterized by distinct properties that find applications in various qualitative analytical techniques. These applications span general laboratory filtration, qualitative air pollution monitoring, soil analysis, food testing, and more. Whatman Filter Paper Grade 1, widely utilized in laboratory filtration, is renowned for its superior fine particle retention and rapid filtration capabilities. Composed of high-quality cellulose fibers, this filter paper basis weighs 88 g/m^2, with a nominal particle retention rating of around 11 μm, making it highly effective in separating very fine particles. With a moderate thickness of 180 μm and a porosity of 10.5 s, it balances quick filtration and efficient particle retention. Whatman Filter Paper Grade 2, another well-established filtration medium, is recognized for its fine particle retention and moderate flow rate. It is manufactured from high-quality cellulose fibers weighing 103 g/m^2 and offers a nominal particle retention of approximately 8 μm. Its balanced construction, with increased thickness (190 μm) compared to Grade 1, ensures efficient particle retention while allowing relatively faster filtration. Whatman Filter Paper Grade 3, made from high-quality alpha cotton cellulose (basis weight of 187 g/m^2), provides a nominal particle retention of approximately 6 μm, with moderate thickness (309 μm) and a porosity of 26 s. Finally, Whatman Filter Paper Grade 4, designed for robust and versatile filtration, is crafted from high-quality alpha cotton cellulose, offering a nominal particle retention of approximately 20–25 μm, with moderate thickness (205 μm) and basis weight of 92 g/m^2.

Nitrocellulose membrane papers, derived from the nitration of cellulose, are integral components in laboratory techniques such as Western blotting and immunoassays. Renowned for their highly porous structure of 0.45 μm and 12 μm pore sizes, these membranes provide efficient protein binding, ensuring a large surface area for immobilization. Their uniform pore size distribution guarantees consistent and reproducible outcomes, while their hydrophobic nature facilitates the transfer of hydrophobic proteins during blotting. With a high binding capacity, compatibility with various immunodetection methods, and versatility for protein and nucleic acid applications, nitrocellulose membrane papers play a crucial role in molecular biology and biochemistry, offering purity and reliability in experimental procedures.

Tissue paper is a versatile and widely used material characterized by its random packing of cellulose microfibers. These microfibers, ranging from several hundred micrometers in length to 50–100 μm in diameter, contribute to tissue paper's softness, absorbency, and strength. Derived from wood pulp or plant-based sources, tissue paper is known for its comfort and durability, making it suitable for facial tissues, toilet paper, and napkins. However, it has been used in research applications such as oil/water separation, wearable sensors, etc. [24].

Glossy paper, traditionally associated with printing and photography, has emerged as a subject of study for paper-based sensors due to its unique composition. Comprising cellulose fibers and inorganic fillers intricately blended into the paper matrix, glossy paper offers distinct advantages in sensor development. For example, Arena et al. [25] explored the use of glossy paper to create a flexible paper-based sensing device specifically designed for detecting ethanol. Unlike traditional filter paper, glossy paper's surface properties are more amenable to modification, providing greater sensor design and customization flexibility. This shift to glossy paper represents a novel approach, capitalizing on its composition to enhance the performance and adaptability of paper-based sensors, thereby expanding the scope of potential applications in analytical and diagnostic fields.

Chromatography paper, composed primarily of high-quality cellulose fibers, is designed with specific technical specifications to facilitate the efficient separation and analysis of substance mixtures. Chromatography paper is available in different formats, such as sheets or rolls, tailored to specific chromatographic techniques, making it an essential tool for diverse analytical applications. Figure 1 shows the morphology of different paper substrates.

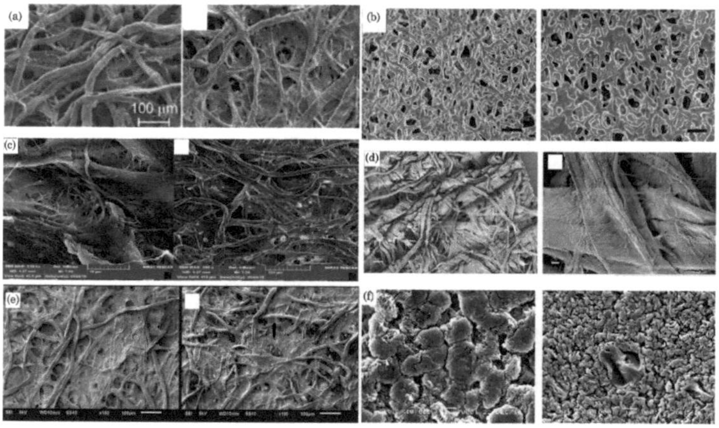

Figure 1. Scanning electron microscopy (SEM) images of different paper substrates. (**a**) Filter paper (Whatman Grade 1) (without treatment and with heat treatment at 600 °C). Reprinted with permission from Jiang et al. [26]. ©2015 The Authors, licensed under a Creative Commons Attribution 4.0 International License (CC BY-NC 4.0). (**b**) NC membrane (untreated and polyurethane acrylate-treated). Reprinted with permission from Lin et al. [27]. ©2022 The Authors, licensed under a Creative Commons Attribution 4.0 International License. Published by Springer Nature Limited. (**c**) Office paper at 10× and 100× magnification. Adapted from Jabar et al. [28] ©2019. Laser. (**d**) Native tissue paper (overview and close view). Reprinted with permission from Cao et al. [24]. ©2017 The Authors, licensed under a Creative Commons Attribution 4.0 International License. (**e**) Chromatographic paper at 150× (untreated and graphene oxide-modified), adapted from Fernandes et al. [29]. ©2019 The Authors, licensed under a Creative Commons Attribution License. Published by Sociedade Brasileira de Química. (**f**) Polyethylene terephthalate (PET) membrane. Adapted from Arahman et al. [30]. ©2017 The Authors, open access article distributed under the terms of the Creative Commons Attribution License.

2.2. Paper Selection Factors

The choice of paper substrate in biosensing relies on several characteristics, including the capillary flow time (the time for the liquid sample to flow through lateral pores), paper thickness, pore size, porosity (percentage of air in the porous structure), and surface quality.

In paper-based microfluidic devices, capillary flow time refers to the duration it takes for a liquid sample to travel through the lateral pores of the paper substrate. The capillary flow time is inversely related to the capillary flow rate and expressed as s cm^{-1}. It plays a vital role in defining the velocity and effectiveness of fluid movement within the microfluidic system. Specifically, the capillary flow time holds significant importance in creating paper-based diagnostic devices, especially in applications like lateral flow assays. Assessing the capillary flow time is instrumental in determining the optimal placement of the test and control lines on the nitrocellulose (NC) membrane.

The thickness of the paper substrate is an important parameter influencing the design and performance of paper-based microfluidic devices. It directly affects the capillary action, fluid flow dynamics, and device functionality. Thicker papers may impede the capillary flow, extending the path length for liquid absorption and affecting the speed of fluid traversal through paper channels. Moreover, paper thickness impacts sample absorption, with thinner papers potentially having a lower sample retention capacity, influencing device sensitivity and detection limits. The mechanical strength and integrity of the paper are also thickness-dependent, where thicker papers contribute to enhanced device durability. However, thicker papers may present challenges in fabrication processes such as cutting, printing, or folding, necessitating consideration of compatibility with chosen techniques. Consequently, optimizing paper thickness involves a delicate balance, requiring careful selection based on the specific needs of the intended microfluidic application.

Pore size is a pivotal parameter in paper microfluidics, exerting a profound impact on the performance and capabilities of microfluidic devices. Acting as conduits for capillary flow, the pores within the paper substrate guide fluid movement throughout the device. The capillary action hinges on pore size, with smaller pores facilitating slower yet controlled fluid flow, while larger pores allow for faster capillary flow. The speed and efficiency of fluid transport within the paper substrate are directly influenced by pore size. Fine-tuning this parameter is critical for optimizing fluid dynamics, ensuring precise sample movement to various device regions. Moreover, pore size governs the volume of sample absorption by the paper, providing customizable control over the sample absorption capacity to meet specific diagnostic or analytical needs. Pore size becomes pivotal in applications necessitating analyte separation, such as chromatographic assays. Varied analytes interact differently with the paper matrix, and adjusting the pore size enables selective separation. The resolution and sensitivity of paper-based assays are intricately tied to pore size, where smaller pores enhance resolution but may impede fluid transport speed. Striking a balance between these factors is essential for attaining paper microfluidic devices' desired sensitivity and resolution.

Porosity refers to the percentage of air present in the porous structure of the paper substrate. It is a crucial parameter influencing the capillary action and fluid flow dynamics within microfluidic devices. The porosity of the paper directly impacts the movement of fluids through its pores. A higher porosity generally allows for better capillary flow, as there is more interconnected space for the fluid to travel. However, excessively high porosity may lead to rapid fluid flow and reduced control over the movement, potentially affecting the performance of the microfluidic device.

Permeability refers to the ability of the paper substrate to allow the flow of fluids through its structure. It is a crucial parameter influencing the capillary action and fluid transport dynamics within microfluidic devices. The permeability of the paper substrate determines how readily and efficiently liquids can traverse through its porous structure. A paper substrate with high permeability allows for rapid capillary flow, enabling the swift movement of fluids along the paper channels. On the other hand, lower permeability may result in slower capillary flow. The permeability of the paper is influenced by factors such as pore size, porosity, and the overall structure of the paper matrix. It is an important consideration in designing and optimizing paper-based microfluidic devices, particularly in applications where the precise control of fluid flow and transport dynamics is essential. For porous materials consisting of fibers with a circular cross-section of radius r_f, permeability can be approximated using the following empirical relation [31]:

$$k = r_f^2 \frac{\pi\varphi(1-\sqrt{1-\varphi})^2}{24(1-\varphi)^{3/2}} \tag{1}$$

For random fibrous media, the permeability can be determined using the following correlation between permeability and porosity [32]:

$$k = C_1 r_f^2 \left(\sqrt{\frac{1-\varphi_c}{1-\varphi}} - 1\right)^{C_2} \tag{2}$$

Here, φ_c represents the critical porosity value required for permeating flow, often referred to as the percolation threshold. The parameters C_1 and C_2 are associated with the network's geometry.

Furthermore, the Kozeny–Carman equation can be employed to predict the permeability of granular isotropic porous materials, such as nitrocellulose membranes [33]:

$$k = \frac{d^2\varphi^3}{180(1-\varphi)^2} \tag{3}$$

Here, d represents the average pore diameter, and porosity φ was determined empirically by measuring the volume of liquid absorbed by the materials [34].

2.3. *Principles of Fluid Transport in Paper*

Paper microfluidics operates on fluid flow without external forces, relying on capillary action to drive passive fluid movement through the paper substrate. The paper and the fluid's contact surface interplay involves cohesive and adhesive forces. Interactions occur within the liquid at the liquid–air interface (cohesion) and between the solid–liquid interfaces (adhesion). The adhesive force facilitates the liquid's spreading onto the porous substrate, while cohesive forces, such as surface tension, work to reduce the liquid–air interface's area. Fluid flow occurs when the effect of adhesion surpasses that of cohesion. The wicking process is influenced by various physical and geometrical properties of porous media, including the paper material, paper structure, pore size, permeability, paper size and shape, and the physical properties of the liquid. Fluid transport can generally be classified into the wet-out process and fully wetted flow [35]. In the wet-out process, the fluid front wicks along the dry porous media and can be modeled using the classical Lucas–Washburn equation. Conversely, fluid transport occurs along the wetted porous media in fully wetted flow and is governed by Darcy's law.

2.3.1. Classical Lucas–Washburn Equation (Capillary Flow)

$$l(t) = 2\sqrt{\frac{k\gamma\cos\theta}{\Phi\mu r_a}}\sqrt{t} \tag{4}$$

where $l(t)$ denotes the length of the wetted region of the paper at time t; k represents the permeability of the paper, reflecting how readily fluid can traverse a specific paper substrate and contingent on pore size and geometry; μ signifies fluid viscosity; γ represents the interfacial surface tension of the liquid; r_a is the average pore radius; and t stands for the liquid penetration time. The Lucas–Washburn equation can be formulated in the following manner by incorporating the influence of tortuosity on capillary flow:

$$l(t) = 2\sqrt{\frac{r_a \gamma \cos\theta}{2\mu\tau^2}}\sqrt{t} \quad (5)$$

In this context, tortuosity is defined as $\tau = (L_e/L)^2$, where L_e represents the effective path length between two intermittent points in the liquid and L is the straight path length. The parameter τ consistently holds a value greater than one.

2.3.2. Darcy's Law for Fluid Flow

Darcy's Law is a fundamental equation describing fluid flow through porous media, and it applies to various contexts, including paper microfluidics. In this context, the imbibition rate of the fluid $\hat{u}$ through the paper substrate can be determined by Darcy's law, as per the following equation:

$$\hat{u} = \frac{k_i \Delta P}{\mu l(t)} \quad (6)$$

In the given expression, $k_i = k/\varphi$ represents the interstitial permeability of the paper strip and $\varphi = 1 - \frac{m}{\rho_c h}$ is the porosity of the medium [36]. Other variables include m, which denotes the basis weight; ρ_c and h representing the density and thickness of the porous substrate, respectively; and ΔP, which stands for the pressure difference over the wetted region, often referred to as Laplace pressure.

For a straight paper strip device, Darcy's law for fluid flow can be modified as follows:

$$Q = -A \cdot k \cdot \frac{\Delta P}{L} \quad (7)$$

Q represents the volumetric flow rate of the liquid; A (equal to $w \times h$) signifies the cross-sectional area of the paper; k stands for the permeability of the paper, which measures the ease of fluid flow through the paper substrate; and $\Delta P = P(0) - P(L)$ indicates the pressure drop across the paper, with $P(0)$ denoting the pressure at $x = 0$ and $P(L)$ representing the average capillary pressure at the fluid front. L is the length of the paper. This equation introduces a negative sign to consider the flow occurring in the direction of decreasing pressure. Additionally, the term $\frac{\mu L}{kwh}$ is defined as the flow resistance (R_{hyd}):

$$Q = -\frac{\Delta P}{R_{hyd}} \quad (8)$$

Equation (8) bears similarity to Ohm's law in an electric circuit, expressed as $\Delta V = RI$, where I denotes the electric current, R is the electrical resistance, and ΔV represents the potential drop. In hydrodynamic systems, the volumetric flow rate Q signifies the volume per unit time, while in an electric system, the current represents the charge per unit time. Moreover, ΔP (energy per volume) draws an analogy to potential drop (energy per charge).

Moreover, beyond the conventional linear channels in paper strips, researchers have introduced a variety of configurations to control fluid transport, each characterized by distinct dynamics attributed to shapes like circular, trapezoidal, sector-shaped, multisection medium, and other arbitrary geometries [37–41]. For an in-depth exploration of these geometries, readers can refer to the comprehensive summary provided by Kumar et al. [42].

2.4. Dimensionless Numbers for Fluid Transport

In paper microfluidics, the intricate phenomena of fluid flow can be effectively characterized and understood by employing a series of dimensionless numbers [43]. These dimensionless numbers play a crucial role in delineating the relative significance of different physical factors governing the behavior of fluids within the porous paper substrate. Dimensionless numbers are fundamental in scaling and normalizing various parameters, enabling researchers and engineers to draw meaningful comparisons and insights across different systems and scales.

2.4.1. Capillary Number (*Ca*)

It is defined as the ratio of viscous forces to capillary forces and is expressed as

$$Ca = \frac{U\mu}{\gamma} \tag{9}$$

Here, U (m/s) represents the velocity of the flow, γ (N/m) denotes the surface tension at the water/paper interface, and μ (kg/(ms)) stands for the viscosity of the fluid.

The capillary number helps to characterize the dominance of capillary forces over viscous forces in a given system. In the context of paper microfluidics, it provides insights into the ability of capillary action to drive fluid flow through the paper substrate. When *Ca* is small, viscous forces dominate, and the flow is slow and controlled by viscosity. On the other hand, when *Ca* is large, capillary forces take precedence, resulting in faster, capillary-driven flow. A low capillary number is often desirable for paper-based devices, ensuring controlled and predictable fluid flow. Understanding and manipulating *Ca* is essential for designing effective paper microfluidic devices, especially in point-of-care diagnostics and environmental monitoring applications.

2.4.2. Reynolds Number (*Re*)

A dimensionless quantity characterizes the relative importance of inertial forces to viscous forces in fluid flow. In paper microfluidics, the Reynolds number helps assess the nature of fluid flow through the porous substrate. It is defined as

$$Re = \frac{\rho UL}{\mu} \tag{10}$$

where ρ is the fluid density, U is the characteristic velocity of the flow, L is a characteristic length (e.g., pore size, channel width), and μ is the dynamic viscosity of the fluid.

The Reynolds number classifies flow regimes into laminar and turbulent. In paper microfluidics, where flow is typically slow and capillary-driven, the flows are often in the laminar regime (low Re). Laminar flow is characterized by smooth and predictable streamlines, making it suitable for the controlled transport of fluids within microchannels or porous media. In a porous medium, if the Re is less than one, the flow is characterized as laminar and linear, and Darcy's Law is applicable. However, when Re exceeds 10, the flow remains laminar but is no longer linear. In this regime, inertial forces become significant, causing a departure from the linear behavior, and consequently, Darcy's Law is no longer valid [43,44].

2.4.3. Weber Number (*We*)

The Weber number (*We*) is a dimensionless parameter that characterizes the ratio of inertial forces to surface tension forces in a fluid flow. It is particularly relevant in understanding the deformation and breakup of liquid droplets. The Weber number is defined as

$$We = \frac{\rho U^2 L}{\sigma} \tag{11}$$

where ρ is the fluid density, U is the characteristic velocity of the flow, L is a characteristic length (e.g., pore size and channel width), and σ is the surface tension at the liquid–air interface.

The Weber number indicates the dominance of inertial forces over surface tension forces. A low Weber number is often desirable in paper microfluidics, where capillary action and surface tension are crucial in fluid transport through porous substrates. A low

We signifies that capillary forces and surface tension are sufficient to overcome inertial forces, allowing for stable and controlled fluid flow.

2.4.4. Schmidt Number (Sc)

The Schmidt number (Sc) is a dimensionless quantity that characterizes the relative importance of momentum and mass transport in the fluid flow. It is defined as the ratio of the kinematic viscosity of the fluid to its mass diffusivity:

$$Sc = \frac{\mu}{\rho\, D} \tag{12}$$

Here, $\nu = \mu/\rho$ is the kinematic viscosity ($m^2\ s^{-1}$) and D is the is the mass diffusivity ($m^2\ s^{-1}$).

Sc plays a crucial role in determining the effectiveness of mass transport processes, especially in cases involving simultaneous flow and diffusion. A high Schmidt number suggests that the diffusional transport of mass is dominant compared to the convective transport by fluid flow. Conversely, a low Schmidt number indicates that convective transport prevails over diffusional processes. Controlling mass transport is essential for applications such as chemical reactions, analyte detection, and other biological or chemical processes in paper microfluidic devices.

2.4.5. *Péclet* Number (Pe_L)

The *Péclet* number (Pe_L) is a dimensionless quantity that characterizes the relative importance of convective transport to diffusive transport in a fluid flow system. It is defined as the ratio of the characteristic time of convective transport to the characteristic time of diffusive transport:

$$Pe_L = Re \times Sc = \frac{U}{D/L} \tag{13}$$

The significance of the *Péclet* number lies in its ability to provide insights into the dominance of convective or diffusive transport mechanisms. A high Pe_L indicates that convective transport is dominant, suggesting that fluid flow is crucial in transporting species within the porous medium. On the other hand, a low Pe_L means that diffusive transport is more significant, indicating that the concentration gradient is the primary driving force for mass transport.

For applications in paper microfluidics, such as chemical reactions, analyte detection, or biological assays, understanding the Pe_L number is crucial for optimizing the design and performance of the devices. Balancing convective and diffusive transport is essential to ensure efficient and controlled mass transport, ultimately influencing the accuracy and reliability of the processes carried out in paper microfluidic systems.

3. Classifications of Paper-Based Assays

Paper-based sensors are versatile diagnostic tools that utilize the properties of paper to detect various analytes. These sensors fall into three main classifications: dipstick tests, lateral flow assays (LFAs), and microfluidic paper-based analytical devices (μPADs).

3.1. Dipstick Assays

Dipstick test strips consist of paper pads with dried capture reagents affixed to a supporting plastic strip. A paper strip is immersed into a liquid sample in the dipstick assay process. The sample traverses the strip, interacting with specific reagents immobilized, resulting in a discernible signal at the test line (Figure 2a). The intensity or shade of the produced color in dipstick assays sometimes enables users to estimate the approximate or semi-quantitative concentration of the analyte. Key advantages of dipstick assays include their simplicity, cost-effectiveness, and the capability to test for multiple analytes simultaneously. However, dipstick assays often exhibit drawbacks such as poor detection limits and limited specificity [45,46].

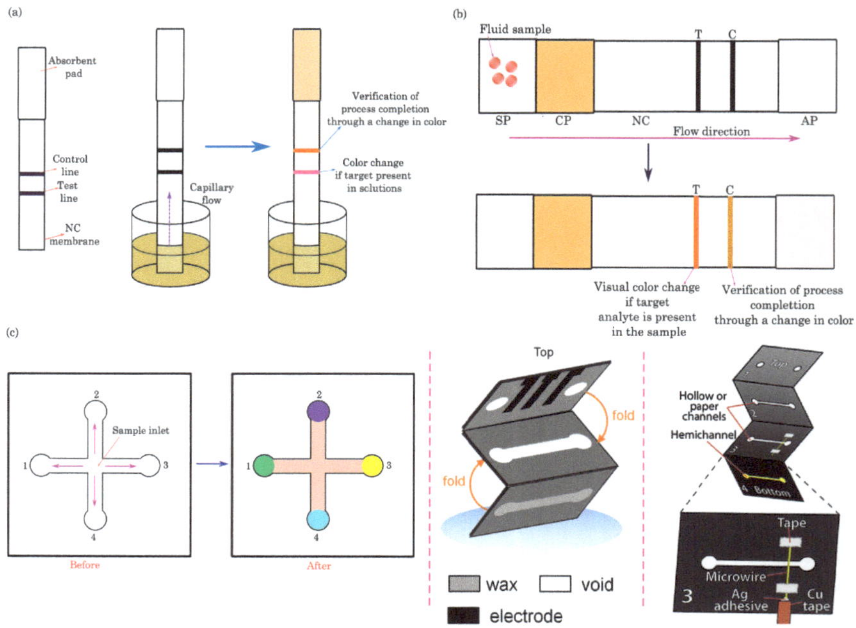

Figure 2. Schematics illustrating various paper-based assays: (**a**) Dipsticks typically comprise the test line and control line printed on the NC membrane, with an absorbent pad (crafted from filter paper) to soak up excess sample fluid. (**b**) Lateral flow assays (LFAs) encompass key components such as SP: sample pad, CP: conjugate pad, NC: nitrocellulose membrane, T: test line, C: control line, and AP: absorbent pad. (**c**) Microfluidic paper-based analytical devices (μPADs): a simple patterned paper device designed for multianalyte detection (left), origami-based 3D μPADs with hollow channels (center and right). Reprinted with permission from Renault et al. [47]. ©2014 American Chemical Society and from Carrell et al. [48]. ©2019 The Authors, under a Creative Commons license CC BY-NC-ND-4.0, Elsevier B.V.

3.2. *Lateral-Flow Assays*

Paper-based lateral flow assays (LFAs) consist of overlapping paper substrates, including a sample pad (SP) for receiving the liquid sample, a conjugate pad (CP) with labeled reagents (e.g., colloidal gold particles), a nitrocellulose (NC) membrane containing test and control lines, and an absorbent pad (AP) to soak up excess sample fluid (Figure 2b). The working principle involves applying the sample to the pad, initiating capillary flow that guides the sample through the conjugate pad, which interacts with the labeled reagents. The sample then traverses the nitrocellulose membrane, binding to immobilized capture agents at the test line if the target analyte is present, forming a visible line. The control line, containing immobilized capture agents for validation, always produces a line. The absorbent pad at the end facilitates liquid flow. This assay allows for rapid on-site detection, with the presence or absence of lines providing a visual interpretation of results, and its versatility makes it valuable for diagnostics and point-of-care testing.

3.3. *Microfluidic Paper-Based Analytical Devices (μPADs)*

Paper-based microfluidic biosensors (μPADs) represent an innovative class of diagnostic tools that integrate microfluidic channels on paper [49]. These biosensors leverage the capillary action of paper to control the flow of liquids through predefined channels. Microfluidic components enable the precise manipulation of samples and reagents, enhancing the sensitivity and specificity of assays. By incorporating various detection zones on the paper, μPADs can be customized to detect multiple analytes simultaneously (Figure 2c).

These paper-based assays are presently employed across diverse applications for detecting diseases, pathogens, toxins, pollutants, food safety, and, most notably, in the recent context, for detecting COVID-19 [50–52]. A detailed discussion on applications of these assays is discussed in Sections 7 and 8.

4. Fabrication Techniques for Paper-Based Devices

Paper-based microfluidic devices have gained popularity due to their simplicity, cost-effectiveness, and ease of fabrication. Several techniques are employed to fabricate these devices, each offering unique advantages. The manufacturing processes for paper microfluidics involve making specific sections of the paper hydrophilic to enable smooth sample flow, while other sections are made hydrophobic to form the channel walls. Broadly, fabrication methods can be categorized into two approaches: the first involves treating hydrophilic paper with hydrophobic materials to shape the desired channels, while the second approach entails cutting the paper using various tools such as knives or lasers to generate the channel pattern. Here are some commonly used fabrication techniques for paper-based microfluidic devices:

4.1. Blade Cutting/Plotting

Blade cutting or plotting is a versatile method providing a straightforward and precise means of creating desired patterns and structures. This technique involves using a cutting or plotting machine equipped with a sharp blade to cut through paper substrates precisely, shaping them according to a predefined digital design.

The process begins with creating a digital design or blueprint of the intended paper-based device using design software. This digital file guides the cutting or plotting machine, detailing the specific features, dimensions, and patterns. The paper substrate, typically selected for its compatibility with blade cutting, is then securely fixed onto the machine's work surface. The machine is calibrated to accommodate the specific properties of the paper and the design requirements. Adjustments to parameters such as blade depth, speed, and cutting force are made to ensure accurate and clean cuts. The digital design file is loaded into the cutting or plotting machine, specifying the desired cutting settings. The cutting or plotting machine, guided by the digital design, moves the sharp blade across the paper substrate, accurately cutting or scoring along the defined lines. The process is precise and repeatable, allowing for the creation of intricate patterns, microfluidic channels, or other features.

Blade cutting/plottings offer advantages such as simplicity, cost-effectiveness, and quick turnaround times. However, it may have limitations in achieving excellent features or complex geometries compared to more advanced fabrication techniques. Nonetheless, it remains a popular choice for rapid prototyping and producing paper-based devices for various applications.

4.2. Laser Cutting

Laser cutting, a meticulous and versatile method for fabricating paper-based devices, utilizes a laser beam to intricately cut or engrave patterns, channels, and features into paper substrates, resulting in well-defined structures. The process involves several key steps, including creating a digital design or blueprint using design software. This digital file serves as a guide for the laser-cutting process, detailing specific features and dimensions. Material selection is crucial, with the chosen paper substrate needing the right thickness and properties to achieve precise cuts without excessive burning. Calibration of the laser cutter ensures alignment with paper and design specifications, adjusting settings such as the laser power, speed, and focus. Once the digital file is loaded into the laser-cutting machine, the process is initiated, and the laser cutter faithfully follows the programmed path to cut through the paper substrate. The high-energy laser beam vaporizes or burns away the material along the designated cutting lines. Figure 3a depicts the conventional laser-cutting method for fabricating paper devices. Laser cutting offers remarkable precision, minimal material wastage, and the capability to craft intricate and personalized designs, making it ideal for applications like paper-based microfluidic devices and sensors requiring precise and detailed structural features.

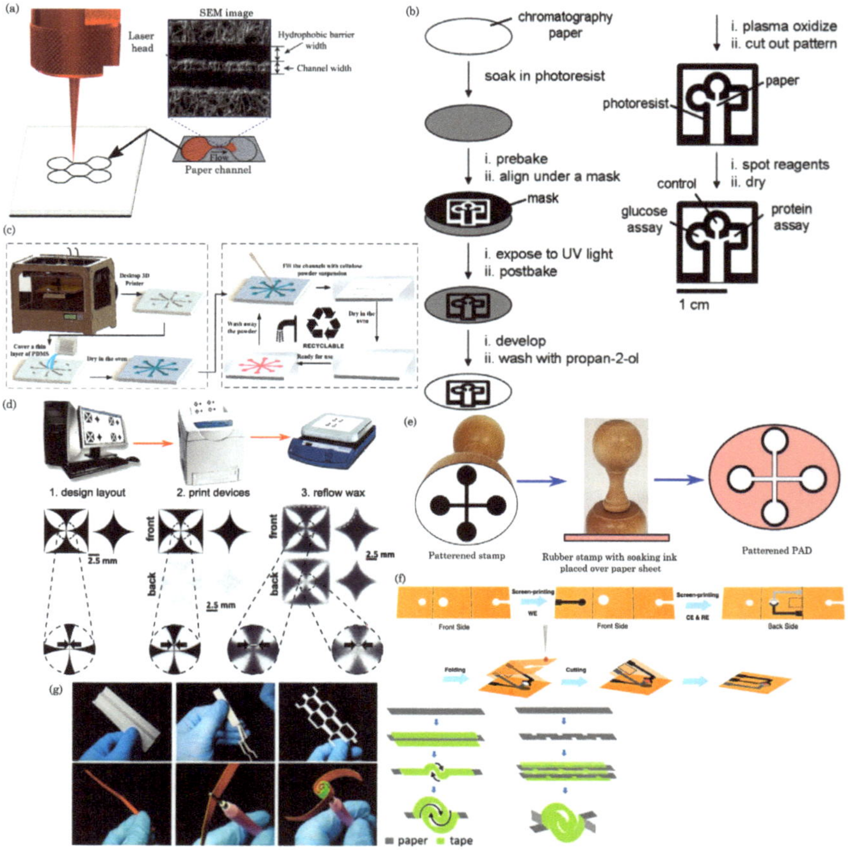

Figure 3. Schematics of various fabrication techniques for paper-based sensors. (**a**) Laser cutting, reproduced with permission from Mahmud et al. [53]. ©2018 The Authors, MDPI (Basel, Switzerland). (**b**) Photolithography, reprinted with permission from Martinez et al. [54]. ©2007 John Wiley & Sons, Inc. (**c**) 3D printing, reprinted with permission from He et al. [55]. ©2016 The Authors, published by MDPI, 2016. (**d**) Wax printing, reprinted with permission from Carrilho et al. [56]. ©2009 American Chemical Society. (**e**) Patterning paper using the stamping method. (**f**) Design and fabrication process of the origami-paper-based device. Reprinted with permission from Wang et al. [57]. Licensed under a Creative Commons Attribution 4.0 International License, ©2024 Springer Nature Limited. (**g**) Images and schematic representation depicting the process of crafting 2D and 3D vertical paper analytical devices (vPADs) through the utilization of quilling and kirigami principles. Reprinted with permission from Gao et al. [58]. ©The Author(s), licensed under a Creative Commons Attribution 4.0 International License.

4.3. Photolithography

Photolithography is a sophisticated technique employed in fabricating paper-based devices, enabling the creation of intricate patterns and microscale features. The process involves several sequential steps, beginning with preparing a photoresist-coated substrate. The chosen paper substrate is coated with a light-sensitive photoresist material, forming a uniform layer. A photomask containing the desired pattern is then placed near the coated substrate. Exposure to ultraviolet (UV) light passes through the transparent regions of the photomask, initiating a chemical reaction in the photoresist. The exposed areas become either more or less soluble, depending on the type of photoresist used. After exposure, the substrate undergoes a development process, where a solvent is applied to remove the soluble portions of the photoresist. This reveals the pattern on the substrate corresponding to the photomask. The developed substrate is subjected to additional treatments, such as baking, to enhance the pattern's stability. The exposed and developed paper substrate can now act as a template for creating hydrophobic barriers, fluidic channels, or other functional elements. Various methods, such as wax printing or plasma treatment, can selectively modify the paper's properties. Figure 3b illustrates the step-by-step photolithography methods for patterning paper devices.

Photolithography offers high precision and resolution, making it suitable for applications that require intricate designs and well-defined microstructures. However, it may involve using specialized equipment and chemicals, adding complexity to the fabrication process.

4.4. 3D Printing

3D printing, or additive manufacturing, is a cutting-edge technique in fabricating paper-based devices. This method enables the creation of three-dimensional structures layer by layer, providing precise control over design and geometry. The 3D printing process involves utilizing digital design software to create a three-dimensional model of the intended paper-based device, serving as a blueprint for the 3D printer. Material selection is crucial, with biodegradable and eco-friendly materials, such as specific polymers, commonly used for paper-based devices compatible with 3D printing.

To ensure precise layer deposition, the 3D printer is calibrated, adjusting parameters like layer thickness, print speed, and temperature based on the chosen material and design specifications. The digital file is loaded into the 3D printer, specifying the desired settings. The printer deposits layers of the selected material, building up the three-dimensional structure according to the digital model. This layer-by-layer approach allows for sophisticated designs and complex geometries, as illustrated in Figure 3c.

After the printing process is complete, any support structures used during printing are removed, and additional post-processing steps, such as sanding or coating, may be performed to refine the surface and enhance specific properties of the 3D-printed paper device. The overall 3D printing method offers advantages such as rapid prototyping, customization, and the ability to produce complex structures that may be challenging with traditional fabrication methods. However, careful consideration of the material selection, printer calibration, and post-processing steps is essential to optimize the performance and quality of 3D-printed paper-based devices.

4.5. Screen Printing

Screen printing is a versatile and economical technique for producing paper-based microfluidic devices. This method facilitates the application of hydrophobic barriers and functional inks onto porous paper substrates, thereby establishing fluidic channels with diverse applications, including diagnostics and chemical analysis. The fabrication begins with developing a digital design or stencil outlining the microfluidic channels, test zones, and additional features. Subsequently, the stencil or design is secured onto the mesh screen, aligning it with the intended microfluidic layout. The screen is then coated with a layer of hydrophobic or wax-based ink. Execution of the printing process involves placing the inked screen onto the paper substrate utilizing a squeegee to distribute the ink evenly across the screen. This action propels the ink through the mesh, deposits it onto the paper, and defines the desired pattern. The screen is lifted to unveil the printed design, and this procedure is iterated for each layer or color in the overall design. Thorough drying of the printed paper is imperative to ensure proper ink adherence and prevent smudging.

Depending on the ink formulation, specific devices may need post-printing treatments. One standard post-printing treatment involves subjecting the printed paper to heat, enhancing the hydrophobic properties of the ink, and ensuring the formation of effective barriers. This step optimizes the paper-based microfluidic device, aligning its performance with specific requirements.

4.6. Wax Printing

Wax printing relies on the hydrophobic properties of wax to create fluidic channels on paper. The process involves selectively depositing wax onto the paper substrate to define the boundaries of channels and hydrophilic zones. The hydrophobic wax barriers prevent liquid flow in certain areas, while the untreated paper remains hydrophilic, facilitating capillary-driven fluid transport.

The printing involves a multistage process. The desired fluidic channel pattern is designed using graphic design software. This layout defines the paper's test zones, channels, and other features. The designed pattern is then printed onto the paper using a wax printer. The wax is typically melted and deposited onto the paper, creating hydrophobic barriers. Commonly used waxes include paraffin or a mixture of paraffin and other additives. After printing, the paper is heated to allow the wax to penetrate the paper fibers, enhancing its hydrophobic properties. This step ensures better control over fluid flow and prevents lateral spreading. The paper may be layered or folded to create three-dimensional structures, and additional materials, such as membranes or reagents, can be integrated at specific locations. Figure 3d describes the step-by-step wax printing methods for patterning paper devices.

4.7. Inkjet Printing

Inkjet printing emerges as a precise and versatile approach for crafting paper-based microfluidic devices, relying on the controlled deposition of liquid inks onto paper substrates. This method facilitates the generation of intricate patterns, microfluidic channels, and functional elements. The process entails several key steps: First, design the desired microfluidic layout, test zones, and other features using digital design software such as AutoCAD and CorelDRAW. The resulting digital file guides the inkjet printer in creating the specified patterns. Next, choose appropriate inks based on application requirements, which may include hydrophobic barriers, conductive materials, or biofunctional agents, depending on the intended purpose of the paper-based device. Calibrate the inkjet printer to ensure accurate and consistent droplet deposition, with crucial parameters such as droplet size, spacing, and positioning. Load the designed digital file into the inkjet printer, specifying the desired settings. The printer then dispenses tiny droplets of ink onto the paper surface based on the digital design, bringing the defined microfluidic features and patterns to life. The inkjet printing process may be repeated for complex designs with multiple layers or colors for each layer, necessitating precise alignment to achieve the intended device structure. Thorough drying of the printed paper is crucial to prevent ink smudging and ensure steadfast adherence to the printed features.

4.8. Embossing

Embossing is a technique that fabricates paper-based devices to create raised patterns or structures on a paper substrate. The process involves designing the desired pattern using digital design software such as AutoCAD and CorelDRAW. This pattern dictates the expanded features of the paper device. A heated metal die is chosen as the embossing material for the embossing process. The die, designed to match the intended pattern, transfers the pattern onto the paper. The paper substrate is prepared on a clean, flat surface. The metal die is heated to the required temperature. Heat is crucial in softening the paper fibers, enabling them to conform to the raised pattern on the die. The embossing process begins by positioning the heated die over the designated area on the paper. Pressure is applied to the die, pressing it onto the paper substrate. The combination of heat and pressure causes the paper fibers to deform, adopting the die pattern. The die is held for a specific duration to ensure proper embossing. After embossing, the paper is allowed to cool and set. This step is essential for the paper to retain the raised pattern effectively. A quality check inspects the embossed paper, ensuring the increased features are well-defined and consistent. Figure 3e shows the step-by-step process for the stamping method.

Depending on the application, additional post-processing steps may be undertaken. For instance, the embossed paper may be coated with hydrophobic substances to modify its fluid-handling properties. Embossing is particularly useful for creating three-dimensional structures on paper, such as microfluidic channels or detection zones. It is a relatively simple and cost-effective method, making it suitable for various applications in paper-based microfluidics and analytical devices.

4.9. Origami, Quilling, and Kirigami

Origami, quilling, and kirigami are innovative methods for fabricating paper-based devices, leveraging folding, quilling, and cutting principles to create intricate structures with diverse functionalities.

Origami, an ancient Japanese art form, involves precisely folding paper to create three-dimensional structures without cutting or adhesive. In paper device fabrication, origami provides an elegant means of constructing complex and functional designs. Researchers and engineers use origami techniques to fold paper into specific shapes, forming containers, channels, or dynamic components. The process typically begins with the design of a flat pattern that, when folded along predetermined lines, transforms into the desired 3D structure. The patterns are often created using computer-aided design (CAD) software. Origami-based paper devices have been developed for μPADS applications, such as creating fluidic channels and reservoirs through folding.

Quilling-based paper device fabrication involves creatively adapting quilling, a paper art form, to construct functional microfluidic devices [59]. This innovative approach utilizes the rolling, shaping, and arranging of paper strips to create intricate structures, including microfluidic channels, reservoirs, and other components essential for analytical or diagnostic purposes. The process includes designing and planning the device layout, selecting suitable paper types, preparing quilling strips, employing quilling techniques to form desired shapes, assembling the components, and integrating functional elements. This method provides a cost-effective and customizable way to prototype simple microfluidic devices, offering accessibility and creativity in fabricating paper-based analytical tools for educational, research, or point-of-care applications.

Kirigami, an extension of origami, introduces the element of cutting into the folding process. This method allows for more intricate and flexible designs by strategically incorporating cuts and folds. In paper device fabrication, kirigami enables the creation of structures that can expand, contract, or exhibit specific movements. Designers use kirigami to craft patterns that, when folded and cut, result in functional and dynamic paper-based devices. This technique is particularly advantageous for applications requiring mechanical actuation or shape-changing capabilities. Kirigami-based devices have been found to be useful in flexible electronics and biomedical devices.

Figure 3f,g illustrates schematic representations of origami, quilling, and kirigami techniques employed in fabricating paper devices. These methods offer simplicity, low cost, and the ability to create complex structures without advanced equipment. However, precision in folding and cutting is crucial to achieving the intended functionalities. These methods have garnered attention for their potential to develop innovative and accessible solutions for various technological applications.

Table 1 provides a comprehensive summary of various fabrication techniques, detailing their specific characteristics and applications for paper-based sensors. The table encompasses a range of methods, highlighting each technique's unique features, advantages, and potential limitations. It serves as a valuable reference for understanding how different fabrication processes can be tailored to enhance the performance and functionality of paper-based sensors in diverse applications.

Table 1. Summary of fabrication techniques for paper-based sensors.

Fabrication Techniques	Equipment and Materials Requirements	Advantages	Limitations	Ref.
Blade cutting/plotting	X-Y plotter, knife	Provides sharp features, no chemical required	Limited to 2D designs	[60,61]
Laser cutting	Laser cutter	Precise, customizable designs, suitable for large-scale production, high resolution (∼60 μm)	Requires specialized equipment and polymer films to protect the paper device from damage, may generate debris	[62–66]
Photolithography	UV light, heating plate, photomask, photoresists (positive/negative), mask aligner, chemicals, oxygen plasma	High resolution (∼200 μm), well-established microfabrication technique	Equipment-intensive, may involve multiple complex steps and chances of channel contamination	[67–69]
3D printing	3D printer, inks	Allows for complex, customized designs	Limited resolution compared to traditional microfabrication	[70–76]
Screen printing	Mesh screen, hot plate, transparency film, wax	Low-cost, scalable for mass production	Resolution may vary, suitable for relatively simple designs, new screens are required for different patterns	[77–83]
Wax printing	Hot plate, wax printer, solid wax	Simple, rapid, cost-effective, and suitable for prototyping	Limited resolution (∼550 μm), wax spread, limited channel height control, temperature sensitivity	[84–89]
Inkjet printing	Customized inkjet printer, hydrophobic ink, hot plate, and chemicals	Noncontact, suitable for rapid prototyping	Resolution may be lower than other techniques, requires multiple steps, and post-printing heating is required for some inks	[90–96]
Embossing	Embossing tools, adhesives, silane	Simple, flexible, suitable for rapid prototyping	Limited resolution, may affect paper integrity, susceptible to contamination	[97–100]
Origami and kirigami	Paper cutting and folding tools, adhesives	Foldable structures, flexible design, enhanced functionality, scalability	Precision challenges, design and assembly complexity, limited material compatibility	[101–108]

5. Detection Techniques

5.1. Colorimetric Sensing

Colorimetric detection is widely used in paper-based microfluidic devices to analyze visual and quantitative data. Colorimetric sensing on paper-based devices operates on the principle of visual color change as an indicator of the presence and concentration of a specific analyte. Immobilized reagents on the paper matrix selectively react with the target substance, leading to a detectable color change upon interaction. The sample application allows the analyte to flow through the paper via capillary action, initiating various biochemical reactions, such as enzymatic reactions, antigen–antibody binding, or pH changes. The resulting color change is proportional to the analyte concentration, providing a simple and cost-effective means of on-site detection without the need for complex instruments. This approach is widely applied in medical diagnostics, environmental monitoring, and food safety, offering a user-friendly solution for rapid analyte quantification. Figure 4a shows schematics of the colorimetric sensing of dengue NS1 using a paper-based lateral flow assay [109]. In this assay, the sample is loaded onto the SP and migrates across the strip. At the CP, the NS1 antigen (Ag) interacts with immobilized Au-rGO-Ab conjugates, forming Au-rGO-Ab-Ag complexes. These complexes travel through the membrane via capillary action. At the test line, they bind to capture antibodies, creating a sandwich (Au-rGO-Ab-Ag-Ab) and producing a colored band. The absence of Ag results in no band. Excess-labeled antibody conjugates bind to secondary antibodies at the control line, creating another colored band and confirming assay completion. The absorbent pad absorbs the excess buffer and unbound nanoparticles.

5.2. Electrochemical Sensing

Electrochemical detection is a widely utilized method in paper-microfluidics-based sensors. It enhances their versatility and efficacy by leveraging electrochemical reactions at the sensor's surface for target analyte detection and quantification. The integration of paper microfluidics, driven by capillary action facilitating fluid flow, seamlessly combines with electrochemical detection, resulting in efficient and portable sensing platforms. Key components of electrochemical sensors, including the working electrode (WE), reference electrode (RE), and counter electrode (CE), play crucial roles. The WE, typically made of conductive materials like carbon or metal, is the primary site for analyte electrochemical reactions, often modified for enhanced selectivity and sensitivity. The RE maintains a stable reference potential, accurately determining the electrochemical reaction at the working electrode. CE completes the electrical circuit by providing a pathway for the flow of electrons during the electrochemical reaction. It is often made of conductive materials such as platinum or graphite and is not directly involved in the analyte reaction.

In paper-based devices, these electrodes are embedded into the paper matrix. These electrodes facilitate electrochemical reactions during the sensing process. Reagents, such as enzymes or antibodies, immobilized on the electrodes selectively interact with the target analyte, initiating an electrochemical response. Upon sample introduction, the immobilized reagents induce electrochemical reactions, with changes in redox states, conductivity, or pH depending on the sensing mechanism. The resulting electrochemical changes are detected using instrumentation like a potentiostat, and the recorded signals indicate the presence and concentration of the target analyte. The quantified electrochemical signals offer a quantitative assessment through digital displays or data analysis software.

In Figure 4b, schematic diagrams of a paper-based electrochemical sensor are depicted. These sensors utilize electrochemical sensing electrodes created by drop-casting a carbon nanotube (CNT) suspension onto paper substrates with varying porosities. The fabrication process involves a combination of laser cutting, CNT solution drop-casting, and origami techniques to produce arrays of diagnostic devices. Laser cutting is employed to delineate the electrode sensing area, facilitating the straightforward drop-casting of the CNT suspension without needing a separate patterning process. Origami techniques are then utilized to establish connections between the working, reference, and counter electrodes with the electrolyte, enhancing the functionality and manufacturability of the device.

5.3. Fluorescence

Fluorescence detection is a robust and widely utilized method in paper-microfluidics-based sensors, offering high sensitivity and specificity for detecting various analytes [110–114]. This approach capitalizes on the innate fluorescence properties of specific molecules known as fluorophores, facilitating target substances' precise identification and quantification.

In the typical configuration of μPADs, specialized reagents or probes contain fluorophores that exhibit selective interactions with the target analyte. When exposed to UV light, these probes undergo distinct fluorescence changes upon introducing a sample containing the target analyte. These changes may manifest as emission intensity or wavelength alterations, which can be detected and measured using a fluorescence imaging system. This imaging system allows for real-time monitoring and quantitative analysis, making it particularly valuable in medical diagnostics, environmental monitoring, and food safety applications. Integrating fluorescence detection into microfluidic paper-based sensors enhances their capabilities, providing a powerful tool for the rapid, on-site, and multiplexed detection of various analytes.

In Figure 4c, a schematic diagram illustrates the fabrication process and detection mechanism of a double-layered, paper-based fluorescent sensor. This sensor comprises an upper reaction layer containing two oxidases (lactate oxidase and choline oxidase) and a bottom fluorescent layer loaded with composite porphine-grafted fluorescent polymer colloids (PF-PDMTP/HQ). The sensor operates by detecting the significant and rapid decrease in fluorescence of porphine resulting from the oxidation reaction between saliva and the oxidases. This reaction is followed by fluorescence resonance energy transfer from oxidized hydroquinone. As a result, the developed fluorescent paper sensor enables the visual detection of oral squamous cell carcinoma (OSCC), which can be further confirmed through grayscale variation analysis using smartphone scanning.

5.4. Chemiluminescence

Chemiluminescence is a detection method commonly employed in paper-microfluidics-based sensors, offering a sensitive and versatile means of analyzing target analytes [115–117]. The fundamental principles of chemiluminescent sensing on paper-based devices involve immobilizing specific chemiluminescent reagents, such as enzymes or light-emitting molecules, onto the paper matrix. These reagents are selected for their ability to produce light upon interacting with the target analyte. Subsequently, the sample containing the analyte is applied to the paper surface, initiating a chemical reaction with the immobilized reagents and resulting in light emission. The chemiluminescent response releases energy through light, which is then detected and quantified using a photodetector or imaging system. The light emission's intensity correlates with the analyte's concentration, providing a quantitative readout that can be visualized through an imaging system or measured using specialized instrumentation like a photodetector.

Figure 4d illustrates a microfluidic paper chip-based multicolor chemiluminescence sensor designed to detect five antioxidants. This paper chip comprises four layers: a polyethylene terephthalate (PET) film, a paper channel, a double-sided adhesive ring, and a round-shaped detection paper. The paper channel includes a large sampling zone connected to a small sampling zone. These components are mass produced using a home craft cutter printer. The detection paper undergoes modification through the sequential addition of 5 μL of 1 mM Co^{2+} solution followed by 5 μL of 5 mM chemiluminescent (CL) reagent, which could be luminol, a mixture of luminol and fluorescein, or a mixture of luminol and rhodamine B. Subsequently, the PET substrate, paper channel, double-sided adhesive ring, and modified detection paper are assembled to create multilayer paper chips suitable for further experimentation.

5.5. Electrochemiluminescence

Electrochemiluminescence (ECL) represents a cutting-edge detection method seamlessly integrated into paper-microfluidics-based sensors, providing a robust and precise analytical tool for detecting target analytes [118,119]. This innovative approach synergistically combines electrochemical and luminescent principles to achieve heightened sensitivity and selectivity.

In an electrochemiluminescence-based paper microfluidic sensor, the device incorporates essential components such as electrodes and chemiluminescent reagents. The electrodes are pivotal in facilitating electrochemical reactions that generate species in excited states. These excited states subsequently release photons during relaxation, resulting in luminescence. The beauty of this method lies in its ability to leverage the controlled electrochemical reactions to induce luminescence, offering a precise and sensitive means of detecting analytes.

The detection mechanism within ECL-based paper microfluidic sensors revolves around measuring the emitted light. The intensity of the emitted light is directly correlated with the concentration of the target analyte present in the sample. This quantitative correlation enables the precise analysis and quantification of analytes, making ECL-based sensors invaluable in various applications, including medical diagnostics, environmental monitoring, and bioanalytical research. Also, integrating electrochemiluminescence into paper microfluidics enhances the analytical capabilities of these sensors and contributes to the development of portable, cost-effective, and efficient platforms for on-site detection. The sensitivity and selectivity achieved through ECL make it a promising technology for advancing point-of-care diagnostics and real-time monitoring in diverse fields.

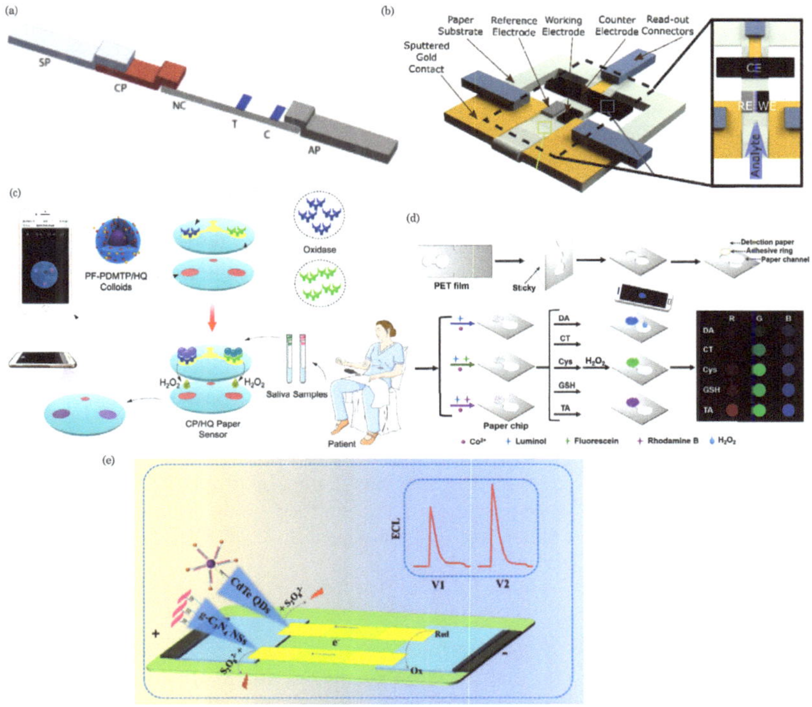

Figure 4. (**a**) Paper-based lateral flow assay for colorimetric sensing of dengue NS1. Reprinted with permission from Kumar et al. [109]. ©2018 AIP Publishing LLC. (**b**) Paper-based electrochemical sensors for glucose sensing, reprinted with permission from Valentine et al. [120]. ©2020 American Chemical Society. (**c**) Schematic diagram of the paper-based fluorescent sensor for rapid early screening of oral squamous cell carcinoma. Reprinted with permission from He et al. [121]. ©2023 American Chemical Society. (**d**) Paper-based chemiluminescence sensing of antioxidants (dopamine, CT, Cys, GSH, and TA): CL spectra (top) and CL images (bottom). Reprinted with permission from Li et al. [122]. ©2023 Elsevier B.V. (**e**) Illustration of a conceptual paper-based bipolar electrode electrochemiluminescence platform for detecting multiple miRNAs. Reprinted with permission from Wang et al. [123]. ©2020 American Chemical Society.

Figure 4e depicts the schematic of a paper-based bipolar electrode electrochemiluminescence platform designed to detect multiple targets, specifically miRNA-155 and miRNA-126. In this setup, the electron transfer process in each bipolar electrode is electrically coupled with the electrochemiluminescence (ECL) reaction of each light-emitting probe due to the connection between the cathode and the anode. The DC power supply is connected to the parallel bipolar electrode sensing platform, executing the most suitable driving voltages for the two light-emitting probes (CdTe QDs and g-C_3N_4 NSs) with their co-reactants. Applying a driving voltage of 9 V to the co-reactant $K_2S_2O_8$ in the hydrophilic unit, which is in close contact with the cathode region of the parallel bipolar electrode, induces an excitation–radiative transition process with the emission of the light signal.

Simultaneously, an oxidation reaction occurs at the anode when in contact with the solution containing H_2O_2, resulting in the bipolar electrode facilitating the electron transfer between the cathode and anode path.

5.6. Surface-Enhanced Raman Spectroscopy (SERS)

Surface-enhanced Raman spectroscopy (SERS) is an analytical technique that enhances the Raman scattering signal of molecules adsorbed on or near metallic nanostructures. SERS exploits the localized surface plasmon resonance (LSPR) phenomenon, where incident light excites collective oscillations of conduction electrons in metal nanostructures. This results in an enhancement of the Raman scattering signal by several orders of magnitude compared to conventional Raman spectroscopy. The enhancement arises from two mechanisms: electromagnetic enhancement due to the strong electromagnetic fields near the metal surface and chemical enhancement due to charge transfer between the molecule and the metal surface.

By incorporating SERS-active substrates onto paper substrates, researchers have created SERS-enhanced paper devices for on-site and point-of-care applications. The fabrication of SERS-active substrates involves the synthesis of noble metal nanoparticles (e.g., gold or silver) and their deposition onto paper substrates. Various methods, such as chemical reduction, physical deposition, inkjet printing, and lithography techniques, are employed to fabricate reproducible and uniform SERS substrates with high enhancement factors [124,125].

These devices have been used for the qualitative and quantitative analysis of various analytes, including chemicals, biomolecules, and pathogens. In healthcare, they can be used for the rapid and sensitive detection of biomarkers for disease diagnosis, monitoring of therapeutic drug levels, and detection of infectious agents. In environmental monitoring, SERS-based paper devices enable the detection of pollutants, toxins, and heavy metals in water, air, and soil. In food safety, they facilitate the identification of contaminants, adulterants, and allergens in food products.

6. Signal Readout Approach

6.1. Qualitative

Qualitative readout methods focus on determining the presence or absence of a particular analyte within a sample through visual inspection or colorimetric assays. Color is one of the most common signals in daily life, and a change in color can be easily distinguished by the naked eye. In the traditional colorimetric detection assay, color changes at the test zone depend on the concentration of the target (i.e., color intensity is proportional to analyte concentration). One common example is a paper-based point-of-care pregnancy kit with a colorimetric signal readout, which offers a convenient and accessible solution for the early detection of pregnancy. It operates on the principle of detecting human chorionic gonadotropin (hCG), a hormone produced during pregnancy, in urine samples. When hCG is present, it triggers a chemical reaction that produces a visible color change on the paper strip. This change serves as a positive indication of pregnancy. These types of paper-based devices provide YES or NO information (i.e., subjective interpretation) and are suitable for point-of-care diagnostics in resource-limited settings.

6.2. Quantitative

Quantitative analysis involves providing numerical data concerning the concentration or quantity of the target analyte in a sample. Meanwhile, sensing techniques such as fluorescence and electrochemical-based sensing offer quantitative signal readouts. As discussed previously, fluorescence-based paper sensors utilize fluorescent molecules that emit light of a specific wavelength upon excitation by an external light source. The presence of the target analyte induces a change in fluorescence intensity directly proportional to the analyte concentration. This alteration can be quantitatively assessed using a fluorescence reader or imaging system.

Furthermore, the results of a colorimetric assay, characterized by a visible color change, can also be quantified through digital image analysis using tools such as digital cameras and smartphones [126].

Quantitative analysis presents several notable advantages, including the precise and accurate quantification of analytes, detection of low concentrations of target molecules, and the ability to monitor dynamic changes in analyte levels. However, it necessitates complex instrumentation and accessories and may involve more intricate sample-preparation procedures.

7. Applications in Health Sensing

7.1. Diagnostic Assays for Infectious Diseases and Others Analytes

Paper-based point-of-care (POC) diagnostic devices have garnered significant attention due to their portability, cost-effectiveness, biodegradability, and ease of use [127]. These devices leverage the unique properties of paper substrates to perform various diagnostic assays, making them promising tools, especially in resource-limited settings, which fulfill the World Health Organization's POC device development guidelines. Paper-based diagnostics typically involve using paper strips or cards that can wick biological samples, such as blood, saliva, or urine, through channels or zones containing reagents for specific assays [128].

Paper-based microfluidic devices have been used for the point-of-care testing of vector-borne and flavivirus families such as malaria [129,130], dengue virus [131–133], and Zika virus [134,135]. For example, Suvanasuthi et al. [136] introduced a paper-based colorimetric biosensor for detecting dengue virus serotypes (DENV1-4). The paper substrate's hydrophobic barriers were fabricated using 3D printing with polylactic acid (PLA) and wax filaments. The developed prototype demonstrated the ability to differentiate between dengue virus serotypes based on subtle nucleotide sequence variations. Figure 5a illustrates the schematics of device assembly and provides photographs showing the visual color changes corresponding to different dengue virus serotypes. Karlikow et al. [137] introduced a paper-based diagnostic platform for detecting Zika and chikungunya viruses in serum samples. The tests achieved high accuracy and sensitivity by utilizing a cell-free expression system, isothermal amplification, toehold-switch reactions, and a custom portable reader and computer vision-enabled image analysis software. Figure 5b depicts the detection mechanism schematics of the paper-based device. In suspected infection cases, the tests demonstrated accuracies of 98.5% for both Zika (95% confidence interval, 96.2–99.6%, 268 serum samples) and chikungunya (95% confidence interval, 91.7–100%, 65 serum samples) viruses, with sensitivities ranging from 2 aM to 5 fM, falling within clinically relevant concentrations. The prototype's performance was successfully validated in field conditions.

Moreover, paper-based devices have been employed to diagnose other diseases and analytes, including influenza virus H5N1 [138], Neisseria meningitides [139], nucleic acid detection [140,141], noncommunicable diseases [142], cancer diagnosis [57,143,144], chronic obstructive pulmonary disease (COPD) biomarkers [145], HIV [146,147], pregnancy, infertility [148,149], and bioanalytes (uric acid, glucose, H_2O_2, and cholesterol) [150–152], etc. In a recent study, Bezdekova et al. [153] introduced a proof-of-concept paper-based device for diagnosing prostate cancer (CaP) from urine samples. Initially, urine samples underwent UV irradiation to induce the formation of fluorescent clusters. Subsequently, a selective molecularly imprinted polymeric layer was prepared on a paper substrate, allowing for the specific capture of these UV-induced fluorescent clusters within the urine sample to be diagnosed. Figure 5c illustrates the process of the formation, capture, and detection of CaP-specific clusters in UV-irradiated urine samples. These clusters, captured using molecular imprinting technology, are then quantified using fluorescence spectroscopy.

Chaiyo et al. [154] introduced a novel 3D electrochemical paper-based analytical device (3D-ePAD) coupled with near-field communication (NFC) potentiostat for the nonenzymatic detection of cholesterol. Figure 5d illustrates the design of the paper device and the strategies employed for cholesterol detection. This integrated platform comprises an origami PAD (oPAD) and an inset PAD (iPAD). β-Cyclodextrin (β-CD) immobilized on the oPAD is the specific material for cholesterol detection without enzymes. The device seamlessly integrates cholesterol detection with a battery-free NFC potentiostat on a smartphone. Cholesterol concentration is assessed through a $[Fe(CN)_6]^{3-/4-}$ current signal, a redox indicator stored in the detection section of the iPAD. The 3D-ePAD/NFC system demonstrates a linear detection range of 1–500 μM and a maximum detection limit of 0.3 μM for cholesterol. Furthermore, the sensor effectively measures cholesterol levels in real human serum samples, yielding results consistent with those obtained from a commercial cholesterol meter.

Most recently, during the COVID-19 pandemic, μPADs have played a pivotal role in point-of-care initial disease screening [155–161]. In one such example, Lee et al. [162] developed a colorimetric lateral flow immunoassay (LFIA) using a recombinant protein linker CBP31-BC to immobilize antibodies on a cellulose membrane in an oriented manner. Figure 5e shows the schematic of the CBP31-BC-based LFIA for detecting SARS-CoV-2. This LFIA demonstrated the sensitive detection of cultured SARS-CoV-2 in 15 min, with a low detection limit of 5×10^4 copies/mL. Clinical evaluation using 19 samples validated by a reverse transcription–polymerase chain reaction (RT-PCR) revealed 100% accuracy in detecting positive and negative samples, even those with low viral loads.

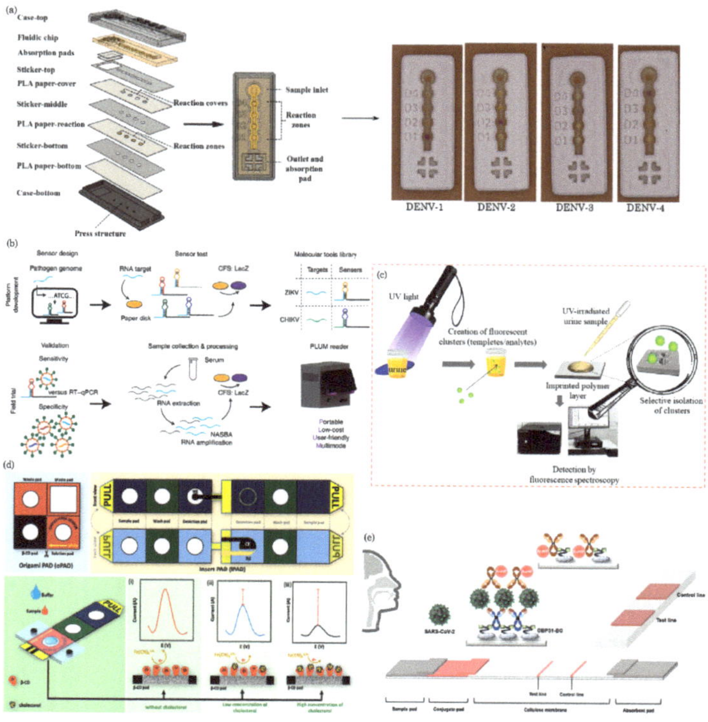

Figure 5. (**a**) Schematics of device assembly and photographs of visual color changes for different dengue virus serotypes. Reprinted with permission from Suvanasuthi et al. [136]. ©2021 Elsevier B.V. (**b**) Schematics of paper-based platforms for detecting the Zika and chikungunya viruses in serum samples. Reprinted with permission from Karlikow et al. [137]. Licensed under a Creative Commons Attribution 4.0 International License, ©2022 The Author(s). (**c**) Illustration of a paper-based analytical device for detecting prostate cancer using UV-irradiated urine samples. Reprinted with permission from Bezdekova et al. [153]. ©2023 Elsevier B.V. (**d**) The design concept of the 3D-ePAD, incorporating origami PAD (oPAD) and insert PAD (iPAD), and the detection mechanism for cholesterol across different concentrations. Reprint with permission from Chaiyo et al. [154]. Licensed under CC-BY-NC-ND 4.0. ©2024 The Authors. (**e**) Schematic of the CBP31-BC-based LFIA for detecting SARS-CoV-2. Reprinted with permission from Lee et al. [162]. ©2022 Elsevier B.V.

These POC devices have been instrumental in enabling rapid and real-time testing for the SARS-CoV-2 virus, facilitating the early identification and containment of infections. Their ease of use, cost-effectiveness, and portability have made them particularly valuable in various settings, including clinics, airports, and resource-limited areas.

7.2. Wearable and Portable Health-Monitoring Devices

Wearable health-monitoring devices utilizing paper-based microfluidic technology represent a cutting-edge application at the intersection of healthcare and materials science. These devices offer a novel approach to continuous health monitoring, leveraging the unique properties of paper microfluidics to create flexible, lightweight, and cost-effective wearable sensors [163–165]. Paper-based biosensors have shown efficacy in detecting specific biomarkers associated with various health conditions. Wearable devices utilizing these biosensors offer the real-time monitoring of conditions such as diabetes, cardiovascular diseases, and infectious diseases, fostering a proactive approach to healthcare. Such wearable health monitors could measure parameters like biophysical features (body temperature, blood pressure, heart rate, and biopotential), sweat biochemicals (pH, uric acid, glucose, cholesterol, cortisol, etc.), lactate, or specific proteins, offering valuable data for individuals managing chronic conditions or athletes optimizing their performance [15,166–170]. For example, Yang et al. [171] designed a paper-based sandwich-structured wearable pH sensor with in situ sebum filtering for reduced interference (Figure 6a). The sensor comprised five layers: a PDMS-based cover layer, a sebum adsorption or ISE top layer (microfluidic snake channel), a filter-paper-based middle microfluidic layer, a sebum adsorption bottom layer (microfluidic snake channel), and an adhesive layer made of double-sided medical adhesive tape for skin fixation. Sweat, introduced through the inlet, moved along the microfluidic layer, allowing sebum adsorption. The sensor effectively adsorbed sebum mixed in sweat, ensuring an accurate pH measurement and facilitating sweat evaporation through the outlet window. Fiore et al. [172] innovated paper-based electrochemical biosensors for cortisol detection in sweat, a stress biomarker (see scheme, Figure 6b). The device uses filter paper for a reagent-free, competitive magnetic-bead-based immunosensor to orchestrate flow and reagent loading. Fabricated with filter paper and solid wax-based printing, the microfluidic pattern features hydrophilic channels defined by hydrophobic wax barriers. Magnetic beads, functionalized with monoclonal antibodies, facilitate specific cortisol measurement in the reaction zone. Integration with a near-field communication wireless module yields a flexible, wearable analytical tool for cortisol detection in sweat. Cheng et al. [173] devised a 3D origami-based µPADs wearable biosensor for multiplexed analyte detection in sweat. Figure 6c illustrates the schematic of a wearable sweat sensor featuring an origami-based 3D paper structure designed for the simultaneous analysis of multiple biomarkers. The square-shaped wearable sweat chip, measuring 36 mm on each side, featured a microfluidic channel with distinct layers for effective analysis. The 3D channel incorporated a collection layer, vertical and horizontal channels, an electrode layer, a colorimetric sensing layer, and a sweat evaporation layer. Screen-printed electrodes were employed for cortisol measurement, while the colorimetric sensing layer utilized cotton-thread-based channels. Sweat absorbed through the collection layer underwent a chromatographic process, reacting at the electrode layer and flowing into the lateral channel for colorimetric analysis. The chip enabled electrochemical and colorimetric sensing, with image analysis conducted using ImageJ and the electrochemical workstation. Recently, Lai et al. [174] presented an ultralight and highly sensitive biological and bioinspired tactile sensation system using printing paper to monitor human wrist pulses, acoustical vibration, and information encryption. The skin's schematic fabrication involves pencil graphite frottage (PGF) for the pressure-sensitive film, creating extended graphite electrodes through pencil writing, and eliminating metal electrodes from the process. A protective ecoflex film is spin-coated onto the printing paper's back, providing a self-adhesive layer. The final e-skin is assembled by placing two graphite-coated printing papers facing each other, with copper wires attached to the graphite electrodes. This innovative approach achieves a versatile and lightweight tactile sensing system. Figure 6d shows a schematic of the skin's tactile function transmitting action potentials to the brain via nerves. In the top right, the photograph shows the successful reproduction of a coin design using the PGF method. At the bottom is the schematic of a graphite-based pressure-sensitive e-skin structure for tactile sensing, comprising a graphite pressure-sensitive layer with an embossed microstructure and graphite electrodes. Niu et al. [175] introduced a pencil-made paper-based hydration sensor for health monitoring, particularly respiratory monitoring, noncontact switching, and skin characterizations. Figure 6e illustrates the fabrication process and response mechanism of the flexible pencil-on-paper hydration sensor, showcasing its potential applications in health monitoring, noncontact switching, and skin characterization. The design and fabrication approaches proposed in this study offer opportunities for the future development of wearable, self-powered, and recyclable sensors and actuators. Karmakar et al. [176] pioneered the development of an origami-inspired conductive paper-based folded pressure sensor tailored for detecting human stimuli. In Figure 6f, the sensor schematics and sensing mechanisms are illustrated, showcasing the intricate design and functionality of the sensor. This innovative sensor design draws inspiration from the principles of origami, leveraging folding techniques to create a flexible and responsive sensor capable of detecting various stimuli. The intricate folding patterns and conductive materials integrated into the paper-based sensor enable the precise detection and measurement of pressure changes, making it suitable for applications in human–computer interaction, wearable technology, and biomedical sensing.

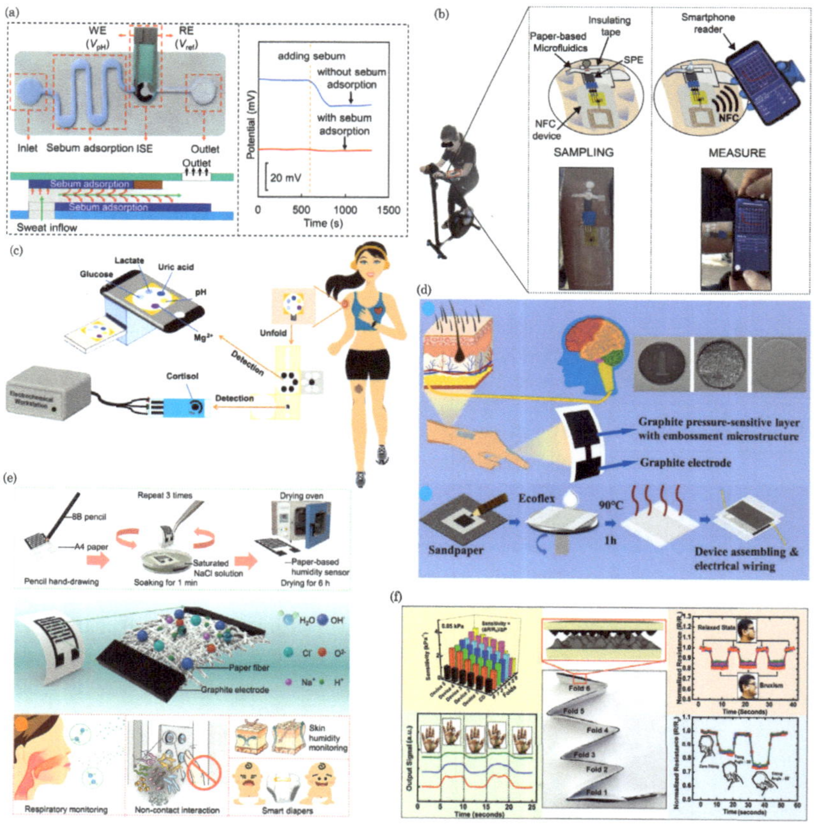

Figure 6. (**a**) Paper-based sandwich-structured sweat sensor with flow schematics and response diagram. Reprinted with permission from Yang et al. [171]. ©2023 American Chemical Society. (**b**) Cortisol

monitoring during physical activity: Illustration of the sampling process, data measurement, and wireless transmission to a smartphone via NFC. Reprinted with permission from Fiore et al. [172]. ©2022 Elsevier B.V. (**c**) Schematics of wearable sweat sensor with an origami-based 3D paper structure for simultaneous analysis of multiple biomarkers (glucose, lactate, uric acid, magnesium ions, and pH value). Reprinted with permission from Cheng et al. [173]. Licensed under a Creative Commons Attribution 4.0 International License. (**d**) Biological and bioinspired tactile sensation system: Schematic of skin's tactile function, a successful coin reproduction using the PGF method, and the graphite-based pressure-sensitive e-skin with embossed microstructure and electrodes. Fabrication process schematic of the PGF-based graphite e-skin. Reprinted with permission from Lai et al. [174]. ©2024 Elsevier B.V. (**e**) Schematics diagram depicting a pencil-on-paper hydration sensor designed to monitor physiological signals and characterize the skin barrier function. Reprinted with permission from Niu et al. [175]. ©2022 American Chemical Society. (**f**) Schematic diagram of the origami-inspired folded tactile sensor for human stimuli detection. Reprinted with permission from Karmakar et al. [176]. ©2023 American Chemical Society.

7.3. Animal Health Screening

Animal health is critical in various sectors, including agriculture, veterinary medicine, and food production. The timely and accurate screening of animal health parameters is essential for disease diagnosis, surveillance, and control. Conventional methods for animal health screening often involve complex and time-consuming laboratory procedures, which may not be suitable for on-site or point-of-care testing. Paper-based microfluidics has emerged as a promising technology that offers a promising alternative due to its portability; low cost; and ability to perform the rapid, sensitive, and specific detection of various analytes.

Several studies have demonstrated the utility of paper-based microfluidics devices in animal health screening, including disease diagnosis, monitoring of biomarkers, and detecting pathogens [177–181]. For example, research by Li et al. [182] demonstrated the utility of paper-based lateral flow biosensors (LFB) for the highly specific, simple, rapid, and visual detection of *Brucella*-specific amplicons (See Figure 7a). Their device utilized *Brucella*-MCDA-functionalized paper strips to capture and detect the *Bscp31* gene (*Brucella* species-specific gene), offering a rapid and cost-effective method for on-site screening. Similarly, Jung et al. [183] showcased the development of a signal-amplifiable nanoprobe-based chemiluminescent lateral flow immunoassay (CL-LFA) for the detection of avian influenza viruses (AIVs) and other viral avian-origin diseases, offering a low-cost alternative to conventional diagnostic methods (See Figure 7b). The nanoprobe allows for the selective immobilization of antibodies and enzymes on sensitive paper-based sensor platforms, enabling enhanced detection sensitivity. Tests conducted with low pathogenicity avian influenza H9N2, H1N1, and high pathogenicity avian influenza H5N9 viruses showed detection limits of $10^{3.5}$ to 10^4 50% egg infective dose (EID_{50})/mL, significantly lower than those of commercial AIV rapid test kits. The CL-LFA also demonstrated high sensitivity and specificity against clinical samples, indicating its potential as a diagnostic tool for sensitive antigen detection in clinical settings.

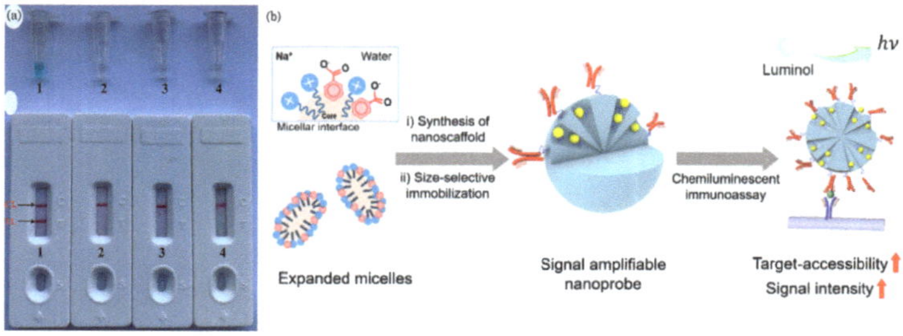

Figure 7. (**a**) Photographs showing the confirmation and verification of *Brucella*-MCDA products: color change in *Brucella*-MCDA tubes (**top**) and visual detection through LFAs (**bottom**) of the presence of *Brucella* species. Tube 1 (biosensor 1): positive amplification; tube 2 (biosensor 2): negative amplification (*Salmonella*), tube 3 (biosensor 3): negative amplification (*Bacillus cereus*), tube 4 (biosensor 4): negative control (DW). Reprinted with permission from [182]. ©2019 The Authors. This open-access article is distributed under the terms of the Creative Commons Attribution License (CC BY). (**b**) Scheme for detection of avian influenza viruses (AIVs) using a signal-amplifiable nanoprobe-based chemiluminescent lateral flow immunoassay (CL-LFA), reprinted with permission from Jung et al. [183], ©2020 American Chemical Society.

8. Environmental Monitoring/Sensing

Environmental pollutants, spanning heavy metals, organic compounds, pathogens, airborne pollutants, and other hazardous substances, pose severe threats to ecosystems and human health [184–186]. Persistent soil and water contamination with heavy metals such as lead, mercury, cadmium, and arsenic is primarily attributed to industrial activities, mining, and improper waste disposal. Contributing to organic pollution in water bodies, organic compounds like pesticides, herbicides, industrial chemicals, and pharmaceuticals impact aquatic life and may enter the human food chain. Waterborne pathogens, encompassing bacteria, viruses, and microorganisms, pose health risks, necessitating effective monitoring systems. Additionally, airborne pollutants, such as particulate matter, volatile organic compounds (VOCs), nitrogen dioxide, and sulfur dioxide, contribute to air pollution, impacting respiratory health and disrupting ecosystem balance. Addressing these multifaceted challenges requires comprehensive monitoring and mitigation strategies.

8.1. Detection of Soil Contaminants

Soil contaminants significantly threaten environmental ecosystems and human health, necessitating efficient early detection and mitigation monitoring systems. Paper-based microfluidic devices have emerged as promising tools in soil contaminant monitoring due to their cost-effectiveness, simplicity, and portability. These devices leverage the capillary action of paper to facilitate the flow of liquids through microchannels, allowing for the detection of various contaminants. They are well-suited for applications in resource-limited settings where sophisticated laboratory equipment may be impractical. These paper-based systems can be designed to detect a range of soil contaminants, including heavy metals, pesticides, and organic pollutants.

Suo et al. [187] developed a high-throughput paper-based fluorescence resonance energy transfer (FRET) aptasensor for the sensitive detection of low concentrations of Pb^{2+}. Figure 8a illustrates the detection methodology. Fabricated on Whatman No. 1 chromatographic paper, the device demonstrated the capability to detect Pb^{2+} in a concentration range spanning from 0.01 to 10 μM, with an impressive limit of detection (LOD) of 6.1 nM. This innovative strategy successfully analyzed various real samples, including water, soil, and food, showcasing its applicability in practical scenarios for environmental and food safety assessments. Integrating FRET technology into a paper-based platform enhances the efficiency and throughput of the aptasensor, offering a versatile and sensitive tool for rapid detection in diverse sample matrices. Yu et al. [188] introduced a fiber-made filter-paper-based device for the real-time monitoring of Cd^{2+} in water, rice, and rice soil (see Figure 8b). The developed test paper exhibited highly sensitive and visible sensing capabilities for Cd^{2+} in water, rice supernatants, and rice soil supernatants. The LODs in these real samples were remarkably low, measuring at 0.0112 ppb for water, 1.1240 ppb for rice supernatants, and 0.1124 ppb for rice soil supernatants. These LODs were found to be lower than the national standards (GB 2762-2022) for food safety in China [189], underscoring the device's potential for precise and reliable detection, with implications for ensuring compliance with stringent safety regulations in diverse environmental and agricultural settings.

Furthermore, pesticides, vital for ensuring food security by controlling pests, weeds, and plant diseases, have significantly increased food availability over the past 50 years. However, their widespread use has led to environmental pollution, adversely affecting ecosystems and human health [190,191]. Implementing efficient management practices and user-friendly point-source monitoring systems accessible to farmers can alleviate pesticides' environmental and health impacts. In this context, paper-microfluidics-based devices have played a significant role in pesticide detection [192]. Zhang et al. [193] developed an innovative paper-based colorimetric sensor for thiacloprid, a commonly used agricultural pesticide, with a low detection limit of 0.04 μM. Figure 8c shows the schematic representation of the principle behind the paper-based colorimetric sensor designed for the real-time monitoring of pesticides. The quantification of the sensor's output was facilitated through RGB analysis, providing a simple and efficient method for detection. Notably, integrating a smartphone app for output reading enhances the accessibility and user-friendliness of the paper-based sensor, offering a promising solution for on-site and real-time monitoring of thiacloprid levels in agricultural settings. Ranveer et al. [194] designed a versatile paper-based dipstick assay for the colorimetric detection of fungicides, organochlorines, organophosphates, carbamates, and herbicides in diverse matrices such as animal feed, water, milk, and soil. Figure 8d presents a schematic illustration depicting the detection of pesticides in dairy samples through the paper-based sensor. The developed dipstick demonstrated versatile applicability with an impressive LOD for different pesticide groups. Specifically, the LOD values ranged from 1 to 10 $\mu g\ L^{-1}$ for fungicides, 1 to 50 $\mu g\ L^{-1}$ for organochlorines, 250 to 500 $\mu g\ L^{-1}$ for organophosphates, 1 to 50 $\mu g\ L^{-1}$ for carbamates, and 1 $\mu g\ L^{-1}$ for herbicides. This paper-based assay showcased sensitivity across a range of pesticide residues. It illustrated its potential as a rapid and cost-effective tool for assessing pesticide contamination in multiple environmental and food matrices. Caratelli et al. [195] introduced a 3D flower-like origami paper-based device designed for the electrochemical detection of pesticides, specifically paraoxon, 2,4-dichloro phenoxy acetic acid, and glyphosate, in the aerosol phase, catering to applications in precision agriculture. Figure 8e illustrates a schematic representation of the electrochemical biosensor for pesticide detection based on an origami-based paper device. The innovative device was seamlessly integrated with a smartphone for convenient output reading. Remarkably, this paper-based system demonstrated the efficient detection of the three classes of pesticides in the aerosol phase, achieving impressive LODs equal to 30 ppb, 10 ppb, and 2 ppb for 2,4-D, glyphosate, and paraoxon, respectively. Integrating electrochemical sensing with a portable paper-based platform enhances accessibility and usability, offering a promising tool for real-time pesticide monitoring in agricultural settings with potential implications for sustainable and precise farming practices.

Paper-based sensors have been found to have a noteworthy application in detecting explosive residues in soil, presenting a valuable forensic-oriented environmental monitoring and security tool [196–200]. The unique attributes of paper microfluidic devices, such as their portability, simplicity, and cost-effectiveness, make them well-suited for the on-site detection of explosive remnants. These sensors can be tailored to detect specific volatile compounds, offering a targeted and efficient approach to soil analysis. The detection mechanism often involves incorporating reactive agents or biomolecules onto the paper substrate, allowing for a rapid and selective response to the presence of explosive residues.

This application is particularly crucial in areas where the remnants of explosives pose environmental and safety concerns, such as former military sites or regions affected by conflict. By leveraging the capabilities of paper-based sensors, ecological professionals and security personnel can conduct real-time, on-site assessments of soil contamination, facilitating prompt remediation efforts and contributing to a safer and more secure environment.

Figure 8. (**a**) Illustration of a high-throughput paper-based FRET aptasensor designed for the detection of Pb^{2+}. Reprinted with permission from Suo et al. [187]. ©2022 Elsevier B.V. (**b**) Schematics of fiber-made filter-paper-based biosensors for the real-time monitoring of Cd^{2+} in water, rice, and rice soil. Reprinted with permission from Yu et al. [188]. ©2023 American Chemical Society. (**c**) Principle illustration of the paper-based colorimetric sensor for real-time monitoring of pesticides. Reprinted with permission from Zhang et al. [193] ©2022 Elsevier B.V. (**d**) Schematic illustration of pesticide detection in dairy samples using the paper-based sensor. Reprinted from Ranveer et al. [194] under a Creative Commons license ©2022 Elsevier B.V. (**e**) Schematic representation of the origami paper-based electrochemical biosensor for pesticide detection. Reprinted with permission from Caratelli et al. [195]. ©2022 Elsevier B.V.

8.2. Water Quality Monitoring

Water quality monitoring ensures clean and safe drinking water access, addressing public health and ecological concerns [201]. Unfortunately, many developing regions face challenges meeting this fundamental need due to inadequate water treatment plans and infrastructure. Countries are grappling with water quality issues due to the rapid growth of human activities like urbanization and industrialization, leading to significant pollution [202]. Thus, ensuring access to clean water has emerged as a significant challenge in recent decades, impacting developing and developed nations. Traditional methods for detecting water contaminants involve chromatographic and spectroscopic techniques, necessitating costly equipment and specialized personnel [203]. Despite the emergence of water toxicity biosensors employing enzymes, antibodies, and microorganisms in recent decades, their specificity limits them to known chemicals, rendering them unsuitable for monitoring unforeseen contaminants in water [204,205]. Addressing the pressing need for on-site and real-time measurements of toxic components in water, there is a demand for a rapid and portable sensor.

Paper-microfluidics-based devices have been instrumental in water quality monitoring in the past decades due to their versatility, cost-effectiveness, and ease of use [206–211]. For example, Da Silva et al. [212] presented an innovative approach by developing a μPAD tailored for the point-of-use colorimetric monitoring of water hardness, phenols, and pH. The fabrication process involved using a cutter printer and 3D printing to create these paper devices. The reading output of the device, specifically the discernible change in visible colors in the presence of analytes, was quantified through captured images using an integrated smartphone (Figure 9a). In a study by Xiong et al. [213], a colorimetric-based μPAD was developed for the simultaneous detection of diverse water quality parameters, including Cu(II), Ni(II), Fe(III), nitrite, and pH. The fabrication involved creating hydrophobic flow patterns on a Whatman Grade 1 filter paper substrate using a wax printer. The visible color changes on the paper device induced by the presence of Cu(II), Ni(II), Fe(III), nitrite, and pH were quantified through RGB analysis using a smartphone app. Figure 9b shows the schematics of the device assembly and detection methods. The device demonstrated impressive detection limits of 0.4 ppm for nitrite, 1.9 ppm for Cu(II), 2.9 ppm for Ni(II), 2.9 ppm for Fe(III), and 5 for pH, with rapid detection achieved within 5 min.

Lin et al. [214] developed a portable paper analytical device modified with nanoclusters and integrated with a syringe for highly sensitive Hg^{2+} detection. The device comprises a paper substrate modified with fluorescent gold nanoclusters (AuNC-paper) enclosed in a reusable cartridge connected to a syringe, facilitating the flow of a large sample volume through the paper for enhanced analyte signal accumulation. The schematic illustration in Figure 9c depicts the AuNC-modified paper device designed for Hg^{2+} ion detection. In the presence of Hg^{2+} ions, the color of the paper substrate changes visibly, enabling naked-eye detection. This technique allows Hg^{2+} ion detection within 30 min, achieving a low detection limit of 1.2 nM.

Aguiar et al. [215] recently presented a μPAD designed for copper detection in natural waters. The μPAD assembly comprises three filter paper discs (R: Whatman 42, B: Whatman 1, E: Whatman 3) with a 9.5 mm diameter, arranged in twenty-four hydrophilic units in each layer (Figure 9d). The R-layer paper discs were prepared by applying 12 μL of Mod-RHOB ligand solution to each disc and were oven-dried at 50 °C for 10 min. The B-layer paper discs were prepared with 10 μL of buffer solution and underwent the same drying process. The E-layer was left untreated. To determine the copper concentration, 20 μL of standard/sample was loaded onto the assembled μPAD through the sample holes, absorbing in approximately 2 min. The reaction between Mod-RHOB and copper produces a pink color complex in the R layer, intensifying with increasing copper concentration.

Uhlikova et al. [216] introduced a μPAD for the colorimetric detection of inorganic nitrogen in water and soil samples. Figure 9e illustrates the detection strategy employed by the device. The developed device demonstrated the capability to detect ammonium and nitrate using bromothymol blue (an acid–base indicator) with quantification limits of 6.5 and 18.2 mg N L^{-1}, respectively. Similarly, using nitrazine yellow (another acid–base indicator), the quantification limits were found to be 2.1 and 4.2 mg N L^{-1} for ammonium and nitrate, respectively. The newly developed μPAD exhibited stability for 62 days when stored in a freezer and 1 day at ambient temperature. Validation with certified reference material confirmed its accuracy, and successful application was demonstrated in determining ammonium and nitrate in spiked environmental water samples and soil extracts.

More recently, Thangjitsirisin et al. [217] introduced a μPAD for the colorimetric determination of ammonium ions in water. The device utilized a superhydrophobic eggshell, an environmentally friendly material, to create a hydrophobic barrier on a circular Whatman No. 1 filter paper substrate. As depicted in Figure 9f, the yellowish color zone on the paper device indicates the presence of the hydrophobic 'eggshell' barrier surrounding a hydrophilic reservoir. The figure also outlines the step-by-step method for the colorimetric detection of ammonium ions in water. The procedure involves pipetting a 3.0 μL aliquot of reagent A (salicylate and nitroprusside) onto the hydrophilic reservoir, followed by the transfer of a 3.0 μL aliquot of the water sample or a series of standard NH_4^+ solutions (5–100 mg N L^{-1}). Subsequently, a 3.0 μL aliquot of reagent B (dichloroisocyanurate and tri-sodium citrate) is added, and after a 5 min reaction period, a visible color change occurs in the hydrophilic reservoir area. The device is then placed in a constant-light illumination studio for image capture and the quantification of color intensity.

Overall, these μPADs offer a promising solution for the on-site and simultaneous monitoring of multiple water quality parameters, showcasing its potential for environmental monitoring and water analysis applications.

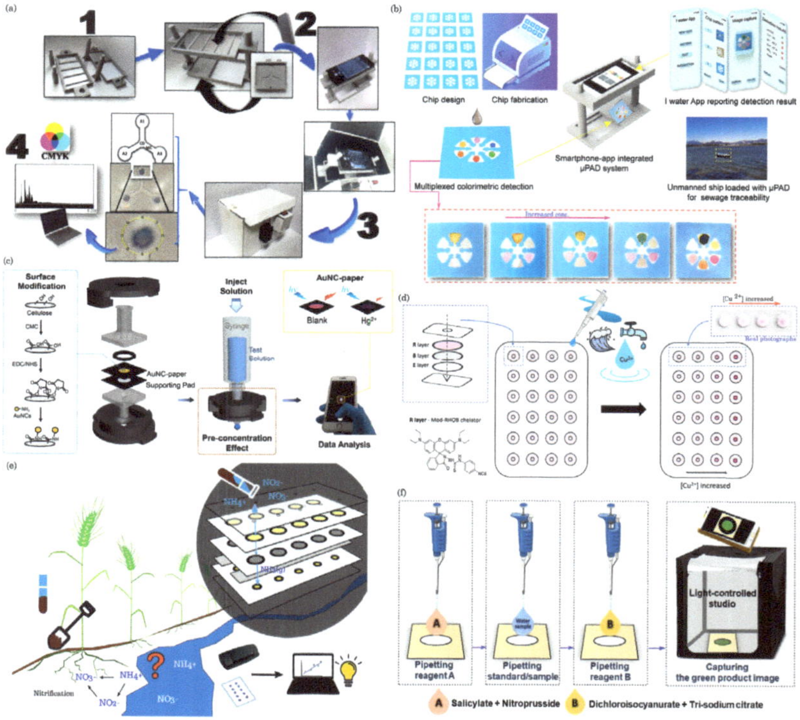

Figure 9. (**a**) Illustration of the μPAD assembly: production of smartphone support parts, coupling smartphone support with 3D-printed μPAD support, assembly of the "closed box" with integrated support and LED white light, and data acquisition through smartphone image capture and conversion of RGB to CMYK color standards using ImageJ® software. Reprinted with permission from Da Silva et al. [212]. ©2020 Elsevier Ltd. (**b**) The schematic diagram illustrates the fabrication process of the μPAD, the multiplexed colorimetric detection strategies, and the integration of a smartphone app for its applications. Reprinted with permission from Xiong et al. [213]. ©2022 The Authors, published by American Chemical Society, licensed under CC BY-NC-ND 4.0. (**c**) Schematic illustration of the gold nanoclusters (AuNC)-modified paper device designed for detecting Hg^{2+} ions. Reprinted with permission from Lin et al. [214]. ©2021 Elsevier B.V. (**d**) Illustration of the μPAD assembly designed for copper determination in water, along with actual photographs of the paper device depicting varying concentrations of Cu^{2+}. The components include laminating pouch sheets L1 and L2, a reagent layer R, a buffer layer B, and an empty layer E. Reproduced with permission from Aguiar et al. [215]. Under a Creative Commons license, ©2024 The Authors. Published by Elsevier B.V. (**e**) Schematic

representation of the μPAD designed for detecting inorganic nitrogen in water and soil samples. Reprinted with permission from Uhlikova et al. [216]. Under a Creative Commons license, ©2024 The Authors. Published by Elsevier B.V. (**f**) Diagram depicting the analytical steps for the straightforward colorimetric determination of NH_4^+ utilizing the proposed paper-based analytical device. Reprinted with permission from Thangjitsirisin et al. [217]. ©2024 Elsevier B.V.

8.3. Air Quality Monitoring/Gas Sensing

Air pollution poses a significant threat to public health and the environment. Conventional air quality monitoring systems are characterized by their high cost, limited portability, and dependency on sophisticated infrastructure. The emergence of paper-based microsystems presents a cost-effective and portable alternative, leveraging the inherent properties of paper substrates for the efficient detection of air pollutants [218–220].

Colorimetric paper-based sensors offer an innovative approach to environmental monitoring, especially in detecting air pollutants. In their work, De Matteis et al. [221] designed a paper-based analytical device (PAD) capable of detecting contaminants such as Fe^{2+}, Cu^{2+} ions in water, and NH_3 and C_2H_4O in the air, even at low concentrations. The researchers employed a wax pen to form a circular hydrophobic barrier on a Whatman filter paper substrate to create distinct sensing zones. These marked spots were utilized to detect the specified analytes at various concentrations. Figure 10a shows the schematics of the detection mechanisms. Notably, the paper sensor displayed a colorimetric response directly correlated with the concentration of the identified pollutant species.

Bordbar et al. [222] developed a paper-based optical nose by depositing bimetallic silver and gold nanoparticles onto a paper substrate, synthesized using both natural and chemical reducing agents. This assay was evaluated for its capability to distinguish between gasoline and five ignitable liquids: diesel, ethanol, methanol, kerosene, and thinner. The interaction between the sensor and sample vapors led to nanoparticle aggregation, resulting in color changes captured by a scanner, producing distinct colorimetric maps for each analyte (Figure 10b). Visual observations were corroborated using multivariate statistical analyses, including principal component analysis and hierarchical clustering analysis. Additionally, partial least-squares regression aided in estimating the quantities of ignitable liquids present as counterfeit substances in gasoline samples, with root mean square errors for prediction ranging from 1.7% to 3.4%. Ultimately, the fabricated sensor demonstrated high efficiency for the onsite detection of pure industrial gasoline samples versus adulterated ones.

Moreover, paper-based devices are extensively utilized for the electrochemical-based detection of air pollutants [223,224]. Mettakoonpitak et al. [225] introduced an innovative electrochemical paper-based device (ePAD) for the multiplexed detection of metals, specifically Cd, Pb, Cu, Fe, and Ni, from a single particulate matter sample. The paper-based device was designed with four independent channels and working electrodes, enabling the implementation of square-wave anodic stripping voltammetry (SWASV) and square-wave cathodic stripping voltammetry (SWCSV) for the simultaneous determination of multiple metals. Figure 10c shows an example of electrochemical-based paper sensors for air pollutant detection. Notably, the device exhibited impressive detection limits, ranging from 0.5 to 400.0 μg L^{-1} for Cd(II), Pb(II), and Fe(II); 1.0 to 400.0 μg L^{-1} for Cu(II); and 0.5 to 200.0 μg L^{-1} for Ni(II). This multiplexed ePAD offers a versatile and efficient solution for sensitively detecting various metals in complex samples, showcasing its potential for environmental monitoring and analytical applications.

Davis et al. [226] engineered a flexible paper-based sensor for acetone detection at room temperature. The paper-based electrodes were crafted through the application of zinc oxide (ZnO)-polyaniline-based conductive inks (Figure 10d). These electrodes exhibited remarkable conductivity (80 S/m) and stability under rigorous mechanical and chemical conditions while demonstrating commendable flexibility (1000 bending cycles).

The acetone sensor displayed a notable sensitivity of 0.02/100 ppm and 0.6/10 μL, with a broad sensitivity range spanning from 260 to >1000 ppm under atmospheric conditions. Moreover, the sensors exhibited an impressive response time of 4 seconds and a recovery time of 15 s for acetone detection at room temperature without necessitating external heaters. The proposed paper device's high sensitivity and long-term stability make it suitable for wearable biosensor applications.

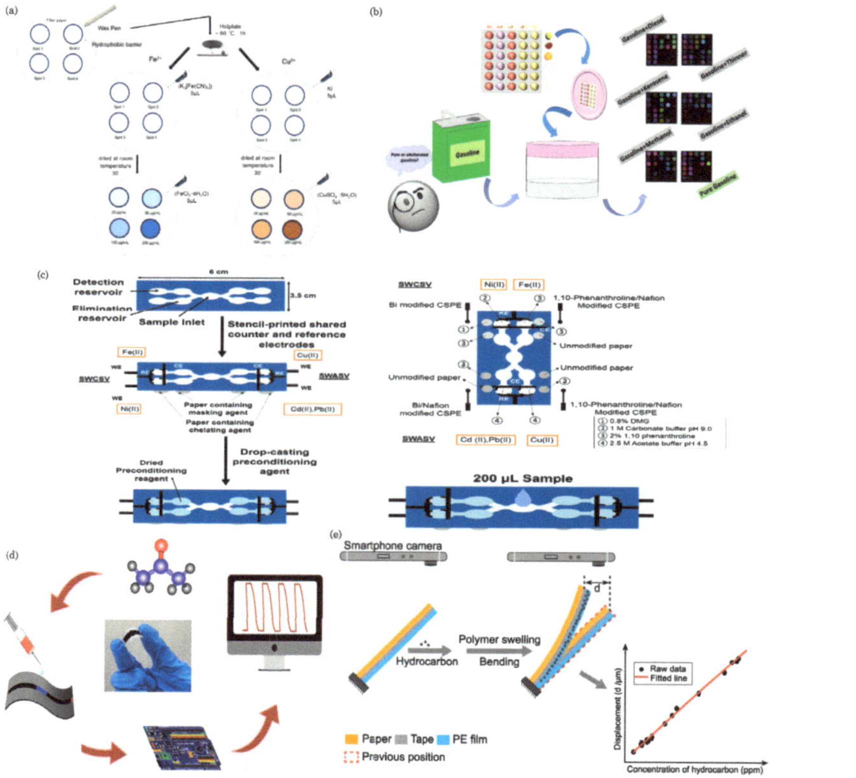

Figure 10. (**a**) Colorimetric detection of Fe^{2+}, Cu^{2+} ions, reprinted with permission from De Matteis et al. [221] under Creative Commons Attribution (CC BY) license. ©2020 The authors, published by MDPI. (**b**) Schematic representation of colorimetric detection process of gasoline utilizing a paper-based optical nose. Reprinted with permission from Bordbar et al. [222]. ©2022 American Chemical Society. (**c**) Electrochemical detection of metals in aerosol samples using paper-based analytical device. Reprinted with permission from Mettakoonpitak et al. [225]. ©2019, American Chemical Society. (**d**) Schematics of the paper-based flexible sensors for detection of acetone at room temperature. Reprinted with permission from Davis et al. [226]. ©2023 American Chemical Society. (**e**) Sensing mechanism of the milli-cantilever. Reprinted with permission from Qin et al. [227]. ©2020 American Chemical Society.

Moreover, a cantilever-based paper-based sensor device was demonstrated by Qin et al. [227]. They developed an inexpensive and lightweight hydrocarbon gas sensor utilizing a smartphone camera for readout. The sensor relies on paper-based milli-cantilever bending induced by polymer swelling. The sensing cantilever comprises three layers: a functional layer of polyethylene film, an adhesive layer of double-sided tape, and a weighing paper substrate. Figure 10e shows schematics of the milli-cantilever. The milli-fabricated sensing cantilever has dimensions of 8 mm length, 0.5 mm width, and 50 μm thickness. The sensor's response is measured as the displacement of the milli-cantilever-free end. Demonstrating its capabilities, the sensor exhibited a linear response to hydrocarbon concentrations, a broad detection range, low detection limits, and rapid response times. For instance, when exposed to xylene, the sensor displayed a detection range of 15–140 ppm, a low detection limit of 15 ppm, and a fast response time of 30 s.

9. Food Safety

Paper-based devices have emerged as valuable tools in ensuring food safety due to their simplicity, cost-effectiveness, and ease of use. These devices are designed to detect various contaminants and ensure the quality of food products. Here are some examples of paper-based devices for food safety applications.

Paper-based devices are widely used to rapidly detect foodborne pathogens such as *Salmonella*, *E. coli*, and *Listeria*. These devices often employ antibodies or DNA probes to capture and identify specific pathogens, providing quick results for on-site testing. For example, Zhuang et al. [228] developed an integrated microfluidic paper-based analytical device, termed RPA-Cas12a-μPAD, combining recombinase polymerase amplification (RPA) with supersensitive surface-enhanced Raman scattering (SERS) detection. Figure 11a illustrates the device schematics, operational steps, and microscopic image of the S. typhi test zone with SERS mapping signals at 1075 cm^{-1}, along with the corresponding Raman spectrum. The successful detection of *Salmonella* in milk and meat samples was achieved with detection limits of 3.72 and 4.04 CFU/mL, respectively.

The detection of toxins in food, such as mycotoxins and chemical contaminants, is critical for ensuring food safety. Paper-based assays can detect specific toxins through colorimetric or electrochemical reactions, providing a visual indication of contamination [229–231]. Dos Santos et al. [232] developed curcumin-modified paper-based sensing platforms for detecting ochratoxin A (OTA) in grape juice and beer samples (Figure 11b). The sensor operates based on specific interactions between curcumin and OTA, involving energy and electron transfer mechanisms in optical detection. Curcumin molecules form complexes with OTA in electrochemical detection, enhancing the binding affinity between OTA and the electrode surface. This results in a greater change in the impedance of the double layer, easily detected by electrochemical impedance spectroscopy (EIS). Sensors exhibit good sensitivity, with limits of detection (LODs) of 0.09 ng/mL and 0.045 ng/mL for optical and electrochemical methods, respectively, remaining effective across various food matrices and in the presence of potential interferents.

Paper-based tests are employed to detect allergens, helping to prevent allergic reactions in individuals with specific sensitivities. These devices can detect the presence of allergenic proteins, allowing for rapid screening in various food products such as ovalbumin and egg white protein [233], milk allergen (β-lactoglobulin) [234–236], histamine in canned tuna [237], and peanut allergen Ara h1 [238]. Recently, Lu et al. [239] developed a paper-based mass spectrometric immunoassay platform for peanut allergen detection. Figure 11c illustrates the microzone paper-based mass spectrometric immunoassay for food allergen detection schematic. They introduced a novel quaternary ammonium-based mass tag and a paper chip with a microzone, resulting in significant signal enhancement. This method could detect Ara h1 with a linear range of 0.1–100 ng/mL and a detection limit of 0.08 ng/mL in milk matrices. Moreover, it accurately quantified Ara h1 in various milk-related beverages, biscuits, and candy bars with complex matrices, demonstrating a capability for low-concentration quantitation.

Paper-based devices are utilized to assess the quality of food products. For example, pH strips on paper can indicate a product's acidity (e.g., carbendazim detection on the skin of apple and cabbage [240] and acidic pH and bisulfite in white wine [241]), ensuring it meets quality standards. Similarly, these devices can monitor the freshness of certain perishable items. Another example is the detection of iodine speciation in seaweed samples. Placer et al. [242] engineered a 3D origami microfluidic paper-based analytical device for quantifying iodide and iodate levels in edible seaweeds via smartphone-based colorimetric detection. The paper device was predesigned to generate hydrophobic patterns on Whatman No. 1 filter paper using wax printing and assembled by folding the paper substrates. Figure 11d illustrates the fabrication of the detection device and plots the analytical signal against analyte concentrations.

The detection of food adulterants and contaminants, such as pesticides or additives, is crucial for maintaining food safety. Paper-based assays can be tailored to identify specific adulterants through selective reactions, quickly assessing food purity. Recent examples include the detection of adulteration in Iranian honey [243] and milk adulteration with melamine [244], starch [245], sugar [246], and urea [247]. Wu et al. [248] developed a surface molecularly imprinted microfluidic paper-based device (SMIPs-μPAD) for detecting butachlor in mung bean samples. When combined with a smartphone, this colorimetric paper chip demonstrated high selectivity and sensitivity to butachlor, with a detection limit of 1.43 ng/g and a detection time of 20 min. Figure 11e depicts the operation steps of SMIPs-μPAD and real photographs of the paper device before and after color development.

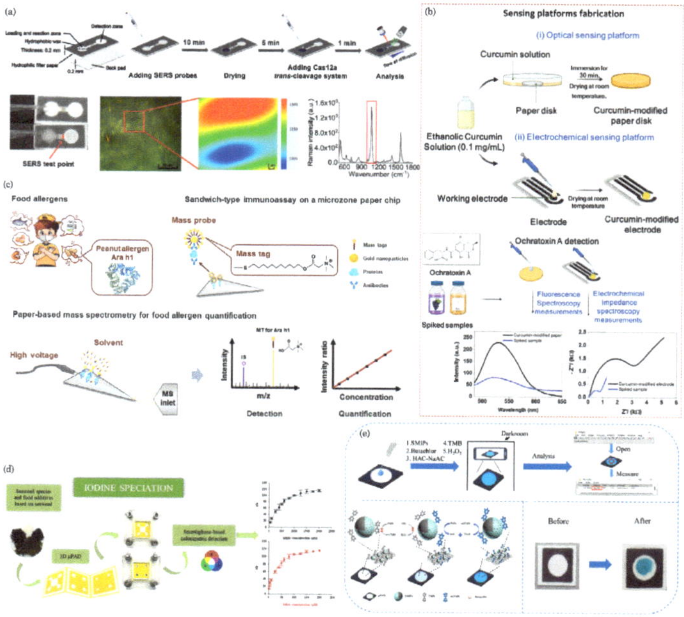

Figure 11. (**a**) Illustration of the device schematics, operational steps, and a microscopic image of the S. typhi test zone, showcasing SERS mapping signals at 1075 cm^{-1}, alongside the corresponding Raman spectrum. Reprinted with permission from Zhuang et al. [228]. ©2022 Elsevier B.V. (**b**) Schematic representation of optical and electrochemical sensing platforms utilizing curcumin-immobilized paper substrates for ochratoxin A detection in grape juice and beer. Reprinted with permission from Dos Santos et al. [232] under a Creative Commons license, ©2023 The Author(s). Published by Elsevier B.V. (**c**) Schematic illustration of microzone paper-based mass spectrometric immunoassay for detecting food allergens (peanut allergen Ara h1). Reprinted with permission from Lu et al. [239]. ©2024 American Chemical Society. (**d**) Schematics of 3D μPAD with colorimetric detection for iodine speciation in seaweed samples. Reprinted with permission from Placer et al. [242] under a Creative Commons license, ©2022 The Authors. Published by Elsevier B.V. (**e**) The procedural steps of SMIPs-μPAD and actual images of the paper device before and after color development. Reprinted with permission from Wu et al. [248]. ©2023 Elsevier Ltd.

10. Biodegradability and Sustainability

10.1. Environmental Impact of Traditional Microfluidic Devices

The environmental impact of traditional microfluidic devices encompasses various aspects, including their fabrication processes, materials, and waste generation. These devices are typically manufactured in cleanroom facilities, requiring controlled environments with stringent conditions. However, maintaining cleanrooms consumes significant energy and entails specialized infrastructure, contributing to environmental concerns.

Traditional microfluidic devices are commonly crafted from silicon, glass, and polymers. These materials' extraction, processing, and manufacturing can have substantial environmental footprints. Silicon wafers, for instance, are resource-intensive to produce, and glass fabrication involves high-temperature processes.

The production of microfluidic devices often involves the use of chemicals, solvents, and photoresists. Disposing of these chemicals and the potential release into the environment pose pollution concerns. Additionally, the energy-intensive processes associated with traditional microfabrication techniques, such as photolithography and etching, contribute to the overall environmental impact.

Waste generation has a significant environmental impact, with manufacturing processes producing unused substrates, chemicals, and contaminated water. The proper disposal and treatment of these wastes are crucial to minimize their ecological impact. Some materials used in traditional microfluidic devices may have limited biodegradability, raising concerns about long-term environmental persistence. Moreover, the single-use nature of many microfluidic devices designed for research and diagnostics contributes to increased waste generation and challenges related to disposal.

10.2. Advantages of Biodegradable Paper Microfluidics

Biodegradable paper microfluidics offers several advantages in alignment with sustainable practices. Derived from renewable resources like wood pulp, biodegradable paper is an eco-friendly alternative to traditional microfluidic materials such as silicon or specific polymers. The production processes for biodegradable paper are generally less energy-intensive, resulting in a lower overall environmental impact throughout the material's life cycle.

Cost-effectiveness is a notable advantage of biodegradable paper microfluidics, making them particularly suitable for applications where cost is a critical consideration. The ease of fabrication is another key feature as paper allows for straightforward manufacturing through cutting, folding, and printing techniques. This simplicity reduces the complexity and cost associated with manufacturing, enhancing accessibility for various applications.

One of the distinctive features of biodegradable paper microfluidics is their inherent biodegradability [249,250]. After disposal, these devices naturally break down over time, minimizing environmental impact and contributing to waste reduction. The customizability and functionalizability of paper microfluidics are additional strengths, allowing researchers to modify surfaces, integrate reagents, and tailor designs for specific assays or diagnostic tests.

The portability and simplicity of paper microfluidic devices make them well-suited for point-of-care applications, especially in remote or resource-limited settings. The reduced usage of chemicals in the fabrication process further adds to their appeal from an environmental standpoint. Biodegradable paper microfluidics offers a sustainable, cost-effective, and customizable solution with reduced ecological impact, promising them for various applications, including eco-friendly and practical diagnostic tools.

11. Challenges and Future Perspectives

11.1. Current Challenges in Paper Microfluidics

In its current state of development, paper microfluidics encounters challenges in creating intricate channels due to limitations in channel design and fluidic pathway complexity [251]. Achieving consistent and reproducible results is hindered by variations in paper properties like thickness and porosity, impacting diagnostic assay reliability. Sensitivity limitations persist, especially compared to advanced lab techniques, posing an ongoing challenge in detecting low analyte concentrations. The paper's susceptibility to environmental conditions, such as humidity, affects reagent stability, raising concerns about long-term stability in resource-limited settings. Multiplexing, integrating multiple tests on a single paper device, is challenging due to the potential for cross-contamination. Achieving uniformity in fabrication processes like printing or cutting proves difficult, introducing variability that affects device performance and reliability. The finite shelf life of paper-based devices, attributed to potential paper and reagent degradation, prompts ongoing research to improve stability for extended storage. While excelling in qualitative analysis, paper devices face challenges in achieving precise quantitative measurements, impacting applications requiring accuracy.

11.2. Prospects and Potential Innovations

The potential innovations of paper microfluidic devices offer significant promise in various fields, particularly in low-cost diagnostics for point-of-care testing in resource-limited settings. The portability of paper devices is well-suited for on-site diagnostics, reducing reliance on centralized laboratories. Advancements in multiplexing capabilities on paper microfluidic devices can revolutionize testing methodologies, simultaneously detecting multiple analytes within a single test. This innovation can significantly impact healthcare and environmental monitoring, particularly in detecting pollutants and contaminants in air and water. Integrating paper microfluidics into wearable devices holds promise for developing flexible and wearable paper-based sensors, providing real-time insights into biomarkers or environmental factors. Advancements in fabrication techniques may enable more customizable designs of paper microfluidic devices, tailoring them to specific applications or user requirements and contributing to their versatility. Integration with smartphones for result readout and data analysis enhances the capabilities of paper microfluidic devices, facilitating remote monitoring and data sharing, aligning with the trend of leveraging smartphones for healthcare and diagnostic applications. Detection methods and sensitivity advancements can broaden paper microfluidics' biological and chemical analysis applications, potentially revolutionizing healthcare, food safety, and environmental monitoring. Exploring hybrid systems that combine paper microfluidics with other technologies, such as electronic sensors or microcontrollers, could lead to sophisticated and versatile platforms with enhanced performance and functionalities, opening up new possibilities across various domains.

Ongoing research and development in paper microfluidics are expected to bring continuous innovations, expanding their applications and impact across diverse fields.

12. Conclusions

In conclusion, the manuscript highlights the significant contributions of paper microfluidics in addressing crucial challenges across various domains. Through innovative applications in healthcare, environmental monitoring, and food safety, paper-based sensing platforms offer versatile, cost-effective, and environmentally friendly solutions. This review underscores the potential of paper microfluidics to revolutionize diagnostics and monitoring, providing accessible tools for health assessments, pollution detection, and food quality assurance. As we move forward, continued research and development in this field promise to unlock further capabilities, paving the way for sustainable sensing solutions with widespread impact on human health and environmental well-being.

Author Contributions: writing—original draft preparation, S.K. and J.B.K.; writing—review and editing, S.K.; visualization, S.K.; supervision, H.P.L. All authors have read and agreed to the published version of the manuscript.

Funding: This research received no external funding.

Data Availability Statement: Not applicable.

Conflicts of Interest: The authors declare no conflicts of interest.

References

1. Peeling, R.W.; Holmes, K.K.; Mabey, D.; Ronald, A. Rapid tests for sexually transmitted infections (STIs): The way forward. *Sex. Transm. Dis.* **2006**, *82*, v1–v6. [CrossRef] [PubMed]
2. Kosack, C.S.; Page, A.L.; Klatser, P.R. A guide to aid the selection of diagnostic tests. *Bull. World Health Organ.* **2017**, *95*, 639–645. [CrossRef] [PubMed]
3. Martinez, A.W.; Phillips, S.T.; Whitesides, G.M.; Carrilho, E. Diagnostics for the developing world: Microfluidic paper-based analytical devices. *Anal. Chem.* **2010**, *82*, 3–10. [CrossRef] [PubMed]
4. Vicente, A.T.; Araújo, A.; Gaspar, D.; Santos, L.; Marques, A.C.; Mendes, M.J.; Pereira, L.; Fortunato, E.; Martins, R. Optoelectronics and bio devices on paper powered by solar cells. In *Nanostructured Solar Cells*; IntechOpen: London, UK , 2017. [CrossRef]
5. Nontawong, N.; Ngaosri, P.; Chunta, S.; Jarujamrus, P.; Nacapricha, D.; Lieberzeit, P.A.; Amatatongchai, M. Smart sensor for assessment of oxidative/nitrative stress biomarkers using a dual-imprinted electrochemical paper-based analytical device. *Anal. Chim. Acta* **2022**, *1191*, 339363. [CrossRef] [PubMed]
6. Patella, B.; Parisi, A.; Moukri, N.; Gitto, F.; Busacca, A.; Aiello, G.; Russo, M.; O'Riordan, A.; Inguanta, R. Phosphate ions detection by using an electrochemical sensor based on laser-scribed graphene oxide on paper. *Electrochim. Acta* **2023**, *461*, 142600. [CrossRef]
7. Lu, Y.; Shi, W.; Qin, J.; Lin, B. Fabrication and characterization of paper-based microfluidics prepared in nitrocellulose membrane by wax printing. *Anal. Chem.* **2010**, *82*, 329–335. [CrossRef] [PubMed]
8. Tang, R.H.; Li, M.; Liu, L.N.; Zhang, S.F.; Alam, N.; You, M.; Ni, Y.H.; Li, Z.D. Chitosan-modified nitrocellulose membrane for paper-based point-of-care testing. *Cellulose* **2020**, *27*, 3835–3846. [CrossRef]
9. Tang, R.; Xie, M.Y.; Li, M.; Cao, L.; Feng, S.; Li, Z.; Xu, F. Nitrocellulose membrane for paper-based biosensor. *Appl. Mater. Today* **2022**, *26*, 101305. [CrossRef]
10. Tang, R.; Xie, M.; Yan, X.; Qian, L.; Giesy, J.P.; Xie, Y. A nitrocellulose/cotton fiber hybrid composite membrane for paper-based biosensor. *Cellulose* **2023**, *30*, 6457–6469. [CrossRef]
11. Boonyasit, Y.; Chailapakul, O.; Laiwattanapaisal, W. A multiplexed three-dimensional paper-based electrochemical impedance device for simultaneous label-free affinity sensing of total and glycated haemoglobin: The potential of using a specific single-frequency value for analysis. *Anal. Chim. Acta* **2016**, *936*, 1–11. [CrossRef]
12. Boonyasit, Y.; Chailapakul, O.; Laiwattanapaisal, W. A folding affinity paper-based electrochemical impedance device for cardiovascular risk assessment. *Biosens. Bioelectron.* **2019**, *130*, 389–396. [CrossRef] [PubMed]
13. Arduini, F.; Cinti, S.; Caratelli, V.; Amendola, L.; Palleschi, G.; Moscone, D. Origami multiple paper-based electrochemical biosensors for pesticide detection. *Biosens. Bioelectron.* **2019**, *126*, 346–354. [CrossRef] [PubMed]
14. Mondal, B.P.; Das, S.; Ranjan, P.; Datta, A. Optimizing Output Performances in Stationery Papers–Based Hybrid Inorganic–Organic Flexible Thermoelectric Generators. *Phys. Status Solidi A* **2023**, *220*, 2300228. [CrossRef]
15. Gong, S.; Schwalb, W.; Wang, Y.; Chen, Y.; Tang, Y.; Si, J.; Shirinzadeh, B.; Cheng, W. A wearable and highly sensitive pressure sensor with ultrathin gold nanowires. *Nat. Commun.* **2014**, *5*, 3132. [CrossRef] [PubMed]
16. Smith, S.; Madzivhandila, P.; Ntuli, L.; Bezuidenhout, P.; Zheng, H.; Land, K. Printed paper–based electrochemical sensors for low-cost point-of-need applications. *Electrocatalysis* **2019**, *10*, 342–351. [CrossRef]
17. Lim, W.Y.; Goh, C.H.; Yap, K.Z.; Ramakrishnan, N. One-step fabrication of paper-based inkjet-printed graphene for breath monitor sensors. *Biosensors* **2023**, *13*, 209. [CrossRef]
18. Camargo, J.R.; Andreotti, I.A.; Kalinke, C.; Henrique, J.M.; Bonacin, J.A.; Janegitz, B.C. Waterproof paper as a new substrate to construct a disposable sensor for the electrochemical determination of paracetamol and melatonin. *Talanta* **2020**, *208*, 120458. [CrossRef] [PubMed]
19. Chen, Z.; Wright, C.; Dincel, O.; Chi, T.Y.; Kameoka, J. A low-cost paper glucose sensor with molecularly imprinted polyaniline electrode. *Sensors* **2020**, *20*, 1098. [CrossRef] [PubMed]
20. Wu, H.; Chiang, S.W.; Lin, W.; Yang, C.; Li, Z.; Liu, J.; Cui, X.; Kang, F.; Wong, C.P. Towards practical application of paper based printed circuits: Capillarity effectively enhances conductivity of the thermoplastic electrically conductive adhesives. *Sci. Rep.* **2014**, *4*, 6275. [CrossRef]
21. Martins, G.V.; Riveiro, A.; Chiussi, S.; Sales, M. Flexible sensing devices integrating molecularly-imprinted polymers for the detection of 3-nitrotyrosine biomarker. *Biosens. Bioelectron. X* **2022**, *10*, 100107. [CrossRef]
22. Qi, J.; Li, B.; Zhou, N.; Wang, X.; Deng, D.; Luo, L.; Chen, L. The strategy of antibody-free biomarker analysis by in-situ synthesized molecularly imprinted polymers on movable valve paper-based device. *Biosens. Bioelectron.* **2019**, *142*, 111533. [CrossRef] [PubMed]
23. Chen, I.H.; You, M.W.; Tsai, J.H.; Chang, J.H.; Cheng, I.C.; Hsu, C.C.; Luo, S.C.; Chen, C.F.; Chen, J.Z. Feasibility study of dielectric barrier discharge jet-patterned perfluorodecyltrichlorosilane-coated paper for biochemical diagnosis. *ECS J. Solid State Sci. Technol.* **2021**, *10*, 037005. [CrossRef]
24. Cao, C.X.; Yuan, J.; Cheng, J.P.; Han, B.H. Synthesis of porous polymer/tissue paper hybrid membranes for switchable oil/water separation. *Sci. Rep.* **2017**, *7*, 3101. [CrossRef] [PubMed]
25. Arena, A.; Donato, N.; Saitta, G.; Bonavita, A.; Rizzo, G.; Neri, G. Flexible ethanol sensors on glossy paper substrates operating at room temperature. *Sens. Actuators B Chem.* **2010**, *145*, 488–494. [CrossRef]
26. Jiang, L.; Nelson, G.W.; Kim, H.; Sim, I.; Han, S.O.; Foord, J.S. Cellulose-derived supercapacitors from the carbonisation of filter paper. *ChemistryOpen* **2015**, *4*, 586–589. [CrossRef] [PubMed]

27. Lin, D.; Li, B.; Fu, L.; Qi, J.; Xia, C.; Zhang, Y.; Chen, J.; Choo, J.; Chen, L. A novel polymer-based nitrocellulose platform for implementing a multiplexed microfluidic paper-based enzyme-linked immunosorbent assay. *Microsyst. Nanoeng.* **2022**, *8*, 53. [CrossRef] [PubMed]
28. Jabar, A.W.; AL-Bawi, Z.F.; Faris, R.A.; Wahhab, H.K. Plasmonic Nanoparticles Decorated Salty Paper Based on SERS Platform for Diagnostic low-Level Contamination: Lab on Paper. *Iraqi J. Laser* **2019**, *18*, 55–61.
29. Fernandes, A.R.; Bernardo, R.A.; Carvalho, T.C.d.; Vaz, B.G.; Chaves, A.R. Graphene oxides coated paper as a substrate to paper spray ionization mass spectrometry for creatinine determination in urine samples. *J. Braz. Chem. Soc.* **2019**, *30*, 1074–1081. [CrossRef]
30. Arahman, N.; Fahrina, A.; Amalia, S.; Sunarya, R.; Mulyati, S. Effect of PVP on the characteristic of modified membranes made from waste PET bottles for humic acid removal. *F1000Research* **2017**, *6*, 668. [CrossRef]
31. Van der Westhuizen, J.; Du Plessis, J.P. An attempt to quantify fibre bed permeability utilizing the phase average Navier-Stokes equation. *Compos. Part A Appl. Sci. Manuf.* **1996**, *27*, 263–269. [CrossRef]
32. Nabovati, A.; Llewellin, E.W.; Sousa, A.C. A general model for the permeability of fibrous porous media based on fluid flow simulations using the lattice Boltzmann method. *Compos. Part A Appl. Sci. Manuf.* **2009**, *40*, 860–869. [CrossRef]
33. Choi, J.R.; Liu, Z.; Hu, J.; Tang, R.; Gong, Y.; Feng, S.; Ren, H.; Wen, T.; Yang, H.; Qu, Z.; et al. Polydimethylsiloxane-paper hybrid lateral flow assay for highly sensitive point-of-care nucleic acid testing. *Anal. Chem.* **2016**, *88*, 6254–6264. [CrossRef] [PubMed]
34. Parolo, C.; Medina-Sánchez, M.; De La Escosura-Muñiz, A.; Merkoçi, A. Simple paper architecture modifications lead to enhanced sensitivity in nanoparticle based lateral flow immunoassays. *Lab Chip* **2013**, *13*, 386–390. [CrossRef] [PubMed]
35. Garnier, G.; Then, W.L. Paper microfluidics: Applications and perspectives. In Proceedings of the 15th Fundamental Research Symposium, Cambridge, UK, 8–13 September 2013; *Advances in Pulp and Paper Research;* Fundamental Research Committee: Manchester, UK, 2013; pp. 541–583. [CrossRef]
36. Hong, S.; Kim, W. Dynamics of water imbibition through paper channels with wax boundaries. *Microfluid. Nanofluid.* **2015**, *19*, 845–853. [CrossRef]
37. Conrath, M.; Fries, N.; Zhang, M.; Dreyer, M.E. Radial capillary transport from an infinite reservoir. *Transp. Porous Med.* **2010**, *84*, 109–132. [CrossRef]
38. Mendez, S.; Fenton, E.M.; Gallegos, G.R.; Petsev, D.N.; Sibbett, S.S.; Stone, H.A.; Zhang, Y.; López, G.P. Imbibition in porous membranes of complex shape: Quasi-stationary flow in thin rectangular segments. *Langmuir* **2010**, *26*, 1380–1385. [CrossRef] [PubMed]
39. Wang, X.; Hagen, J.A.; Papautsky, I. Paper pump for passive and programmable transport. *Biomicrofluidics* **2013**, *7*, 014107. [CrossRef]
40. Shou, D.; Ye, L.; Fan, J.; Fu, K.; Mei, M.; Wang, H.; Chen, Q. Geometry-induced asymmetric capillary flow. *Langmuir* **2014**, *30*, 5448–5454. [CrossRef] [PubMed]
41. Elizalde, E.; Urteaga, R.; Berli, C.L. Rational design of capillary-driven flows for paper-based microfluidics. *Lab Chip* **2015**, *15*, 2173–2180. [CrossRef]
42. Kumar, S.; Bhushan, P.; Bhattacharya, S. Fluid Transport Mechanisms in Paper-Based Microfluidic Devices. In *Paper Microfluidics Theory and Applications*; Springer: Singapore, 2019; Chapter 2, pp. 7–28.
43. Zhong, Z.; Wu, R.; Wang, Z.; Tan, H. An investigation of paper based microfluidic devices for size based separation and extraction applications. *J. Chromatogr. B* **2015**, *1000*, 41–48. [CrossRef]
44. Garra, R.; Salusti, E. Application of the nonlocal Darcy law to the propagation of nonlinear thermoelastic waves in fluid saturated porous media. *Phys. D Nonlinear Phenom.* **2013**, *250*, 52–57. [CrossRef]
45. Mambatta, A.K.; Jayarajan, J.; Rashme, V.L.; Harini, S.; Menon, S.; Kuppusamy, J. Reliability of dipstick assay in predicting urinary tract infection. *J. Family Med. Prim. Care* **2015**, *4*, 265–268. [CrossRef]
46. Dadzie, I.; Quansah, E.; Puopelle Dakorah, M.; Abiade, V.; Takyi-Amuah, E.; Adusei, R. The effectiveness of dipstick for the detection of urinary tract infection. *Can. J. Infect. Dis. Med. Microbiol.* **2019**, *2019*, 8642628. [CrossRef] [PubMed]
47. Renault, C.; Anderson, M.J.; Crooks, R.M. Electrochemistry in hollow-channel paper analytical devices. *J. Am. Chem. Soc.* **2014**, *136*, 4616–4623. [CrossRef] [PubMed]
48. Carrell, C.; Kava, A.; Nguyen, M.; Menger, R.; Munshi, Z.; Call, Z.; Nussbaum, M.; Henry, C. Beyond the lateral flow assay: A review of paper-based microfluidics. *Microelectron. Eng.* **2019**, *206*, 45–54. [CrossRef]
49. Noviana, E.; Ozer, T.; Carrell, C.S.; Link, J.S.; McMahon, C.; Jang, I.; Henry, C.S. Microfluidic paper-based analytical devices: From design to applications. *Chem. Rev.* **2021**, *121*, 11835–11885. [CrossRef] [PubMed]
50. Peto, T.; Affron, D.; Afrough, B.; Agasu, A.; Ainsworth, M.; Allanson, A.; Allen, K.; Allen, C.; Archer, L.; Ashbridge, N.; et al. COVID-19: Rapid antigen detection for SARS-CoV-2 by lateral flow assay: A national systematic evaluation of sensitivity and specificity for mass-testing. *EClinicalMedicine* **2021**, *36*, 100924. [CrossRef] [PubMed]
51. Davies, B.; Araghi, M.; Moshe, M.; Gao, H.; Bennet, K.; Jenkins, J.; Atchison, C.; Darzi, A.; Ashby, D.; Riley, S.; et al. Acceptability, usability, and performance of lateral flow immunoassay tests for severe acute respiratory syndrome coronavirus 2 antibodies: REACT-2 study of self-testing in nonhealthcare key workers. *Open Forum Infect. Dis.* **2021**, *8*, ofab496. [CrossRef] [PubMed]
52. Budd, J.; Miller, B.S.; Weckman, N.E.; Cherkaoui, D.; Huang, D.; Decruz, A.T.; Fongwen, N.; Han, G.R.; Broto, M.; Estcourt, C.S.; et al. Lateral flow test engineering and lessons learned from COVID-19. *Nat. Rev. Bioeng.* **2023**, *1*, 13–31. [CrossRef]

53. Mahmud, M.A.; Blondeel, E.J.; Kaddoura, M.; MacDonald, B.D. Features in microfluidic paper-based devices made by laser cutting: How small can they be? *Micromachines* **2018**, *9*, 220. [CrossRef]
54. Martinez, A.W.; Phillips, S.T.; Butte, M.J.; Whitesides, G.M. Patterned paper as a platform for inexpensive, low-volume, portable bioassays. *Angew. Chem.* **2007**, *119*, 1340–1342. [CrossRef] [PubMed]
55. He, Y.; Gao, Q.; Wu, W.B.; Nie, J.; Fu, J.Z. 3D printed paper-based microfluidic analytical devices. *Micromachines* **2016**, *7*, 108. [CrossRef]
56. Carrilho, E.; Martinez, A.W.; Whitesides, G.M. Understanding wax printing: A simple micropatterning process for paper-based microfluidics. *Anal. Chem.* **2009**, *81*, 7091–7095. [CrossRef]
57. Wang, Y.; Sun, S.; Luo, J.; Xiong, Y.; Ming, T.; Liu, J.; Ma, Y.; Yan, S.; Yang, Y.; Yang, Z.; et al. Low sample volume origami-paper-based graphene-modified aptasensors for label-free electrochemical detection of cancer biomarker-EGFR. *Microsyst. Nanoeng.* **2020**, *6*, 32. [CrossRef]
58. Gao, B.; Chi, J.; Liu, H.; Gu, Z. Vertical paper analytical devices fabricated using the principles of quilling and kirigami. *Sci. Rep.* **2017**, *7*, 7255. [CrossRef] [PubMed]
59. Johnston, M. *The Book of Paper Quilling: Techniques & Projects for Paper Filigree*; Sterling Publishing Company, Inc.: New York, NY, USA, 1995.
60. Fenton, E.M.; Mascarenas, M.R.; López, G.P.; Sibbett, S.S. Multiplex lateral-flow test strips fabricated by two-dimensional shaping. *ACS Appl. Mater. Interfaces* **2009**, *1*, 124–129. [CrossRef] [PubMed]
61. Nie, J.; Zhang, Y.; Lin, L.; Zhou, C.; Li, S.; Zhang, L.; Li, J. Low-cost fabrication of paper-based microfluidic devices by one-step plotting. *Anal. Chem.* **2012**, *84*, 6331–6335. [CrossRef] [PubMed]
62. Mahmud, M.A.; Blondeel, E.J.; Kaddoura, M.; MacDonald, B.D. Creating compact and microscale features in paper-based devices by laser cutting. *Analyst* **2016**, *141*, 6449–6454. [CrossRef]
63. de Araujo, W.R.; Frasson, C.M.; Ameku, W.A.; Silva, J.R.; Angnes, L.; Paixão, T.R. Single-step reagentless laser scribing fabrication of electrochemical paper-based analytical devices. *Angew. Chem.* **2017**, *129*, 15309–15313. [CrossRef]
64. Theillet, G.; Rubens, A.; Foucault, F.; Dalbon, P.; Rozand, C.; Leparc-Goffart, I.; Bedin, F. Laser-cut paper-based device for the detection of dengue non-structural NS1 protein and specific IgM in human samples. *Arch. Virol.* **2018**, *163*, 1757–1767. [CrossRef]
65. Zhang, R.; Cai, S.; Wu, Q.; Zhu, Y.; Yin, X.; Xu, Y.; Yang, Y.; Chang, H. Laser cutting assisted fabrication of assembled solid-state supercapacitors based on polypyrrole coated paper. *J. Electroanal. Chem.* **2022**, *919*, 116522. [CrossRef]
66. Li, X.; Su, D.; Gu, Y.; Zhang, J.; Li, S.; Xiao, Y.; He, J.; Wang, W.; Li, D. Laser fabrication of epidermal paper-based graphene sensors. *Appl. Mater. Today* **2024**, *36*, 102051. [CrossRef]
67. Martinez, A.W.; Phillips, S.T.; Wiley, B.J.; Gupta, M.; Whitesides, G.M. FLASH: A rapid method for prototyping paper-based microfluidic devices. *Lab Chip* **2008**, *8*, 2146–2150. [CrossRef] [PubMed]
68. Kakoti, A.; Siddiqui, M.F.; Goswami, P. A low cost design and fabrication method for developing a leak proof paper based microfluidic device with customized test zone. *Biomicrofluidics* **2015**, *9*, 026502. [CrossRef] [PubMed]
69. Yu, L.; Shi, Z.Z. Microfluidic paper-based analytical devices fabricated by low-cost photolithography and embossing of Parafilm®. *Lab Chip* **2015**, *15*, 1642–1645. [CrossRef]
70. Park, C.; Han, Y.D.; Kim, H.V.; Lee, J.; Yoon, H.C.; Park, S. Double-sided 3D printing on paper towards mass production of three-dimensional paper-based microfluidic analytical devices (3D-μPADs). *Lab Chip* **2018**, *18*, 1533–1538. [CrossRef] [PubMed]
71. Fu, X.; Xia, B.; Ji, B.; Lei, S.; Zhou, Y. Flow controllable three-dimensional paper-based microfluidic analytical devices fabricated by 3D printing technology. *Anal. Chim. Acta* **2019**, *1065*, 64–70. [CrossRef]
72. He, Y.; Wu, W.b.; Fu, J.z. Rapid fabrication of paper-based microfluidic analytical devices with desktop stereolithography 3D printer. *RSC Adv.* **2015**, *5*, 2694–2701. [CrossRef]
73. Puneeth, S.; Salve, M.; Akshatha, R.; Goel, S. Realization of microfluidic paper-based analytical devices using a 3-D printer: Characterization and optimization. *IEEE T. Device Mat. Re.* **2019**, *19*, 529–536. [CrossRef]
74. Chiang, C.K.; Kurniawan, A.; Kao, C.Y.; Wang, M.J. Single step and mask-free 3D wax printing of microfluidic paper-based analytical devices for glucose and nitrite assays. *Talanta* **2019**, *194*, 837–845. [CrossRef]
75. Silva-Neto, H.A.; Duarte-Junior, G.F.; Rocha, D.S.; Bedioui, F.; Varenne, A.; Coltro, W.K. Recycling 3D Printed Residues for the Development of Disposable Paper-Based Electrochemical Sensors. *ACS Appl. Mater. Interfaces* **2023**, *15*, 14111–14121. [CrossRef] [PubMed]
76. Zaki, M.F.; Wu, Y.H.; Chen, P.C.; Chen, P.S. Determination of psychoactive substances in one microliter plasma using a novel 3D printing microfluidic paper-based column coupled to liquid chromatography-mass spectrometry. *Sens. Actuators B Chem.* **2023**, *393*, 134243. [CrossRef]
77. Dungchai, W.; Chailapakul, O.; Henry, C.S. A low-cost, simple, and rapid fabrication method for paper-based microfluidics using wax screen-printing. *Analyst* **2011**, *136*, 77–82. [CrossRef] [PubMed]
78. Sameenoi, Y.; Nongkai, P.N.; Nouanthavong, S.; Henry, C.S.; Nacapricha, D. One-step polymer screen-printing for microfluidic paper-based analytical device (μPAD) fabrication. *Analyst* **2014**, *139*, 6580–6588. [CrossRef] [PubMed]
79. Lamas-Ardisana, P.J.; Casuso, P.; Fernandez-Gauna, I.; Martínez-Paredes, G.; Jubete, E.; Añorga, L.; Cabañero, G.; Grande, H.J. Disposable electrochemical paper-based devices fully fabricated by screen-printing technique. *Electrochem. Commun.* **2017**, *75*, 25–28. [CrossRef]

80. Tasaengtong, B.; Sameenoi, Y. A one-step polymer screen-printing method for fabrication of microfluidic cloth-based analytical devices. *Microchem. J.* **2020**, *158*, 105078. [CrossRef]
81. Mettakoonpitak, J.; Khongsoun, K.; Wongwan, N.; Kaewbutdee, S.; Siripinyanond, A.; Kuharuk, A.; Henry, C.S. Simple biodegradable plastic screen-printing for microfluidic paper-based analytical devices. *Sens. Actuators B Chem.* **2021**, *331*, 129463. [CrossRef]
82. Xiong, C.; Li, M.; Han, Q.; Zhao, W.; Dai, L.; Ni, Y. Screen printing fabricating patterned and customized full paper-based energy storage devices with excellent photothermal, self-healing, high energy density and good electromagnetic shielding performances. *J. Mater. Sci. Technol.* **2022**, *97*, 190–200. [CrossRef]
83. Oliveira, A.E.; Pereira, A.C.; de Resende, M.A. Fabrication of Low-cost Screen-printed Electrode in Paper Using Conductive Inks of Graphite and Silver/Silver Chloride. *Electroanalysis* **2023**, *35*, e202200093. [CrossRef]
84. Liu, J.; Kong, X.; Wang, H.; Zhang, Y.; Fan, Y. Roll-to-roll wax transfer for rapid and batch fabrication of paper-based microfluidics. *Microfluid. Nanofluidics* **2020**, *24*, 6. [CrossRef]
85. Walia, S.; Bhatnagar, I.; Liu, J.; Mitra, S.K.; Asthana, A. A novel method for fabrication of paper-based microfluidic devices using BSA-ink. *Int. J. Biol. Macromol.* **2021**, *193*, 1617–1622. [CrossRef] [PubMed]
86. Espinosa, A.; Diaz, J.; Vazquez, E.; Acosta, L.; Santiago, A.; Cunci, L. Fabrication of paper-based microfluidic devices using a 3D printer and a commercially-available wax filament. *Talanta Open* **2022**, *6*, 100142. [CrossRef] [PubMed]
87. Tran, B.T.; Rijiravanich, P.; Puttaraksa, N.; Surareungchai, W. Wax gates in laminated microfluidic paper-based immunosensors. *Microchem. J.* **2022**, *178*, 107343. [CrossRef]
88. Brito-Pereira, R.; Ribeiro, C.; Costa, P.; Correia, V.; Cardoso, V.F.; Lanceros-Mendez, S. Graphene Based Printable Conductive Wax for Low-Power Thermal Actuation in Microfluidic Paper-Based Analytical Devices. *Adv. Mater. Technol.* **2023**, *8*, 2300051. [CrossRef]
89. Monju, T.; Hirakawa, M.; Kuboyama, S.; Saiki, R.; Ishida, A. A fabrication technique for paper-based analytical devices via two-sided patterning with thermal-transfer printer and laminator. *Sens. Actuators B Chem.* **2023**, *375*, 132886. [CrossRef]
90. Yamada, K.; Henares, T.G.; Suzuki, K.; Citterio, D. Paper-based inkjet-printed microfluidic analytical devices. *Angew. Chem. Int. Ed.* **2015**, *54*, 5294–5310. [CrossRef]
91. Paschoalino, W.J.; Kogikoski Jr, S.; Barragan, J.T.; Giarola, J.F.; Cantelli, L.; Rabelo, T.M.; Pessanha, T.M.; Kubota, L.T. Emerging considerations for the future development of electrochemical paper-based analytical devices. *ChemElectroChem* **2019**, *6*, 10–30. [CrossRef]
92. Ahmad, M.; Costa Angeli, M.A.; Ibba, P.; Vasquez, S.; Shkodra, B.; Lugli, P.; Petti, L. Paper-Based Printed Antenna: Investigation of Process-Induced and Climatic-Induced Performance Variability. *Adv. Eng. Mater.* **2023**, *25*, 2201703. [CrossRef]
93. Galliani, M.; Ferrari, L.M.; Bouet, G.; Eglin, D.; Ismailova, E. Tailoring inkjet-printed PEDOT: PSS composition toward green, wearable device fabrication. *APL Bioeng.* **2023**, *7*, 016101. [CrossRef]
94. Le, N.N.; Dinh, D.M.T.; Lam, P.H.; Le, A.V.T.; Le, M.T.; Pham, M.D.; Dang, D.M.T.; Dang, C.M. Fabrication of microfluidic paper-based channels by inkjet printing process for analytical applications. *Adv. Nat. Sci. Nanosci. Nanotechnol.* **2023**, *14*, 015015. [CrossRef]
95. Ray, A.; Mohan, J.M.; Amreen, K.; Dubey, S.K.; Javed, A.; Ponnalagu, R.; Goel, S. Ink-jet-printed CuO nanoparticle-enhanced miniaturized paper-based electrochemical platform for hypochlorite sensing. *Appl. Nanosci.* **2023**, *13*, 1855–1861. [CrossRef]
96. Silvestri, A.; Vázquez-Díaz, S.; Misia, G.; Poletti, F.; López-Domene, R.; Pavlov, V.; Zanardi, C.; Cortajarena, A.L.; Prato, M. An Electroactive and Self-Assembling Bio-Ink, based on Protein-Stabilized Nanoclusters and Graphene, for the Manufacture of Fully Inkjet-Printed Paper-Based Analytical Devices. *Small* **2023**, *19*, 2300163. [CrossRef] [PubMed]
97. de Tarso Garcia, P.; Cardoso, T.M.G.; Garcia, C.D.; Carrilho, E.; Coltro, W.K.T. A handheld stamping process to fabricate microfluidic paper-based analytical devices with chemically modified surface for clinical assays. *RSC Adv.* **2014**, *4*, 37637–37644. [CrossRef]
98. He, Y.; Wu, Y.; Xiao, X.; Fu, J.; Xue, G. A low-cost and rapid microfluidic paper-based analytical device fabrication method: Flash foam stamp lithography. *RSC Adv.* **2014**, *4*, 63860–63865. [CrossRef]
99. Guan, Y.; Sun, B. Detection and extraction of heavy metal ions using paper-based analytical devices fabricated via atom stamp printing. *Microsyst. Nanoeng.* **2020**, *6*, 14. [CrossRef]
100. de Araujo, T.A.; de Moraes, N.C.; Petroni, J.M.; Ferreira, V.S.; Lucca, B.G. Simple, fast, and instrumentless fabrication of paper analytical devices by novel contact stamping method based on acrylic varnish and 3D printing. *Microchim. Acta* **2021**, *188*, 437. [CrossRef] [PubMed]
101. Kim, M.; Lee, C.; Jeon, K.; Lee, J.Y.; Kim, Y.J.; Lee, J.G.; Kim, H.; Cho, M.; Kim, D.N. Harnessing a paper-folding mechanism for reconfigurable DNA origami. *Nature* **2023**, *619*, 78–86. [CrossRef] [PubMed]
102. Chen, C.A.; Yeh, W.S.; Tsai, T.T.; Chen, C.F. Three-dimensional origami paper-based device for portable immunoassay applications. *Lab Chip* **2019**, *19*, 598–607. [CrossRef] [PubMed]
103. Tian, T.; An, Y.; Wu, Y.; Song, Y.; Zhu, Z.; Yang, C. Integrated distance-based origami paper analytical device for one-step visualized analysis. *ACS Appl. Mater. Interfaces* **2017**, *9*, 30480–30487. [CrossRef]
104. Ding, J.; Li, B.; Chen, L.; Qin, W. A three-dimensional Origami paper-based device for potentiometric biosensing. *Angew. Chem. Int. Ed.* **2016**, *55*, 13033–13037. [CrossRef]

105. Scida, K.; Li, B.; Ellington, A.D.; Crooks, R.M. DNA detection using origami paper analytical devices. *Anal. Chem.* **2013**, *85*, 9713–9720. [CrossRef]
106. Liu, H.; Xiang, Y.; Lu, Y.; Crooks, R.M. Aptamer-based origami paper analytical device for electrochemical detection of adenosine. *Angew. Chem.* **2012**, *124*, 7031–7034. [CrossRef]
107. Liu, X.; Sun, J.; Tong, Y.; Zhang, M.; Wang, X.; Guo, S.; Han, X.; Zhao, X.; Tang, Q.; Liu, Y. Calligraphy and Kirigami/Origami-Inspired All-Paper Touch–Temperature Sensor with Stimulus Discriminability. *ACS Appl. Mater. Interfaces* **2023**, *15*, 1726–1735. [CrossRef] [PubMed]
108. Chen, X.; Li, Y.; Wang, X.; Yu, H. Origami paper-based stretchable humidity sensor for textile-attachable wearable electronics. *ACS Appl. Mater. Interfaces* **2022**, *14*, 36227–36237. [CrossRef] [PubMed]
109. Kumar, S.; Bhushan, P.; Krishna, V.; Bhattacharya, S. Tapered lateral flow immunoassay based point-of-care diagnostic device for ultrasensitive colorimetric detection of dengue NS1. *Biomicrofluidics* **2018**, *12*, 034104. [CrossRef] [PubMed]
110. Assi, N.; Rypar, T.; Macka, M.; Adam, V.; Vaculovicova, M. Microfluidic paper-based fluorescence sensor for L-homocysteine using a molecularly imprinted polymer and in situ-formed fluorescent quantum dots. *Talanta* **2023**, *255*, 124185. [CrossRef]
111. Goncharov, A.; Joung, H.A.; Ghosh, R.; Han, G.R.; Ballard, Z.S.; Maloney, Q.; Bell, A.; Aung, C.T.Z.; Garner, O.B.; Carlo, D.D.; et al. Deep Learning-Enabled Multiplexed Point-of-Care Sensor using a Paper-Based Fluorescence Vertical Flow Assay. *Small* **2023**, *19*, 2300617. [CrossRef]
112. Ren, Y.; Cao, L.; Zhang, X.; Jiao, R.; Ou, D.; Wang, Y.; Zhang, D.; Shen, Y.; Ling, N.; Ye, Y. A novel fluorescence resonance energy transfer (FRET)-based paper sensor with smartphone for quantitative detection of Vibrio parahaemolyticus. *Food Control* **2023**, *145*, 109412. [CrossRef]
113. Tong, X.; Cai, G.; Xie, L.; Wang, T.; Zhu, Y.; Peng, Y.; Tong, C.; Shi, S.; Guo, Y. Threaded 3D microfluidic paper analytical device-based ratiometric fluorescent sensor for background-free and visual detection of organophosphorus pesticides. *Biosens. Bioelectron.* **2023**, *222*, 114981. [CrossRef]
114. Yuan, M.; Li, C.; Zheng, Y.; Cao, H.; Ye, T.; Wu, X.; Hao, L.; Yin, F.; Yu, J.; Xu, F. A portable multi-channel fluorescent paper-based microfluidic chip based on smartphone imaging for simultaneous detection of four heavy metals. *Talanta* **2024**, *266*, 125112. [CrossRef]
115. Liu, W.; Guo, Y.; Li, H.; Zhao, M.; Lai, Z.; Li, B. A paper-based chemiluminescence device for the determination of ofloxacin. *Spectrochim. Acta Part A Mol. Biomol. Spectrosc.* **2015**, *137*, 1298–1303. [CrossRef]
116. Liu, D.; Ju, C.; Han, C.; Shi, R.; Chen, X.; Duan, D.; Yan, J.; Yan, X. Nanozyme chemiluminescence paper test for rapid and sensitive detection of SARS-CoV-2 antigen. *Biosens. Bioelectron.* **2021**, *173*, 112817. [CrossRef] [PubMed]
117. Al Yahyai, I.; Hassanzadeh, J.; Al-Lawati, H.A. A novel and selective multi-emission chemiluminescence system for the quantification of deltamethrin in food samples. *Sens. Actuators B Chem.* **2021**, *327*, 128927. [CrossRef]
118. Delaney, J.L.; Hogan, C.F.; Tian, J.; Shen, W. Electrogenerated chemiluminescence detection in paper-based microfluidic sensors. *Anal. Chem.* **2011**, *83*, 1300–1306. [CrossRef] [PubMed]
119. Liu, H.; Zhou, X.; Liu, W.; Yang, X.; Xing, D. Paper-based bipolar electrode electrochemiluminescence switch for label-free and sensitive genetic detection of pathogenic bacteria. *Anal. Chem.* **2016**, *88*, 10191–10197. [CrossRef] [PubMed]
120. Valentine, C.J.; Takagishi, K.; Umezu, S.; Daly, R.; De Volder, M. Paper-based electrochemical sensors using paper as a scaffold to create porous carbon nanotube electrodes. *ACS Appl. Mater. Interfaces* **2020**, *12*, 30680–30685. [CrossRef] [PubMed]
121. He, Z.; Huang, J.; Shen, W.; Lei, X.; Zhang, Y.; Zhu, L.; Shen, X.; Zhang, D.; Yu, D.; Zhou, M. A Paper-Based Fluorescent Sensor for Rapid Early Screening of Oral Squamous Cell Carcinoma. *ACS Appl. Mater. Interfaces* **2023**, *15*, 24913–24922. [CrossRef]
122. Li, Z.; Zhu, M.; Li, F.; Li, Z.; Zhao, A.; Haghighatbin, M.A.; Cui, H. Microfluidic paper chip based multicolor chemiluminescence sensing strategy for discrimination of antioxidants. *Sens. Actuators B Chem.* **2023**, *393*, 134166. [CrossRef]
123. Wang, F.; Liu, Y.; Fu, C.; Li, N.; Du, M.; Zhang, L.; Ge, S.; Yu, J. Paper-based bipolar electrode electrochemiluminescence platform for detection of multiple miRNAs. *Anal. Chem.* **2020**, *93*, 1702–1708. [CrossRef]
124. Tay, L.L.; Poirier, S.; Ghaemi, A.; Hulse, J.; Wang, S. Paper-based surface-enhanced Raman spectroscopy sensors for field applications. *J. Raman Spectrosc.* **2021**, *52*, 563–572. [CrossRef]
125. Eskandari, V.; Sahbafar, H.; Karooby, E.; Heris, M.H.; Mehmandoust, S.; Razmjoue, D.; Hadi, A. Surface-Enhanced Raman scattering (SERS) filter paper substrates decorated with silver nanoparticles for the detection of molecular vibrations of Acyclovir drug. *Spectrochim. Acta Part A Mol. Biomol. Spectrosc.* **2023**, *298*, 122762. [CrossRef] [PubMed]
126. Cate, D.M.; Adkins, J.A.; Mettakoonpitak, J.; Henry, C.S. Recent developments in paper-based microfluidic devices. *Anal. Chem.* **2015**, *87*, 19–41. [CrossRef] [PubMed]
127. Magro, L.; Escadafal, C.; Garneret, P.; Jacquelin, B.; Kwasiborski, A.; Manuguerra, J.C.; Monti, F.; Sakuntabhai, A.; Vanhomwegen, J.; Lafaye, P.; et al. Paper microfluidics for nucleic acid amplification testing (NAAT) of infectious diseases. *Lab Chip* **2017**, *17*, 2347–2371. [CrossRef] [PubMed]
128. Li, J.; Macdonald, J. Multiplexed lateral flow biosensors: Technological advances for radically improving point-of-care diagnoses. *Biosens. Bioelectron.* **2016**, *83*, 177–192. [CrossRef]
129. Arias-Alpízar, K.; Sánchez-Cano, A.; Prat-Trunas, J.; de la Serna Serna, E.; Alonso, O.; Sulleiro, E.; Sánchez-Montalvá, A.; Diéguez, A.; Baldrich, E. Malaria quantitative POC testing using magnetic particles, a paper microfluidic device and a hand-held fluorescence reader. *Biosens. Bioelectron.* **2022**, *215*, 114513. [CrossRef] [PubMed]

130. Prat-Trunas, J.; Arias-Alpizar, K.; Álvarez-Carulla, A.; Orio-Tejada, J.; Molina, I.; Sánchez-Montalvá, A.; Colomer-Farrarons, J.; Del Campo, F.; Miribel-Català, P.L.; Baldrich, E. Paper-based microfluidic electro-analytical device (PMED) for magneto-assay automation: Towards generic point-of-care diagnostic devices. *Biosens. Bioelectron.* **2024**, *246*, 115875. [CrossRef] [PubMed]
131. Biswas, P.; Mukunthan Sulochana, G.N.; Banuprasad, T.N.; Goyal, P.; Modak, D.; Ghosh, A.K.; Chakraborty, S. All-Serotype Dengue Virus Detection through Multilayered Origami-Based Paper/Polymer Microfluidics. *ACS Sens.* **2022**, *7*, 3720–3729. [CrossRef] [PubMed]
132. Le, T.N.; Hsiao, W.W.W.; Cheng, Y.Y.; Lee, C.C.; Huynh, T.T.; Pham, D.M.; Chen, M.; Jen, M.W.; Chang, H.C.; Chiang, W.H. Spin-Enhanced Lateral Flow Immunoassay for High-Sensitivity Detection of Nonstructural Protein NS1 Serotypes of the Dengue Virus. *Anal. Chem.* **2022**, *94*, 17819–17826. [CrossRef]
133. Trakoolwilaiwan, T.; Takeuchi, Y.; Leung, T.S.; Sebek, M.; Storozhuk, L.; Nguyen, L.; Thanh, N.T.K.; et al. Development of a thermochromic lateral flow assay to improve sensitivity for dengue virus serotype 2 NS1 detection. *Nanoscale* **2023**, *15*, 12915–12925. [CrossRef]
134. Kaarj, K.; Akarapipad, P.; Yoon, J.Y. Simpler, faster, and sensitive Zika virus assay using smartphone detection of loop-mediated isothermal amplification on paper microfluidic chips. *Sci. Rep.* **2018**, *8*, 12438. [CrossRef]
135. Seok, Y.; Batule, B.S.; Kim, M.G. Lab-on-paper for all-in-one molecular diagnostics (LAMDA) of zika, dengue, and chikungunya virus from human serum. *Biosens. Bioelectron.* **2020**, *165*, 112400. [CrossRef]
136. Suvanasuthi, R.; Chimnaronk, S.; Promptmas, C. 3D printed hydrophobic barriers in a paper-based biosensor for point-of-care detection of dengue virus serotypes. *Talanta* **2022**, *237*, 122962. [CrossRef] [PubMed]
137. Karlikow, M.; da Silva, S.J.R.; Guo, Y.; Cicek, S.; Krokovsky, L.; Homme, P.; Xiong, Y.; Xu, T.; Calderón-Peláez, M.A.; Camacho-Ortega, S.; et al. Field validation of the performance of paper-based tests for the detection of the Zika and chikungunya viruses in serum samples. *Nat. Biomed. Eng.* **2022**, *6*, 246–256. [CrossRef] [PubMed]
138. Gong, H.; Tang, L.; Chen, C.; Chen, F.; Cai, C. Portable paper-based molecularly imprinted sensor for visual real-time detection of influenza virus H5N1. *Chem. Eng. J.* **2023**, *477*, 146990. [CrossRef]
139. Tavakoli, H.; Hirth, E.; Luo, M.; Timilsina, S.S.; Dou, M.; Dominguez, D.C.; Li, X. A microfluidic fully paper-based analytical device integrated with loop-mediated isothermal amplification and nano-biosensors for rapid, sensitive, and specific quantitative detection of infectious diseases. *Lab Chip* **2022**, *22*, 4693–4704. [CrossRef] [PubMed]
140. Choi, J.R.; Hu, J.; Tang, R.; Gong, Y.; Feng, S.; Ren, H.; Wen, T.; Li, X.; Abas, W.A.B.W.; Pingguan-Murphy, B.; et al. An integrated paper-based sample-to-answer biosensor for nucleic acid testing at the point of care. *Lab Chip* **2016**, *16*, 611–621. [CrossRef] [PubMed]
141. Tang, R.; Yang, H.; Gong, Y.; You, M.; Liu, Z.; Choi, J.R.; Wen, T.; Qu, Z.; Mei, Q.; Xu, F. A fully disposable and integrated paper-based device for nucleic acid extraction, amplification and detection. *Lab Chip* **2017**, *17*, 1270–1279. [CrossRef]
142. Natarajan, S.; Su, F.; Jayaraj, J.; Shah, M.I.I.; Huang, Y. A paper microfluidics-based fluorescent lateral flow immunoassay for point-of-care diagnostics of non-communicable diseases. *Analyst* **2019**, *144*, 6291–6303. [CrossRef] [PubMed]
143. Shalaby, A.A.; Tsao, C.W.; Ishida, A.; Maeki, M.; Tokeshi, M. Microfluidic paper-based analytical devices for cancer diagnosis. *Sens. Actuators B Chem.* **2023**, *379*, 133243. [CrossRef]
144. Sana, T.; Sharma, P.; Khanuja, M.; Narang, J. Detection of Uterine Cancer Biomarker EGFR through an Aptasensor Utilizing a Carbon Electrode Modified with Silver Nanowires. *Mater. Chem. Phys.* **2024**, *319*, 129412. [CrossRef]
145. Gutiérrez-Capitán, M.; Sanchís, A.; Carvalho, E.O.; Baldi, A.; Vilaplana, L.; Cardoso, V.F.; Calleja, Á.; Wei, M.; de la Rica, R.; Hoyo, J.; et al. Engineering a Point-of-Care Paper-Microfluidic Electrochemical Device Applied to the Multiplexed Quantitative Detection of Biomarkers in Sputum. *ACS Sens.* **2023**, *8*, 3032–3042. [CrossRef] [PubMed]
146. Rohrman, B.A.; Richards-Kortum, R.R. A paper and plastic device for performing recombinase polymerase amplification of HIV DNA. *Lab Chip* **2012**, *12*, 3082–3088. [CrossRef] [PubMed]
147. Fu, X.; Cheng, Z.; Yu, J.; Choo, P.; Chen, L.; Choo, J. A SERS-based lateral flow assay biosensor for highly sensitive detection of HIV-1 DNA. *Biosens. Bioelectron.* **2016**, *78*, 530–537. [CrossRef] [PubMed]
148. Tsao, Y.T.; Yang, C.Y.; Wen, Y.C.; Chang, T.C.; Matsuura, K.; Chen, Y.; Cheng, C.M. Point-of-care semen analysis of patients with infertility via smartphone and colorimetric paper-based diagnostic device. *Bioeng. Transl. Med.* **2021**, *6*, e10176. [CrossRef] [PubMed]
149. Sarabi, M.R.; Yigci, D.; Alseed, M.M.; Mathyk, B.A.; Ata, B.; Halicigil, C.; Tasoglu, S. Disposable paper-based microfluidics for fertility testing. *iScience* **2022**, *25*, 104986. [CrossRef] [PubMed]
150. Kumar, S.; Bhushan, P.; Bhattacharya, S. Development of a paper-based analytical device for colorimetric detection of uric acid using gold nanoparticles–graphene oxide (AuNPs–GO) conjugates. *Anal. Methods* **2016**, *8*, 6965–6973. [CrossRef]
151. Kumar, S.; Bhushan, P.; Bhattacharya, S. Facile synthesis of Au@ Ag–hemin decorated reduced graphene oxide sheets: A novel peroxidase mimetic for ultrasensitive colorimetric detection of hydrogen peroxide and glucose. *RSC Adv.* **2017**, *7*, 37568–37577. [CrossRef]
152. Pérez-Rodríguez, M.; del Pilar Cañizares-Macías, M. A prototype microfluidic paper-based chromatic device for simultaneous determination of copper (II) and zinc (II) in urine. *Talanta Open* **2023**, *7*, 100178. [CrossRef]
153. Bezdekova, J.; Plevova, M.; Nejdl, L.; Macka, M.; Masarik, M.; Pacik, D.; Adam, V.; Vaculovicova, M. Prostate cancer diagnosed and staged using UV-irradiated urine samples and a paper-based analytical device. *Sens. Actuators B Chem.* **2024**, *403*, 135146. [CrossRef]

154. Chaiyo, S.; Kunpatee, K.; Kalcher, K.; Yakoh, A.; Pungjunun, K. 3D Paper-Based Device Integrated with a Battery-Less NFC Potentiostat for Nonenzymatic Detection of Cholesterol. *ACS Meas. Sci. Au* **2024**. [CrossRef]
155. Ozer, T.; Henry, C.S. Paper-based analytical devices for virus detection: Recent strategies for current and future pandemics. *TrAC Trends Anal. Chem.* **2021**, *144*, 116424. [CrossRef] [PubMed]
156. Adrover-Jaume, C.; Alba-Patino, A.; Clemente, A.; Santopolo, G.; Vaquer, A.; Russell, S.M.; Baron, E.; Del Campo, M.D.M.G.; Ferrer, J.M.; Berman-Riu, M.; et al. Paper biosensors for detecting elevated IL-6 levels in blood and respiratory samples from COVID-19 patients. *Sens. Actuators B Chem.* **2021**, *330*, 129333. [CrossRef]
157. Song, Q.; Sun, X.; Dai, Z.; Gao, Y.; Gong, X.; Zhou, B.; Wu, J.; Wen, W. Point-of-care testing detection methods for COVID-19. *Lab Chip* **2021**, *21*, 1634–1660. [CrossRef] [PubMed]
158. Zhang, T.; Deng, R.; Wang, Y.; Wu, C.; Zhang, K.; Wang, C.; Gong, N.; Ledesma-Amaro, R.; Teng, X.; Yang, C.; et al. A paper-based assay for the colorimetric detection of SARS-CoV-2 variants at single-nucleotide resolution. *Nat. Biomed. Eng.* **2022**, *6*, 957–967. [CrossRef] [PubMed]
159. Fabiani, L.; Mazzaracchio, V.; Moscone, D.; Fillo, S.; De Santis, R.; Monte, A.; Amatore, D.; Lista, F.; Arduini, F. based immunoassay based on 96-well wax-printed paper plate combined with magnetic beads and colorimetric smartphone-assisted measure for reliable detection of SARS-CoV-2 in saliva. *Biosens. Bioelectron.* **2022**, *200*, 113909. [CrossRef] [PubMed]
160. Wang, Q.; Chen, Z.; Yang, H. Colorimetric detection of SARS-CoV-2 variants with paper-based analytical devices. *MedComm–Biomater. Appl.* **2023**, *2*, e35. [CrossRef]
161. Sen, A.; Masetty, M.; Weerakoon, S.; Morris, C.; Yadav, J.S.; Apewokin, S.; Trannguyen, J.; Broom, M.; Priye, A. Paper-based loop-mediated isothermal amplification and CRISPR integrated platform for on-site nucleic acid testing of pathogens. *Biosens. Bioelectron.* **2024**, *257*, 116292. [CrossRef] [PubMed]
162. Lee, A.S.; Kim, S.M.; Kim, K.R.; Park, C.; Lee, D.G.; Heo, H.R.; Cha, H.J.; Kim, C.S. A colorimetric lateral flow immunoassay based on oriented antibody immobilization for sensitive detection of SARS-CoV-2. *Sens. Actuators B Chem.* **2023**, *379*, 133245. [CrossRef]
163. Jin, H.; Abu-Raya, Y.S.; Haick, H. Advanced materials for health monitoring with skin-based wearable devices. *Adv. Healthc. Mater.* **2017**, *6*, 1700024. [CrossRef]
164. Xu, Y.; Fei, Q.; Page, M.; Zhao, G.; Ling, Y.; Stoll, S.B.; Yan, Z. Paper-based wearable electronics. *iScience* **2021**, *24*, 102736. [CrossRef]
165. Deroco, P.B.; Wachholz Junior, D.; Kubota, L.T. Paper-based Wearable Electrochemical Sensors: A New Generation of Analytical Devices. *Electroanalysis* **2023**, *35*, e202200177. [CrossRef]
166. Bandodkar, A.J.; Hung, V.W.; Jia, W.; Valdés-Ramírez, G.; Windmiller, J.R.; Martinez, A.G.; Ramírez, J.; Chan, G.; Kerman, K.; Wang, J. Tattoo-based potentiometric ion-selective sensors for epidermal pH monitoring. *Analyst* **2013**, *138*, 123–128. [CrossRef] [PubMed]
167. Nassar, J.M.; Mishra, K.; Lau, K.; Aguirre-Pablo, A.A.; Hussain, M.M. Recyclable nonfunctionalized paper-based ultralow-cost wearable health monitoring system. *Adv. Mater. Technol.* **2017**, *2*, 1600228. [CrossRef]
168. Pal, A.; Goswami, D.; Cuellar, H.E.; Castro, B.; Kuang, S.; Martinez, R.V. Early detection and monitoring of chronic wounds using low-cost, omniphobic paper-based smart bandages. *Biosens. Bioelectron.* **2018**, *117*, 696–705. [CrossRef] [PubMed]
169. Xu, Y.; Zhao, G.; Zhu, L.; Fei, Q.; Zhang, Z.; Chen, Z.; An, F.; Chen, Y.; Ling, Y.; Guo, P.; et al. Pencil–paper on-skin electronics. *Proc. Natl. Acad. Sci. USA* **2020**, *117*, 18292–18301. [CrossRef]
170. Mogera, U.; Guo, H.; Namkoong, M.; Rahman, M.S.; Nguyen, T.; Tian, L. Wearable plasmonic paper–based microfluidics for continuous sweat analysis. *Sci. Adv.* **2022**, *8*, eabn1736. [CrossRef]
171. Yang, M.; Sun, N.; Lai, X.; Wu, J.; Wu, L.; Zhao, X.; Feng, L. Paper-Based Sandwich-Structured Wearable Sensor with Sebum Filtering for Continuous Detection of Sweat pH. *ACS Sens.* **2023**, *8*, 176–186. [CrossRef]
172. Fiore, L.; Mazzaracchio, V.; Serani, A.; Fabiani, G.; Fabiani, L.; Volpe, G.; Moscone, D.; Bianco, G.M.; Occhiuzzi, C.; Marrocco, G.; et al. Microfluidic paper-based wearable electrochemical biosensor for reliable cortisol detection in sweat. *Sens. Actuators B Chem.* **2023**, *379*, 133258. [CrossRef]
173. Cheng, Y.; Feng, S.; Ning, Q.; Li, T.; Xu, H.; Sun, Q.; Cui, D.; Wang, K. Dual-signal readout paper-based wearable biosensor with a 3D origami structure for multiplexed analyte detection in sweat. *Microsyst. Nanoeng.* **2023**, *9*, 36. [CrossRef]
174. Lai, Q.T.; Liang, H.Q.; Tang, X.G.; Zhang, D.; Roy, V.A.; Sun, Q.J. Printing Paper-Derived Ultralight and Highly Sensitive E-Skin for Health Monitoring and Information Encryption. *J. Alloys Compd.* **2024**, *976*, 173411. [CrossRef]
175. Niu, G.; Wang, Z.; Xue, Y.; Yan, J.; Dutta, A.; Chen, X.; Wang, Y.; Liu, C.; Du, S.; Guo, L.; et al. Pencil-on-paper humidity sensor treated with NaCl solution for health monitoring and skin characterization. *Nano Lett.* **2023**, *23*, 1252–1260. [CrossRef]
176. Karmakar, R.S.; Huang, J.F.; Chu, C.P.; Mai, M.H.; Chao, J.I.; Liao, Y.C.; Lu, Y.W. Origami-Inspired Conductive Paper-Based Folded Pressure Sensor with Interconnection Scaling at the Crease for Novel Wearable Applications. *ACS Appl. Mater. Interfaces* **2024**, *16*, 4231–4241. [CrossRef]
177. Busin, V.; Burgess, S.; Shu, W. A hybrid paper-based microfluidic platform toward veterinary P-ELISA. *Sens. Actuators B Chem.* **2018**, *273*, 536–542. [CrossRef]
178. Zhao, G.; Wang, H.; Hou, P.; He, C.; He, H. Rapid visual detection of *Mycobacterium avium* subsp. *paratuberculosis* by recombinase polymerase amplification combined with a lateral flow dipstick. *J. Vet. Sci.* **2018**, *19*, 242–250. [CrossRef]

179. Yang, Z.; Xu, G.; Reboud, J.; Ali, S.A.; Kaur, G.; McGiven, J.; Boby, N.; Gupta, P.K.; Chaudhuri, P.; Cooper, J.M. Rapid veterinary diagnosis of bovine reproductive infectious diseases from semen using paper-origami DNA microfluidics. *ACS Sens.* **2018**, *3*, 403–409. [CrossRef]
180. Huang, L.; Xiao, W.; Xu, T.; Chen, H.; Jin, Z.; Zhang, Z.; Song, Q.; Tang, Y. Miniaturized paper-based smartphone biosensor for differential diagnosis of wild-type pseudorabies virus infection versus vaccination immunization. *Sens. Actuators B Chem.* **2021**, *327*, 128893. [CrossRef]
181. Feng, S.; Hua, M.Z.; Roopesh, M.; Lu, X. Rapid detection of three mycotoxins in animal feed materials using competitive ELISA-based origami microfluidic paper analytical device (μPAD). *Anal. Bioanal. Chem.* **2023**, *415*, 1943–1951. [CrossRef]
182. Li, S.; Liu, Y.; Wang, Y.; Wang, M.; Liu, C.; Wang, Y. Rapid detection of *Brucella* spp. and elimination of carryover using multiple cross displacement amplification coupled with nanoparticles-based lateral flow biosensor. *Front. Cell. Infect. Microbiol.* **2019**, *9*, 78. [CrossRef]
183. Jung, H.; Park, S.H.; Lee, J.; Lee, B.; Park, J.; Seok, Y.; Choi, J.H.; Kim, M.G.; Song, C.S.; Lee, J. A size-selectively biomolecule-immobilized nanoprobe-based chemiluminescent lateral flow immunoassay for detection of avian-origin viruses. *Anal. Chem.* **2020**, *93*, 792–800. [CrossRef]
184. Meredith, N.A.; Quinn, C.; Cate, D.M.; Reilly, T.H.; Volckens, J.; Henry, C.S. Paper-based analytical devices for environmental analysis. *Analyst* **2016**, *141*, 1874–1887. [CrossRef]
185. Colozza, N.; Caratelli, V.; Moscone, D.; Arduini, F. Paper-based devices as new smart analytical tools for sustainable detection of environmental pollutants. *Case Stud. Chem. Environ. Eng.* **2021**, *4*, 100167. [CrossRef]
186. Alahmad, W.; Cetinkaya, A.; Kaya, S.I.; Varanusupakul, P.; Ozkan, S.A. Electrochemical paper-based analytical devices for environmental analysis: Current trends and perspectives. *Trends Environ. Anal. Chem.* **2023**, *40*, e00220. [CrossRef]
187. Suo, Z.; Liang, R.; Liu, R.; Wei, M.; He, B.; Jiang, L.; Sun, X.; Jin, H. A convenient paper-based fluorescent aptasensor for high-throughput detection of Pb^{2+} in multiple real samples (water-soil-food). *Anal. Chim. Acta* **2023**, *1239*, 340714. [CrossRef] [PubMed]
188. Yu, X.; Chang, W.; Zhang, H.; Cai, Z.; Yang, Y.; Zeng, C. Visual and Real-Time Monitoring of Cd^{2+} in Water, Rice, and Rice Soil with Test Paper Based on [2 + 2] Lanthanide Clusters. *Inorg. Chem.* **2023**, *62*, 6387–6396. [CrossRef]
189. *GB 2762-2022*; National Food Safety Standard Maximum Levels of Contaminants in Foods. State Administration for Market Regulation, National Health Commission: Beijing, China, 2022. Available online: https://food.chemlinked.com/database/view/3048 (accessed on 12 May 2024).
190. Wang, S.; Ge, L.; Li, L.; Yan, M.; Ge, S.; Yu, J. Molecularly imprinted polymer grafted paper-based multi-disk micro-disk plate for chemiluminescence detection of pesticide. *Biosens. Bioelectron.* **2013**, *50*, 262–268. [CrossRef] [PubMed]
191. Ma, Z.; Li, Y.; Lu, C.; Li, M. On-site Screening Method for Bioavailability Assessment of the Organophosphorus Pesticide, Methyl Parathion, and Its Primary Metabolite in Soils by Paper Strip Biosensor. *J. Hazard. Mater.* **2023**, *457*, 131725. [CrossRef] [PubMed]
192. Cioffi, A.; Mancini, M.; Gioia, V.; Cinti, S. Office paper-based electrochemical strips for organophosphorus pesticide monitoring in agricultural soil. *Environ. Sci. Technol.* **2021**, *55*, 8859–8865. [CrossRef] [PubMed]
193. Zhang, X.; Wang, Z.; Huang, X.; Huang, Q.; Wen, Y.; Li, B.; Holmes, M.; Shi, J.; Zou, X. Uniform stain pattern of robust MOF-mediated probe for flexible paper-based colorimetric sensing toward environmental pesticide exposure. *Chem. Eng. J.* **2023**, *451*, 138928. [CrossRef]
194. Ranveer, S.A.; Harshitha, C.; Dasriya, V.; Tehri, N.; Kumar, N.; Raghu, H. Assessment of developed paper strip based sensor with pesticide residues in different dairy environmental samples. *Curr. Res. Food Sci.* **2023**, *6*, 100416. [CrossRef]
195. Caratelli, V.; Fegatelli, G.; Moscone, D.; Arduini, F. A paper-based electrochemical device for the detection of pesticides in aerosol phase inspired by nature: A flower-like origami biosensor for precision agriculture. *Biosens. Bioelectron.* **2022**, *205*, 114119. [CrossRef]
196. Taudte, R.V.; Beavis, A.; Wilson-Wilde, L.; Roux, C.; Doble, P.; Blanes, L. A portable explosive detector based on fluorescence quenching of pyrene deposited on coloured wax-printed μPADs. *Lab Chip* **2013**, *13*, 4164–4172. [CrossRef]
197. Wang, J.; Yang, L.; Liu, B.; Jiang, H.; Liu, R.; Yang, J.; Han, G.; Mei, Q.; Zhang, Z. Inkjet-printed silver nanoparticle paper detects airborne species from crystalline explosives and their ultratrace residues in open environment. *Anal. Chem.* **2014**, *86*, 3338–3345. [CrossRef]
198. Ueland, M.; Blanes, L.; Taudte, R.V.; Stuart, B.H.; Cole, N.; Willis, P.; Roux, C.; Doble, P. Capillary-driven microfluidic paper-based analytical devices for Lab on a Chip screening of explosive residues in soil. *J. Chromatogr. A* **2016**, *1436*, 28–33. [CrossRef]
199. de Araujo, W.R.; Cardoso, T.M.; da Rocha, R.G.; Santana, M.H.; Munoz, R.A.; Richter, E.M.; Paixão, T.R.; Coltro, W.K. Portable analytical platforms for forensic chemistry: A review. *Anal. Chim. Acta* **2018**, *1034*, 1–21. [CrossRef]
200. Raucci, A.; Miglione, A.; Cimmino, W.; Cioffi, A.; Singh, S.; Spinelli, M.; Amoresano, A.; Musile, G.; Cinti, S. Technical Evaluation of a Paper-Based Electrochemical Strip to Measure Nitrite Ions in the Forensic Field. *ACS Meas. Sci. Au* **2023**, *4*, 136–143. [CrossRef]
201. Hunter, P.R.; Zmirou-Navier, D.; Hartemann, P. Estimating the impact on health of poor reliability of drinking water interventions in developing countries. *Sci. Total. Environ.* **2009**, *407*, 2621–2624. [CrossRef]
202. Adu-Manu, K.S.; Tapparello, C.; Heinzelman, W.; Katsriku, F.A.; Abdulai, J.D. Water quality monitoring using wireless sensor networks: Current trends and future research directions. *ACM Trans. Sens. Netw.* **2017**, *13*, 1–41. [CrossRef]
203. Vikesland, P.J. Nanosensors for water quality monitoring. *Nat. Nanotec.* **2018**, *13*, 651–660. [CrossRef]

204. Liu, F.; Nordin, A.; Li, F.; Voiculescu, I. A lab-on-chip cell-based biosensor for label-free sensing of water toxicants. *Lab Chip* **2014**, *14*, 1270–1280. [CrossRef]
205. Liu, B.; Lei, Y.; Li, B. A batch-mode cube microbial fuel cell based "shock" biosensor for wastewater quality monitoring. *Biosens. Bioelectron.* **2014**, *62*, 308–314. [CrossRef]
206. Sicard, C.; Glen, C.; Aubie, B.; Wallace, D.; Jahanshahi-Anbuhi, S.; Pennings, K.; Daigger, G.T.; Pelton, R.; Brennan, J.D.; Filipe, C.D. Tools for water quality monitoring and mapping using paper-based sensors and cell phones. *Water Res.* **2015**, *70*, 360–369. [CrossRef] [PubMed]
207. İncel, A.; Akın, O.; Çağır, A.; Yıldız, Ü.H.; Demir, M.M. Smart phone assisted detection and quantification of cyanide in drinking water by paper based sensing platform. *Sens. Actuators B Chem.* **2017**, *252*, 886–893. [CrossRef]
208. Cho, J.H.; Gao, Y.; Choi, S. A portable, single-use, paper-based microbial fuel cell sensor for rapid, on-site water quality monitoring. *Sensors* **2019**, *19*, 5452. [CrossRef]
209. Cho, J.H.; Gao, Y.; Ryu, J.; Choi, S. Portable, disposable, paper-based microbial fuel cell sensor utilizing freeze-dried bacteria for in situ water quality monitoring. *ACS Omega* **2020**, *5*, 13940–13947. [CrossRef] [PubMed]
210. Charbaji, A.; Heidari-Bafroui, H.; Anagnostopoulos, C.; Faghri, M. A new paper-based microfluidic device for improved detection of nitrate in water. *Sensors* **2020**, *21*, 102. [CrossRef] [PubMed]
211. Jaballah, M.B.; Karrat, A.; Amine, A.; Dridi, C. Immobilization of diphenylcarbazide on paper-based analytical devices for the pre-concentration and detection of chromium VI in water samples. *Talanta* **2023**, *265*, 124889. [CrossRef] [PubMed]
212. da Silva, V.A.O.P.; de Freitas, R.C.; de Oliveira, P.R.; Moreira, R.C.; Marcolino-Júnior, L.H.; Bergamini, M.F.; Coltro, W.K.; Janegitz, B.C. Microfluidic paper-based device integrated with smartphone for point-of-use colorimetric monitoring of water quality index. *Measurement* **2020**, *164*, 108085. [CrossRef]
213. Xiong, X.; Guo, C.; Yan, G.; Han, B.; Wu, Z.; Chen, Y.; Xu, S.; Shao, P.; Song, H.; Xu, X.; et al. Simultaneous Cross-type Detection of Water Quality Indexes via a Smartphone-App Integrated Microfluidic Paper-Based Platform. *ACS Omega* **2022**, *7*, 44338–44345. [CrossRef] [PubMed]
214. Lin, J.H.; Chen, S.J.; Lee, J.E.; Chu, W.Y.; Yu, C.J.; Chang, C.C.; Chen, C.F. The detection of Mercury (II) ions using fluorescent gold nanoclusters on a portable paper-based device. *Chem. Eng. J.* **2022**, *430*, 133070. [CrossRef]
215. Aguiar, J.I.; Ribeiro, S.O.; Leite, A.; Rangel, M.; Rangel, A.O.; Mesquita, R.B. Use of a rhodamine-based chelator in a microfluidic paper-based analytical device for the in-situ copper quantification in natural waters. *Talanta* **2024**, *271*, 125683. [CrossRef]
216. Uhlikova, N.; Almeida, M.I.G.; McKelvie, I.D.; Kolev, S.D. Microfluidic paper-based analytical device for the speciation of inorganic nitrogen species. *Talanta* **2024**, *271*, 125671. [CrossRef] [PubMed]
217. Thangjitsirisin, K.; Seeharaj, P.; Choengchan, N. Superhydrophobic eggshell for fabrication of hydrophobic barrier of paper-based analytical device for colorimetric determination of ammonium ion in water. *Microchem. J.* **2024**, *200*, 110464. [CrossRef]
218. Sameenoi, Y.; Panymeesamer, P.; Supalakorn, N.; Koehler, K.; Chailapakul, O.; Henry, C.S.; Volckens, J. Microfluidic paper-based analytical device for aerosol oxidative activity. *Environ. Sci. Technol.* **2013**, *47*, 932–940. [CrossRef] [PubMed]
219. Khachornsakkul, K.; Hung, K.H.; Chang, J.J.; Dungchai, W.; Chen, C.H. A rapid and highly sensitive paper-based colorimetric device for the on-site screening of ammonia gas. *Analyst* **2021**, *146*, 2919–2927. [CrossRef] [PubMed]
220. Mettakoonpitak, J.; Sawatdichai, N.; Thepnuan, D.; Siripinyanond, A.; Henry, C.S.; Chantara, S. Microfluidic paper-based analytical devices for simultaneous detection of oxidative potential and copper in aerosol samples. *Microchim. Acta* **2023**, *190*, 241. [CrossRef] [PubMed]
221. De Matteis, V.; Cascione, M.; Fella, G.; Mazzotta, L.; Rinaldi, R. Colorimetric paper-based device for hazardous compounds detection in air and water: A proof of concept. *Sensors* **2020**, *20*, 5502. [CrossRef] [PubMed]
222. Bordbar, M.M.; Tashkhourian, J.; Hemmateenejad, B. Paper-Based Optical Nose Made with Bimetallic Nanoparticles for Monitoring Ignitable Liquids in Gasoline. *ACS Appl. Mater. Interfaces* **2022**, *14*, 8333–8342. [CrossRef] [PubMed]
223. Arroyo, P.; Gómez-Suárez, J.; Suárez, J.I.; Lozano, J. Low-Cost Air Quality Measurement System Based on Electrochemical and PM Sensors with Cloud Connection. *Sensors* **2021**, *21*, 6228. [CrossRef]
224. Li, G.; Yuan, H.; Mou, J.; Dai, E.; Zhang, H.; Li, Z.; Zhao, Y.; Dai, Y.; Zhang, X. Electrochemical detection of nitrate with carbon nanofibers and copper co-modified carbon fiber electrodes. *Compos. Commun.* **2022**, *29*, 101043. [CrossRef]
225. Mettakoonpitak, J.; Volckens, J.; Henry, C.S. Janus electrochemical paper-based analytical devices for metals detection in aerosol samples. *Anal. Chem.* **2019**, *92*, 1439–1446. [CrossRef]
226. Davis, D.; Narayanan, S.K.; Ajeev, A.; Nair, J.; Jeeji, J.; Vijayan, A.; Viyyur Kuttyadi, M.; Nelliparambil Sathian, A.; Arulraj, A.K. Flexible Paper-Based Room-Temperature Acetone Sensors with Ultrafast Regeneration. *ACS Appl. Mater. Interfaces* **2023**, *15*, 25734–25743. [CrossRef]
227. Qin, X.; Wu, T.; Zhu, Y.; Shan, X.; Liu, C.; Tao, N. A paper based milli-cantilever sensor for detecting hydrocarbon gases via smartphone camera. *Anal. Chem.* **2020**, *92*, 8480–8486. [CrossRef] [PubMed]
228. Zhuang, J.; Zhao, Z.; Lian, K.; Yin, L.; Wang, J.; Man, S.; Liu, G.; Ma, L. SERS-based CRISPR/Cas assay on microfluidic paper analytical devices for supersensitive detection of pathogenic bacteria in foods. *Biosens. Bioelectron.* **2022**, *207*, 114167. [CrossRef]
229. Lin, X.; Li, C.; Tong, X.; Duan, N.; Wang, Z.; Wu, S. A portable paper-based aptasensor for simultaneous visual detection of two mycotoxins in corn flour using dual-color upconversion nanoparticles and Cu-TCPP nanosheets. *Food Chem.* **2023**, *404*, 134750. [CrossRef] [PubMed]

230. He, Y.; Wang, H.; Yu, Z.; Tang, X.; Zhou, M.; Guo, Y.; Xiong, B. A disposable immunosensor array using cellulose paper assembled chemiresistive biosensor for simultaneous monitoring of mycotoxins AFB1 and FB1. *Talanta* **2024**, *276*, 126145. [CrossRef] [PubMed]
231. Xiang, J.; Qi, J.; Hu, D.; Wang, C.; Wang, L.; Wu, Y.; Chen, J.; Zhang, Z.; Wang, X.; Li, B.; et al. Molecularly imprinted metal-organic frameworks assisted cloth and paper hybrid microfluidic devices for visual detection of gonyautoxin. *J. Hazard. Mater.* **2024**, *469*, 133969. [CrossRef] [PubMed]
232. dos Santos, D.M.; Migliorini, F.L.; Coatrini-Soares, A.; Soares, J.C.; Mattoso, L.H.; Oliveira, O.N.; Correa, D.S. Low-cost paper-based sensors modified with curcumin for the detection of ochratoxin a in beverages. *Sens. Actuators Rep.* **2024**, *7*, 100184. [CrossRef]
233. Hua, M.Z.; Lu, X. Development of a microfluidic paper-based immunoassay for rapid detection of allergic protein in foods. *ACS Sens.* **2020**, *5*, 4048–4056. [CrossRef] [PubMed]
234. Kunpatee, K.; Panphloi, M.; Charoenkitamorn, K.; Pimpitak, U.; Puthong, S.; Buakeaw, A.; Komolpis, K.; Sain, M.M.; Yakoh, A.; Chaiyo, S. Electrochemical lateral flow immunosensor with enhanced reproducibility for milk allergen detection. *Sens. Actuators B Chem.* **2024**, *401*, 135042. [CrossRef]
235. Han, X.; Zhang, D.; Xie, M.; Yang, J.; Wang, Y.; Li, H.; Wang, S.; Pan, M. Microfluidic paper-based analysis device applying black phosphorus nanosheets@ MWCNTs-COOH: A portable and efficient strategy for detection of β-Lactoglobulin in dairy products. *Food Chem.* **2024**, *446*, 138844. [CrossRef]
236. Ortiz-Gómez, I.; Ipatov, A.; Barreiro-Docío, E.; Salinas-Castillo, A.; de Orbe-Payá, I.; Capitán-Vallvey, L.F.; Prado, M. Microfluidic paper-based analytical aptasensor for fluorometric β-lactoglobulin determination. *Microchem. J.* **2024**, *198*, 110121. [CrossRef]
237. He, Y.; Hua, M.Z.; Feng, S.; Lu, X. Development of a smartphone-integrated microfluidic paper-based optosensing platform coupled with molecular imprinting technique for in-situ determination of histamine in canned tuna. *Food Chem.* **2024**, *451*, 139446. [CrossRef]
238. Pan, M.; Han, X.; Chen, S.; Yang, J.; Wang, Y.; Li, H.; Wang, S. Paper-based microfluidic device for selective detection of peanut allergen Ara h1 applying black phosphorus-Au nanocomposites for signal amplification. *Talanta* **2024**, *267*, 125188. [CrossRef] [PubMed]
239. Lu, C.; Xu, S.; Wang, S.; Wang, T.; Wang, W.L.; Yang, C.; Zhang, Y. Facile and Ultrasensitive Food Allergen Quantification Using Microzone Paper-Based Mass Spectrometric Immunoassay. *Anal. Chem.* **2024**, *96*, 2387—2395. [CrossRef] [PubMed]
240. Martins, T.S.; Machado, S.A.; Oliveira Jr, O.N.; Bott-Neto, J.L. Optimized paper-based electrochemical sensors treated in acidic media to detect carbendazim on the skin of apple and cabbage. *Food Chem.* **2023**, *410*, 135429. [CrossRef] [PubMed]
241. Zhang, D.; Wang, S.; Yang, F.; Li, Z.; Huang, W. Visual inspection of acidic pH and bisulfite in white wine using a colorimetric and fluorescent probe. *Food Chem.* **2023**, *408*, 135200. [CrossRef] [PubMed]
242. Placer, L.; Lavilla, I.; Pena-Pereira, F.; Bendicho, C. A 3D microfluidic paper-based analytical device with smartphone-assisted colorimetric detection for iodine speciation in seaweed samples. *Sens. Actuators B Chem.* **2023**, *377*, 133109. [CrossRef]
243. Masoomi, S.; Sharifi, H.; Hemmateenejad, B. A paper-based optical tongue for characterization of Iranian honey: Identification of geographical/botanical origins and adulteration detection. *Food Control* **2024**, *155*, 110052. [CrossRef]
244. Selvaraj, S.; Ravi Shankaran, D. Nano Enabled Plasmonic Strips For Colorimetric Detection Of Food Adulterants. *ChemistrySelect* **2023**, *8*, e202302027. [CrossRef]
245. Andrade, L.M.; Romanholo, P.V.; Ananias, A.C.A.; Venancio, K.P.; Silva-Neto, H.A.; Coltro, W.K.; Sgobbi, L.F. Pocket test for instantaneous quantification of starch adulterant in milk using a counterfeit banknote detection pen. *Food Chem.* **2023**, *405*, 134844. [CrossRef]
246. Dortez, S.; Crevillen, A.G.; Escarpa, A.; Cinti, S. Electroanalytical paper-based device for reliable detection and quantification of sugars in milk. *Sens. Actuators B Chem.* **2024**, *398*, 134704. [CrossRef]
247. Shalileh, F.; Sabahi, H.; Dadmehr, M.; Hosseini, M. Sensing approaches toward detection of urea adulteration in milk. *Microchem. J.* **2023**, *193*, 108990. [CrossRef]
248. Wu, Y.; Zhang, L.; Zhang, D.; Yu, R. A surface molecularly imprinted microfluidic paper based device with smartphone assisted colorimetric detection for butachlor in mung bean. *Food Chem.* **2024**, *435*, 137659. [CrossRef] [PubMed]
249. Qin, X.; Liu, J.; Zhang, Z.; Li, J.; Yuan, L.; Zhang, Z.; Chen, L. Microfluidic paper-based chips in rapid detection: Current status, challenges, and perspectives. *TrAC Trends Anal. Chem.* **2021**, *143*, 116371. [CrossRef]
250. Gautam, N.; Verma, R.; Ram, R.; Singh, J.; Sarkar, A. Development of a biodegradable microfluidic paper-based device for blood-plasma separation integrated with non-enzymatic electrochemical detection of ascorbic acid. *Talanta* **2024**, *266*, 125019. [CrossRef]
251. Yamada, K.; Shibata, H.; Suzuki, K.; Citterio, D. Toward practical application of paper-based microfluidics for medical diagnostics: State-of-the-art and challenges. *Lab Chip* **2017**, *17*, 1206–1249. [CrossRef]

biosensors

MDPI

Article

Modeling of Paper-Based Bi-Material Cantilever Actuator for Microfluidic Biosensors

Ashutosh Kumar *, Hojat Heidari-Bafroui, Nassim Rahmani, Constantine Anagnostopoulos and Mohammad Faghri *

Microfluidics Laboratory, Department of Mechanical, Industrial and Systems Engineering, University of Rhode Island, 2 East Alumni Avenue, Kingston, RI 02881, USA; nara7@uri.edu (N.R.); anagnostopoulos@uri.edu (C.A.)
* Correspondence: ashutosh@uri.edu (A.K.); faghrim@uri.edu (M.F.)

Abstract: This research explores the dynamics of a fluidically loaded Bi-Material cantilever (B-MaC), a critical component of μPADs (microfluidic paper-based analytical devices) used in point-of-care diagnostics. Constructed from Scotch Tape and Whatman Grade 41 filter paper strips, the B-MaC's behavior under fluid imbibition is examined. A capillary fluid flow model is formulated for the B-MaC, adhering to the Lucas–Washburn (LW) equation, and supported by empirical data. This paper further investigates the stress–strain relationship to estimate the modulus of the B-MaC at various saturation levels and to predict the behavior of the fluidically loaded cantilever. The study shows that the Young's modulus of Whatman Grade 41 filter paper drastically decreases to approximately 20 MPa (about 7% of its dry-state value) upon full saturation. This significant decrease in flexural rigidity, in conjunction with the hygroexpansive strain and coefficient of hygroexpansion (empirically deduced to be 0.008), is essential in determining the B-MaC's deflection. The proposed moderate deflection formulation effectively predicts the B-MaC's behavior under fluidic loading, emphasizing the measurement of maximum (tip) deflection using interfacial boundary conditions for the B-MaC's wet and dry regions. This knowledge of tip deflection will prove instrumental in optimizing the design parameters of B-MaCs.

Keywords: paper-based sensor; Bi-Material cantilever; paper-based valve; fluid imbibition; hygroexpansion coefficient; hygroexpansion strain; Whatman Grade 41 filter paper; modulus of paper

Citation: Kumar, A.; Heidari-Bafroui, H.; Rahmani, N.; Anagnostopoulos, C.; Faghri, M. Modeling of Paper-Based Bi-Material Cantilever Actuator for Microfluidic Biosensors. *Biosensors* **2023**, *13*, 580. https://doi.org/10.3390/bios13060580

Received: 27 April 2023
Revised: 23 May 2023
Accepted: 24 May 2023
Published: 26 May 2023

1. Introduction

Lab-on-a-chip technology, employing innovative materials and components, has made significant strides in diagnosing diseases and detecting a broad range of phenomena [1]. A crucial breakthrough is the development of the Bi-Material cantilever (B-MaC) valve in microfluidic paper-based analytical devices (μPADs), which allows for autonomous control of multiple fluid reagents. This paper-based cantilever, consisting of a sensing and an actuating layer, responds to changes in moisture levels, initiating mechanical motion, a concept inspired by microcantilever sensors in atomic force microscopy [2,3].

B-MaC valves are fundamental to the operation of μPADs for biosensing [4]. Their self-actuation results from hygroexpansion, much like thermal expansion in thermostats. Composite bilayers have found wide-ranging applications in electronics, biomimetics, and biomedical applications [5–9]. Silicon-based μPADs have evolved to develop microactuators and lab-on-a-chip devices [10–14], and various polymers have also been explored [15–18]. However, nonbiodegradable materials have significant drawbacks, rendering paper-based μPADs attractive due to their biodegradability. Various valving configurations for μPADs have been developed [19–25]. Paper deformation upon fluid imbibition is a common feature across different applications [26–29]. Fluid imbibition leads to the expansion of cellulose fibers and subsequent B-MaC bending. This behavior is influenced by solution properties, the paper material, and environmental conditions [30,31], and has been the focus of many studies [32–37]. Furthermore, the characteristics of micro check valves have been extensively studied, highlighting their importance in microfluidic systems [38]. The bimaterial cantilever (B-MaC) actuator primarily emphasizes its use in single-use, disposable applications. The design of the paper-based sensor aligns well with low-cost, lightweight, and biodegradable requirements, making it a particularly effective solution in scenarios where the sensor's reusability might compromise the accuracy of subsequent readings. Such applications could include certain biosensing or environmental monitoring tasks. Furthermore, using the B-MaC design for one-time applications can help avoid potential complications such as contamination, degradation, or material property alterations over multiple uses, which could adversely affect the sensor's reliability.

This research builds on and extends the recent work on the bending behavior and modeling of these cantilevers [39]. Our research intensively explores the behavior of B-MaCs under fluidic loading, with a specific focus on the impact of material properties, particularly the Young's modulus. Compared to our previous work, where the wetted length and spatial coordinates were determined assuming the arc length for the radius of curvature of bilayer beam, we now adopt a more nuanced approach. In our current formulation, we utilize the classical beam theory to establish the curvature, factoring in the deflection and curvature of the bilayer beam. We delve deeply into the behavior of B-MaCs under fluidic loading, emphasizing the influence of material properties such as the Young's modulus. Moreover, we propose a mathematical model that accounts for the hygroexpansive response of the paper [40,41].

One of the significant advancements in our study is the determination of the hygroexpansion coefficient of Whatman 41 paper and the Young's modulus of both Whatman 41 paper and tape at different moisture levels. These values, previously unreported in the literature, are crucial to accurately understanding the behavior of B-MaCs constructed with these materials. Using arbitrary values does not predict real case deflections accurately; thus, our research fills a critical gap in the existing literature. Despite progress in the field, the current literature lacks a comprehensive model that considers the behavior of bilayers under fluidic loading. Our study introduces a comprehensive model of a fluidically activated B-MaC for the automation of a paper-based assay [42], implemented in a fluidic circuit to sequentially load multiple reagents for analyte detection. By considering various geometric and material properties, our model, which is validated using experimental results, makes a novel and significant contribution to the field.

2. Materials and Methods

A borosilicate capillary is utilized to load a certain amount of fluid onto the paper-based bimaterial cantilever (B-MaC). The sample fluid transfers from the fixed to the free end of the B-MaC due to the capillary action. The fluidic loading of paper-based B-MaC results in the hygroexpansion of cellulose fibers, and the B-MaC starts to deflect over several seconds. In addition, an Instron pull test was performed on a paper-based bimaterial Cantilever (B-MaC) to estimate the Young's modulus of Whatman Grade 41 filter paper. For this purpose, dog bone samples were utilized with iterations of wet and dry Whatman Grade 41 filter paper and Scotch Tape.

2.1. Materials

The following materials were used in preparing, fabricating, and testing the paper-based bimaterial cantilever (B-MaC) valves used in this study: Whatman filter papers grade 41 (GE Healthcare Whatman 41-1441866) purchased from Thermo Fisher Scientific (Waltham, MA, USA); Scotch® Tape 600 (3M, St. Paul, MN, USA); food coloring (Wilton Icing Colors, Illinois, USA) for visual aid; ASTM Type 1 deionized water (resistivity > 18 MΩ/cm, (LabChem-LC267405, Pennsylvania, USA). The dimensions of the cantilevers were cut using Vector 13 graphics software (CorelDraw X6 2022 v24. 1). The cantilevers were then cut out from paper, in a cross-machine direction, using a laser engraver (Epilog mini 40 W 800 Laser System). For material testing, Shimadzu EZ-LX Instron and SCG 1kNA grips were utilized. An 8-megapixel video camera with 30-frames-per-second capability and media player (Avidemux 2.8.1) was used to record and play back the recording and collect the data for the actuation of the cantilever valve.

Experiment Flow

A picture of the experiment model is shown in Figure 1, consisting of a stationary component, paper-based Bi-Material cantilever (B-MaC) valve and capillary tube to load the fluid and obtain the response deflection.

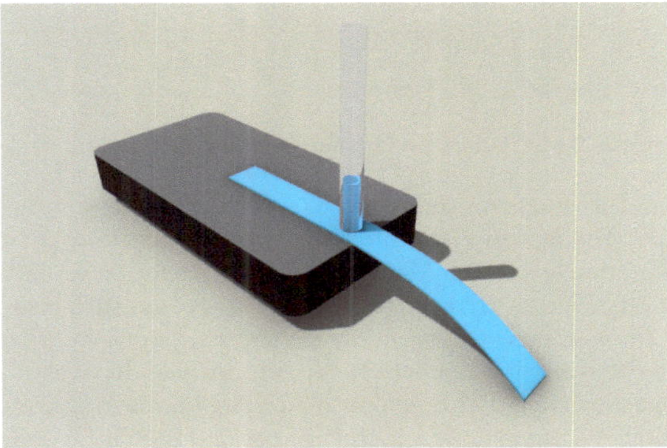

Figure 1. Experimental model representation for paper-based Bi-Material cantilever (B-MaC) valve.

The Whatman Grade 41 filter paper with one side laminated with tape was cut in a cross-machine direction with a 4 mm width and a 20 mm length using an Epilog Mini laser engraver. The picture of samples for B-MaC can be seen in Figure 2. A 2 mm diameter capillary tube was used to introduce fluid into the paper-based Bi-Material cantilever (B-MaC) valve. The fixture for the positioning of the paper-based cantilever and capillary was designed and utilized to reduce the uncontrollable error of running experiments. Figure 3 shows (on the left) the tape-side-down (normally closed) B-MaC positioned on the fixture in the unloaded condition and (on the right) actuated B-MaC in a loaded condition.

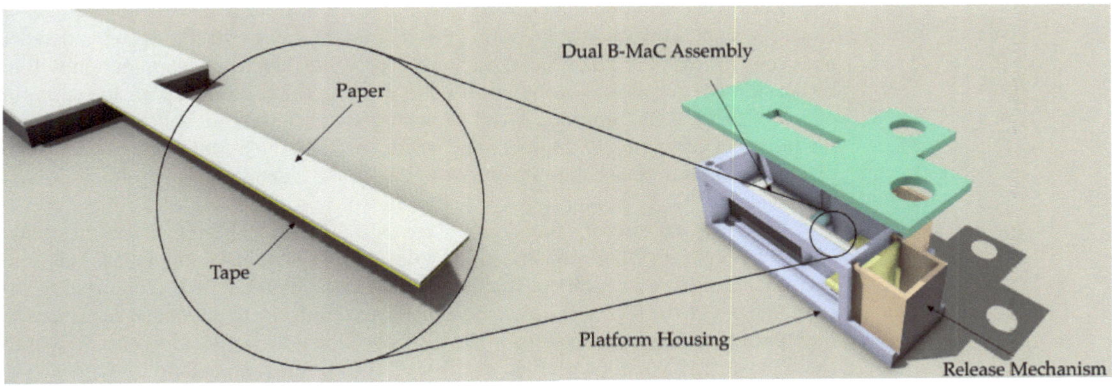

Figure 2. Paper-based Bi-Material cantilever (B-MaC) valve (on **left**); An exploded view of the platform housing [42] (on **right**).

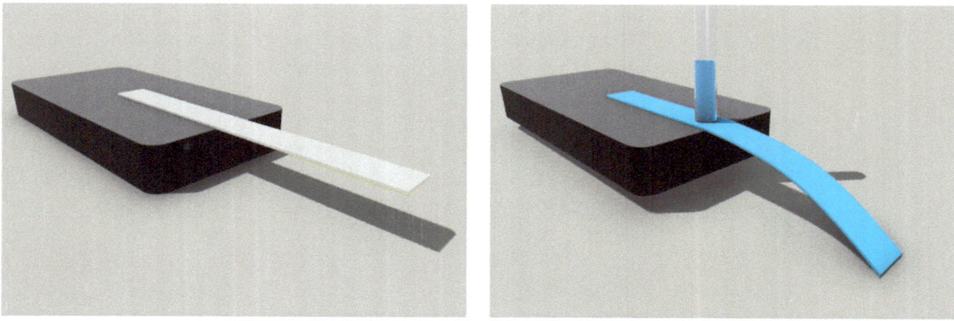

Figure 3. B-MaC before (on **left**) and after (on **right**) fluidic loading.

The study was conducted to experimentally assess the Young's modulus of paper and tape. The modulus for different saturation levels of paper, obtained using six different saturation levels of B-MaC ranging from 0% to 100%, was arbitrarily chosen with steps of 20% increments to cover the possible range of moisture content that a B-MaC can experience in autonomous assays. Moreover, test specimens were loaded onto the Instron machine after a dwell of 30 s to allow the specimens to have even distribution of fluid along the gage dimensions. The paper and tape dog bone test specimens were prepared for the study using Whatman Grade 41 filter paper and Scotch Tape. The filter paper and tape with a 4 mm gage width and a 10 mm gage length were cut using an Epilog Mini laser engraver. A similar setting for moisture content and sample for W-41 filter paper with gauge lengths 10 mm, 20 mm, and 30 mm were utilized to determine the empirical hygroexpansion strain.

2.2. Modeling—Bimaterial Cantilever (B-MaC)

For this study, a 2D quasi-static model is adapted with the wetted length as a dependent variable. The experiments were conducted under controlled laboratory conditions, and the effect of temperature and humidity of surroundings are not considered to approach a simpler model. It is important that the model takes dynamic deflection for the B-MaC into consideration; this is handled by using a moving (variable) boundary condition for the wetted length of the cantilever in a 2D model. B-MaC consists of a layer of paper, laminated with tape on one side, and fluidic loading of B-MaC leads to the possibility of delamination. However, significant observation during the experiment assures that the delamination does not occur for the given time for the actuation process; therefore, delamination will not be considered. A classical beam was used to develop the relationship for the curvature of B-MaC, with the assumption that the thickness, h, of B-MaC is small in comparison to the radius of curvature, R; the stress and strain profile in the B-MaC is homogenous; the plane of remains normal before and after bending; and the deflection is the only function of wetted length.

The B-MaC consists of a thin layer of tape (Scotch® Tape 600) laminated on the thinker layer of filter paper (Whatman Grade 41). Figure 4 displays the behavior of the B-MaC upon fluidic loading in two conditions (bonded and unbonded). The B-MaC remains in neutral condition before the fluid is loaded onto the bilayer cantilever. In unbonded conditions the filter paper exhibits hygroexpansion, but no deflection is achieved. However, in bonded conditions, upon loading with fluid, the B-MaC absorbs the sample fluid through wicking and actuates, resulting in bending. The paper layer exhibits hygroexpansion and tape being hydrophobic in nature does not expand. Since the paper and tape layers are bonded together, the hygroexpansion of paper shall be compensated with the inextensible tape layer. Therefore, the generated inconsistency in the strain is responsible for tensile force in the paper layer and compressive strain in the tape layer; these forces and moments generated are equal and opposite in nature to maintain equilibrium.

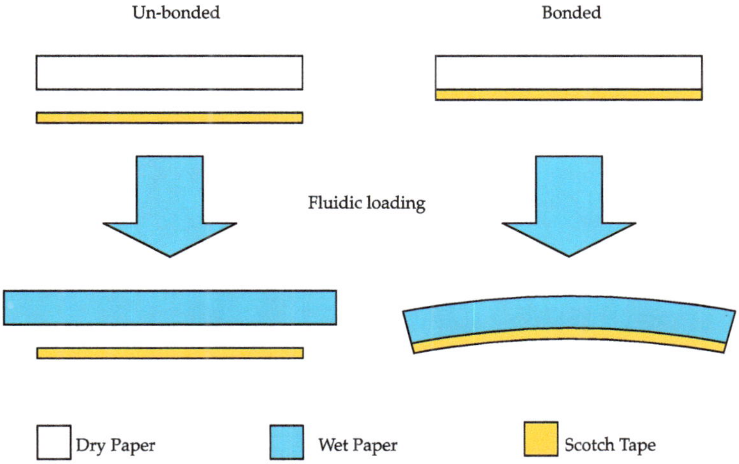

Figure 4. Paper-based bimaterial cantilever (B-MaC) valve upon fluidic loading.

2.2.1. Strain in B-MaC

Filter paper generally exhibits anisotropic and nonlinear mechanical behavior [43]. The hygroexpansive strain field is locally defined for the filter paper as the relative displacement of the length of the filter paper before and after deformation due to fluid imbibition. On the other hand, the hydrophobic nature of tape does not allow for axial deformation of tape on fluid imbibition.

A schematic representation of the homogenous strain field due to the hygroexpansion of the B-MaC element at 100% saturation is provided in Figure 5. The hygroexpansive strain can be expressed as:

$$\epsilon_h = \frac{\Delta l_h}{l} \tag{1}$$

where Δl_h is the change in length due to hygroexpansion and l is the original length.

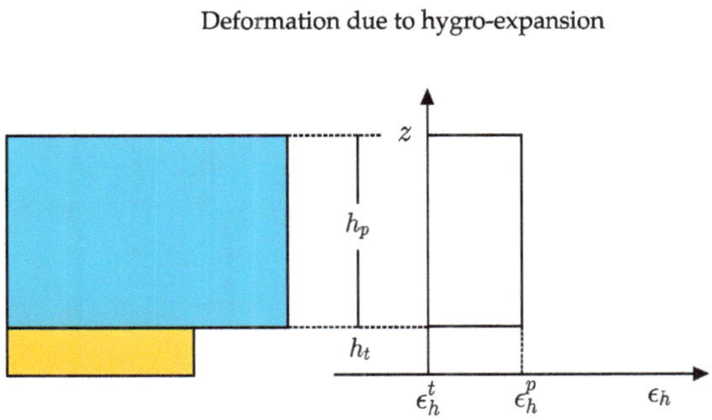

Figure 5. Hygroexpansive strain in B-MaC element.

Figure 5 represents the hygroexpansion strain in the paper layer and tape layer as ϵ_h^p and ϵ_h^t. Tape being hydrophobic in nature does not exhibit hygroexpansion, resulting in zero value.

The paper and tape layer of B-MaC upon bending restore the bending strain that is responsible for the curvature of B-MaC. In the pure bending state of B-MaC, the bending strain is given in Figure 6:

$$\epsilon_b = \epsilon_o - z\kappa \tag{2}$$

where ϵ_o is the reference plane strain and κ is the bending curvature.

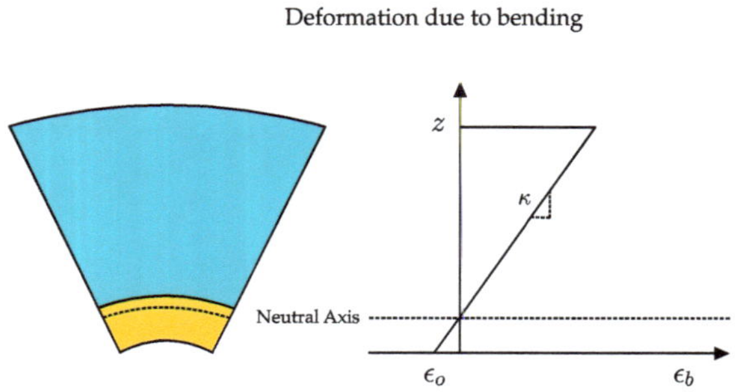

Figure 6. Bending strain in B-MaC element.

The total strain in B-MaC under no external load is given by the difference in strain due to bending (Equation (2)) and hygroexpansion (Equation (1)), where positive and negative strain values are the result of tension and compression, respectively.

$$\epsilon = \epsilon_o - z\kappa - \epsilon_h \quad (3)$$

2.2.2. Stress of B-MaC

The paper upon fluidic loading is subjected to tensile force due to hygroexpansion; the relation for stress and strain for anisotropic paper exhibiting elastic–plastic behavior, established by Ramberg–Osgood form (1943) in slightly modified form, is given by $\epsilon = (\sigma/E) + (\sigma/E_0)^n$, where E is Young's modulus, E_0 is the Hardening modulus, n is the Hardening exponent, σ is the 1D axial stress, and ϵ is corresponding strain. Filter paper in elastic range $(\sigma/E_0)^n$ vanishes from the relation, and future modification of the relation for B-MaC results in

$$\sigma = \epsilon E \quad (4)$$

2.2.3. Fluid Flow in B-MaC

Fluid imbibition phenomena in paper-like porous material are carried out due to capillary action at the microscale; this is known as pore-level transport. The schematic representation of the samples utilized for the study is provided in Figure 7. Fluid imbibition in B-MaC is an ambiguous phenomenon, and to better understand the fluid flow in B-MaC, we must model the fluid flow for the filter paper layer. The fluid flow in a porous system driven by capillary action is exemplary of fluid flow in filter paper. In this study, the capillary model is adapted for fluid flow into filter paper that obeys the Lucas–Washburn relationship. According to the LW equation [44]:

$$l_w = \sqrt{\frac{r\gamma\cos\phi}{2\eta}t} \quad (5)$$

where l_w—wetted length, r—average pore radius, γ—surface tension of liquid, t—time taken for fluid imbibition, ϕ—contact angle of the liquid on capillary walls and η—viscosity of the fluid. The equation can be modified by squaring both sides and using the diffusivity coefficient ψ as $\frac{r\gamma\cos\phi}{2\eta}$.

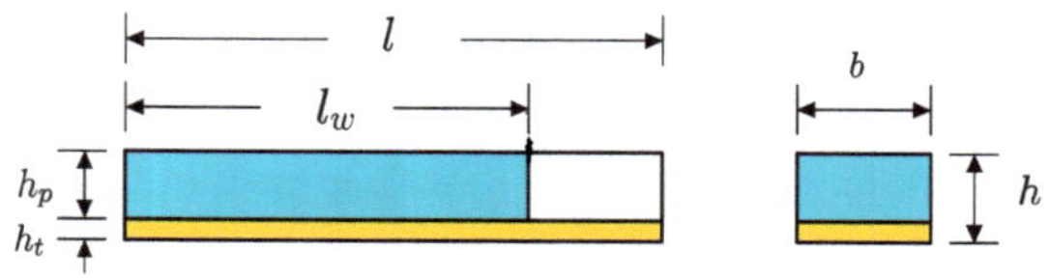

Figure 7. Schematic representation of 2D wetted B-MaC at an arbitrary time.

Simplifying,

$$l_w{}^2 = \psi t \tag{6}$$

2.2.4. Material Properties of B-MaC

In relation to the material properties of B-MaC, one of the critical aspects is the Young's modulus of the paper and tape layer of B-MaC. A schematic representation of the fluidically loaded paper and tape layer of B-MaC is provided in Figure 8, where E_p and E_t are the Young's modulus of fluidically saturated paper and tape respectively. The Young's modulus is the measure of elastic property defining the ability to withstand the change in length under tension or compressive load before failure. Mathematically, the Young's modulus of B-MaC can be defined as the ratio of internal axial stress induced to the hygroexpansion strain in the material due to fluidic loading, given by Equation (4).

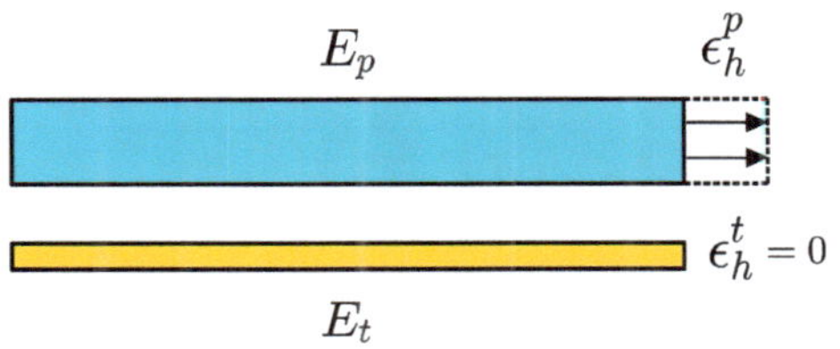

Figure 8. Young's modulus of fluidically saturated filter paper and tape.

In conjunction with studying the behavior of B-MaC on fluidic loading, the material properties of paper-based cantilevers were obtained experimentally. For this purpose, an Instron tensile test was performed in laboratory conditions to attain values for the paper-based cantilever's Young's modulus.

2.2.5. Linear Coefficient of Hygroexpansion

The tendency of matter to change its shape, area, volume, and density in response to changes in moisture content results in the hygroexpansion of the paper-based cantilever. The ratio of the hygroexpansion strain to the water imbibition content can be defined as the linear coefficient of hygroexpansion. This paper later presents the empirical value of the linear coefficient of hygroexpansion (β_h) for the Whatman Grade 41 paper.

$$\beta_h = \frac{\epsilon_h}{\Delta M} \tag{7}$$

where β_h is the coefficient of linear hygroexpansion and ΔM is the change in moisture content.

2.3. Modeling of B-MaC

This paper models the response of B-MaC on fluidic loading of the paper-based cantilever. Our previous work modeled the curvature of the bilayer cantilever utilizing the average intralayer force and moment [39]. This paper presents the model for response deflection considering interfacial conditions of continuous strain and slope between the bilayers and the wet-dry zone, respectively. The wet zone determined by Washburn on fluidic loading models the curvature for coupled fluid and structure bilayer model and utilizes the classical beam relationship to obtain the deflection for a given wetted length. The dry zone is perpetuated as a straight line and utilizes the slope at the interface to obtain the tip deflection of the bilayer cantilever. The modeling of B-MaC determines the response deflection over the period of actuation. Table 1 provides the details of the parameters required for modeling,

Table 1. B-MaC parameters.

Variables	Description	Dimension
l	Length of B-MaC	L
w	Deflection in y	L
u	Deformation in x	L
R	Bending radius	L
l_w	Wetted length	L
b	B-MaC width	L
h_t	Tape layer thickness	L
h_p	Paper layer thickness	L
t_s	Time required for saturation	T
σ	Stress per unit length	$ML^{-2}T^{-2}$
F_h	Hygroexpansion force	MLT^{-2}
M_h	Bending moment due to actuation	ML^2T^{-2}
E_t	Young's modulus of tape	$ML^{-1}T^{-2}$
E_p	Young's modulus of saturated paper	$ML^{-1}T^{-2}$

Geometry is inspired by the bending of a paper-based cantilever when exposed to fluid [1]. B-MaC actuation is considered as the system output defined by fluid loading, please refer to Table 1 for parameters. A schematic representation of stress, force, and moment in the paper and tape layer that evolved in B-MaC upon fluidic loading is provided in Figure 9.

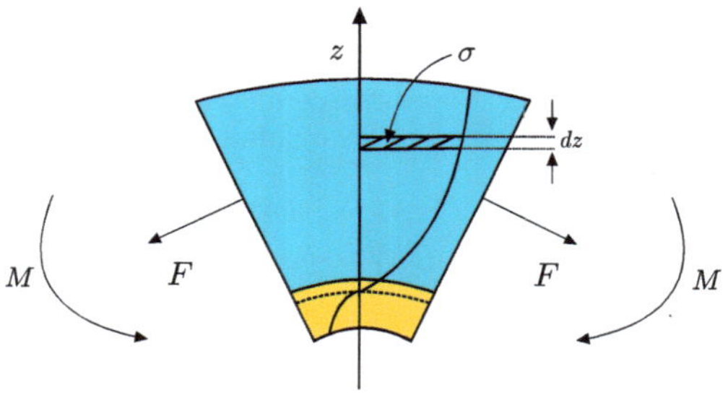

Figure 9. B-MaC element subjected to force and moment due to stress developed in layers.

The force and moment can be obtained by integrating the stress along the cross section of B-MaC.

Force in B-MaC:

$$F = \int_0^h \sigma\, dz = \int_0^h E(\epsilon_o - z\kappa - \epsilon_h)dz \tag{8}$$

Moment in B-MaC:

$$M = \int_0^h \sigma\, z\, dz = \int_0^h E(\epsilon_o - z\kappa - \epsilon_h)z\, dz \tag{9}$$

The B-MaC undergoes bending until the hygroexpansion strain attains its maximum value; at equilibrium, the net force and net moment in the cross section of B-MaC is

$$F = 0; \cdots\cdots\cdots\cdots M = 0; \tag{10}$$

on combining Equations (8)–(10) and writing in matrix form,

$$\begin{bmatrix} P & -Q \\ Q & -R \end{bmatrix} \begin{bmatrix} \epsilon_o \\ \kappa \end{bmatrix} = \begin{bmatrix} F_h \\ M_h \end{bmatrix} \tag{11}$$

where

$$P = \int_0^h E\,dz;\; Q = \int_0^h E\,z\,dz;\; R = \int_0^h E\,z^2\,dz; \tag{12}$$

$$F_h = \int_0^h E\,\epsilon_h\,dz;\; M_h = \int_0^h E\,\epsilon_h\,z\,dz \tag{13}$$

The reference strain and bending curvature can be obtained by solving Equation (11):

$$\epsilon_o = \frac{RF_h - QM_h}{Q^2 - PR} \tag{14}$$

$$\kappa = \frac{PM_h - QF_h}{Q^2 - PR} \tag{15}$$

To approach a static bilayer bending, Young's modulus of paper on saturation is considered to be constant. Moreover, the hygroexpansion strain attained at full saturation of paper corresponds to a constant value. The above Equations (8)–(15) were solved using MATLAB, provided as supplementary information. Substituting values from Equations (12) and (13) in Equations (14) and (15), we obtain

$$\epsilon_o = \frac{E_t\epsilon_h^t h_t\left(E_t h_t^3 + 6E_p h_p h_t^2 + 4E_p h_p^3 + 9E_p h_p^2 h_t\right) - E_p\epsilon_h^p h_p\left(E_p h_p^3 - 2E_t h_t^3 - 3E_t h_p h_t^2\right)}{\left(E_t^2 h_t^4 + 4E_t E_p h_t^3 h_p + 6E_t E_p h_t^2 h_p^2 + 4E_t E_p h_t h_p^3 + E_p^2 h_p^4\right)} \tag{16}$$

$$\kappa = \frac{6(E_p E_t)(h_p h_t)(h_p + h_t)\left(\epsilon_h^t - \epsilon_h^p\right)}{\left(E_t^2 h_t^4 + 4E_t E_p h_t^3 h_p + 6E_t E_p h_t^2 h_p^2 + 4E_t E_p h_t h_p^3 + E_p^2 h_p^4\right)} \tag{17}$$

where $\left(\epsilon_h^t - \epsilon_h^p\right)$ is the actuation strain in B-MaC. Since the tape is hydrophobic in nature, the hygroexpansion strain in tape, $\epsilon_h^t = 0$.

On simplifying,

$$\epsilon_o = \left(-\epsilon_h^p\right)\left[\frac{\left(1 - 2E_r h_r^3 - 3E_r h_r^2\right)}{\left(1 + 4E_r h_r^3 + 6E_r h_r^2 + 4E_r h_r + E_r^2 h_r^4\right)}\right] \tag{18}$$

$$\kappa = -\left(\frac{\epsilon_h^p}{h_p}\right)\left[\frac{6E_r h_r(1 + h_r)}{\left(1 + 4E_r h_r^3 + 6E_r h_r^2 + 4E_r h_r + E_r^2 h_r^4\right)}\right] \tag{19}$$

$$E_r = \frac{E_t}{E_p};\; h_r = \frac{h_t}{h_p} \tag{20}$$

$$\left[\frac{6E_r h_r(1 + h_r)}{\left(1 + 4E_r h_r^3 + 6E_r h_r^2 + 4E_r h_r + E_r^2 h_r^4\right)}\right] = \kappa_{Bilayer} \tag{21}$$

where $\kappa_{Bilayer}$ is the static curvature of bilayer beam.

The bending force and moment contribute to the deflection of B-MaC on fluid imbibition. The placement of tape on B-MaC plays an important role in deciding the curvature of bending; in our case, the bottom surface of filter paper is laminated. A bottom-laminated B-MaC will correspond to negative curvature of bending for B-MaC. From Figure 10, the deflection of B-MaC is given by Equation (A5) (Appendix A):

$$\frac{d^2 w}{dx^2} = \kappa \tag{22}$$

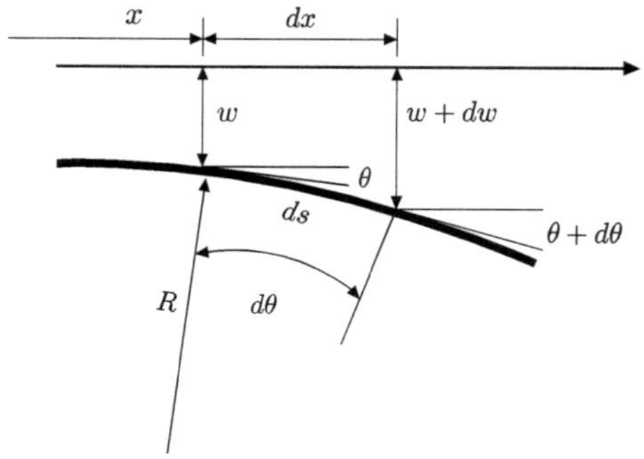

Figure 10. Curvature of B-MaC at any time instance upon fluidic loading.

Substituting above expression with Equation (19), we obtain

$$\frac{d^2w}{dx^2} = -\left(\frac{\epsilon_h^p}{h_p}\right)\kappa_{Bilayer} \tag{23}$$

Further, the following boundary conditions are accompanied with the governing equations to obtain specific solutions for the modeling of B-MaC. For fixed end (x = 0): deflection $w(x)$ = 0 and slope $w'(x)$ = 0.

To synthesize the analysis of the modeling, a nondimensionless form is presented in Table 2 below:

Table 2. B-MaC nondimensional parameters.

Variables	Expression	Description
L_w	$\frac{l_w}{l}$	Characteristic Wetted Length
W	$\frac{w}{l}$	Characteristic Deflection
H	$\frac{h}{l}$	Characteristic Thickness
T	$\frac{t}{t_s}$	Characteristic Time

Nondimensional governing equations:

Wet zone $(0 < X < L_w)$

$$W'' = -\xi\,\kappa_{Bilayer} \tag{24}$$

where $\xi = \left(\frac{\epsilon_h^p}{h_p}\right)/l$.

Dry zone $(L_w < X < 1)$

The dry length is the portion of the B-MaC that the wicking fluid front has not yet reached. Since it is not wetted, it remains straight. As seen in Figure 11, the dry zone length is the difference between the B-MaC total length and the wetted length:

$$(y - W) = m\left(x - L'_W\right) \tag{25}$$

where $m = \left(\frac{dw}{dx}\right)_{l'_w} = W'(L'_w)$.

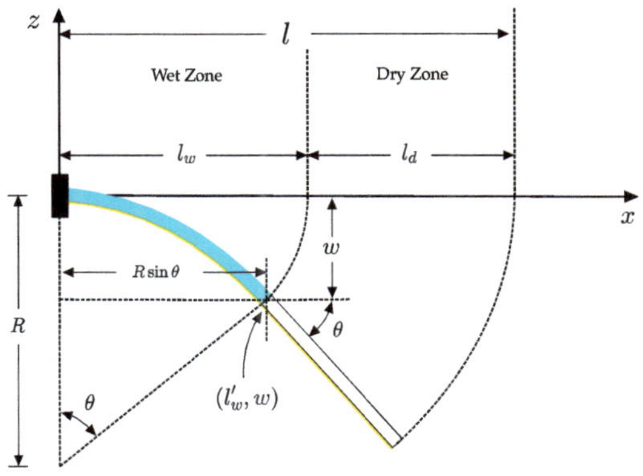

Figure 11. Wet and dry zones of B-MaC.

Boundary conditions:

Fixed-end ($x = 0$):

$$W(0) = 0 \tag{26}$$

$$W'(0) = 0 \tag{27}$$

Interface ($x = L_w$):

$$W(L_w{}')_{Wet\ Zone} = W(L_w{}')_{Dry\ Zone} \tag{28}$$

Specific solution for the modeling of B-MaC is evaluated for wet and dry zones using the set of nondimensional equations formulated above with the boundary conditions.

Solution for Equation (24) uses boundary conditions in Equations (26) and (27):

$$W(x) = -\xi\, \kappa_{Bilayer}\frac{x^2}{2} \tag{29}$$

To obtain the numerical values for deflection in future, the above solution requires empirical values for modulus for bilayers, hygroexpansion strain, and other geometrical parameters. Details for these parameters are discussed in the following sections.

3. Results

This study brings results for two important aspects of B-MaCs, i.e., the material properties and the behavior of the B-MaC on fluidic loading. There are other factors resulting from the experiment conducted for the B-MaC modeling, which will be discussed in detail below. All the results were obtained using the variables in Table 3.

Table 3. Variables utilized for the study.

Variables	Description	Value
l	Length of B-MaC	20 mm
b	B-MaC width	4 mm
h_t	Tape layer height	58×10^{-3} mm
h_p	Paper layer wetted height	246×10^{-3} mm
E_t	Young's modulus of tape	300 MPa
E_p	Young's modulus of saturated paper	20 MPa
ϵ_h^p	Hygroexpansion strain in paper	0.008

3.1. Fluid Flow in B-MaC

The results for fluid flow in the B-MaC that are adapted to the capillary flow model are compared to the experimental results. The two-dimensional rectangular geometry is inspired by the previous cantilever design utilized in µPADs [1]. The rectangular channel model provides good accuracy using the LW theory (Figure 12). The flow is prominently considered only in the lengthwise direction.

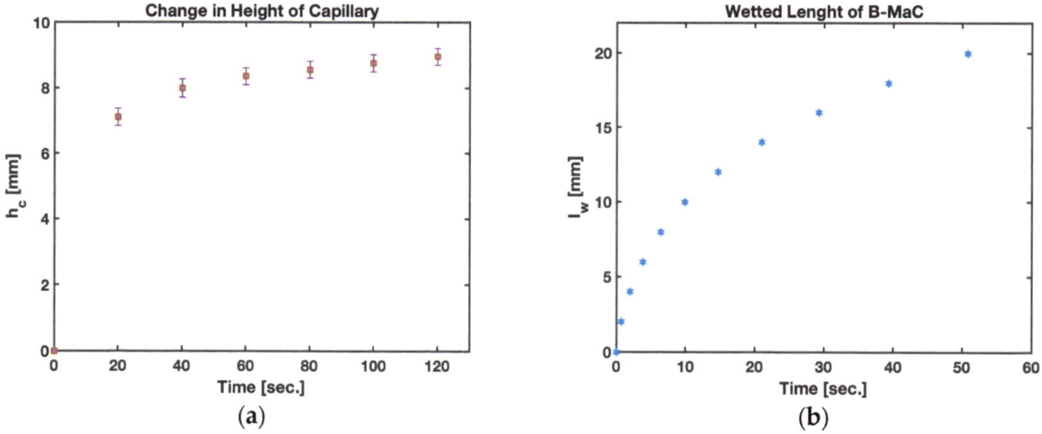

Figure 12. Fluid imbibition in B-MaC (cross-machined direction). (**a**) Change in height of capillary water column h_c (mm) vs. time t (sec.); (**b**) wetted length l_w (mm) vs. time t (sec.).

The results for fluid imbibition in the B-MaC (cross machined direction) for the change in height of the capillary water column and wetted length vs. time are presented in Figure 13. In case (a), the change in height of the capillary is inversely proposal to the unsaturated area on the B-MaC; with time, the available unsaturated area of the B-MaC decreases by the relation given in Equation (7). On fluid imbibition, the average change in height of the capillary for the given B-MaC (20 mm × 4 mm) is found to be around 9 mm for 100% saturation. Upon wetting of the B-MaC, the fluid front travels at a rate given by the LW equation. This relation is validated by experimental results obtained for this study, and the plot is presented in Figure 13. The wetted length of the B-MaC is found to be proportional to the sq. root of time, endorsed by the experiment results of this paper. The velocity of fluid imbibition in B-MaC is maximum at the start of fluidic loading and gradually reduced to zero with time, indicating the high volumetric water ingress upon loading B-MaC with fluid at the start and diminishing the volumetric water ingress by the end of loading (or 100% saturation of B-MaC). The experimental results obtained for capillary height and wetted length vs. time are in good agreement with the numerical predictions, shown in Figure 13.

3.2. Young's Modulus of B-MaC

The modulus of elasticity is one of the important material properties of B-MaCs that help us predict the behavior of paper-based cantilevers on fluidic loading. Experiments were conducted to obtain stress and strain relationships for the Whatman Grade 41 filter paper in CMD fiber orientation, as well as for Scotch Tape, as shown in Figure 14.

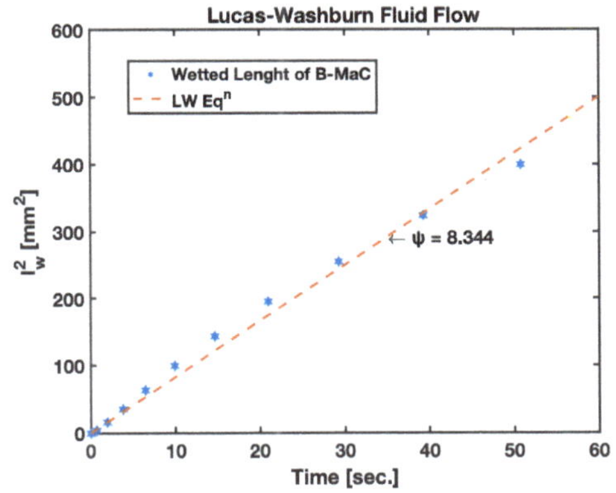

Figure 13. Fluid imbibition in B-MaC (cross-machined direction).

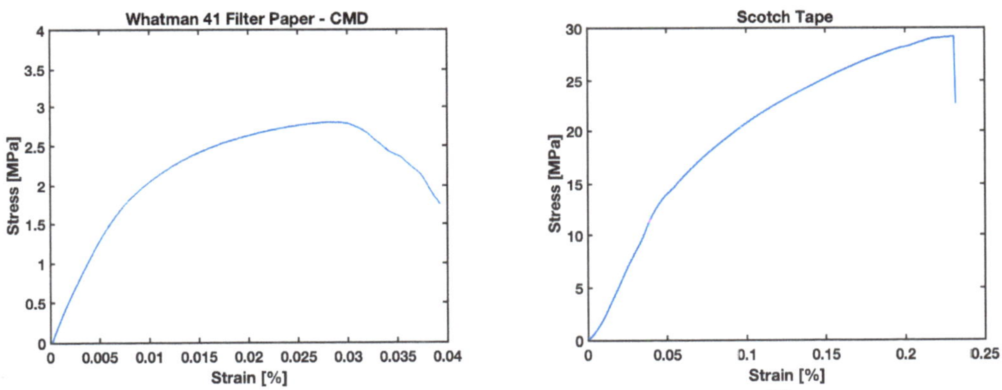

(a) (b)

Figure 14. Stress vs. strain relationship: (**a**) Whatman 41 filter paper in CMD; (**b**) Scotch Tape.

The trend illustrated in Figure 14 provides a relation for different zones, i.e., elastic, plastic, stress-hardening, and failure zones. The proportional limit of the stress–strain relationship is utilized to obtain the Young's modulus of W41 filter paper and Scotch Tape. The experimental values of the Young's Modulus are found to be ~300 MPa for filter paper in the cross-machine direction and Scotch Tape. Additionally, it is important for the study to obtain the Young's modulus for saturated filter paper. For this purpose, the study was conducted to acquire values of wetted cantilevers for different saturations (moisture content), and the results are presented in Figure 15. It is evident from the data obtained for the study that the value of the Young's modulus is drastically affected by the saturation levels of filter paper. The value obtained for the Young's modulus of saturated filter paper in a cross-machine direction was found to be ~20 MPa. The Young's modulus drops to ~7% of the respective value for fully (100%) saturated samples. From Figure 15, these drop values of the modulus are very evident for the slight wetting of cantilevers; the instant that fluid is loaded onto the cantilever, the cellulose fibers of the paper start swelling and disintegrating, causing them to lose their rigidity. This justifies the sudden fall in values for the Young's modulus of the B-MaC and PBC on wetting. These experimental values are utilized for modeling and predicting the behavior of B-MaCs on fluidic loading.

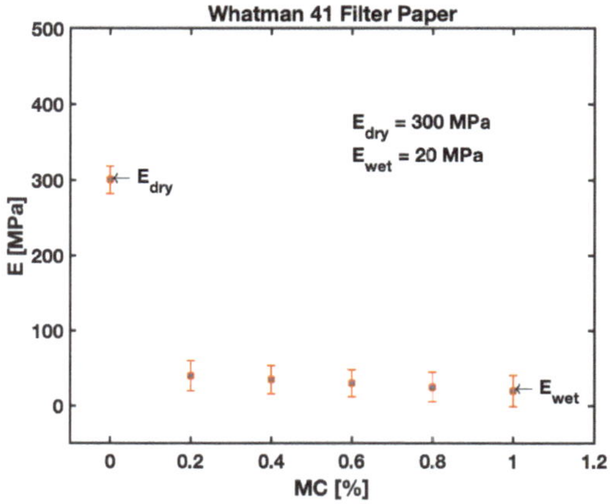

Figure 15. Young's modulus E (MPa) of Whatman Grade 41 filter paper with different saturation levels MC (%).

3.3. Stress and Strain in B-MaC

The results of this section primarily focus on the behaviors of filter paper on fluidic loading; therefore, further on, we will be discussing results obtained for the important parameters concerning the modeling of B-MaCs. The stress and strain relationships for paper-based cantilevers were discussed in the previous paper [1] and provided a detailed discussion of the considerate factors for modeling PBC on fluidic loading. Unlike PBC, B-MaCs consist of a layer of laminated tape that encourages us to study the effect of a bi-layer cantilever on fluidic loading. Relationships established in the previous sections for effective strain (ϵ) in B-MaCs can be obtained by Equation (3), i.e., the effective strain is the result of the bending and hygroexpansive strain of the B-MaC. The values of bending strain (ϵ_b) and hygroexpansion strain (ϵ_h) for the B-MaC are carefully deduced empirically and found to be around 0.2% (at tape and paper interface) and 0.8%, respectively.

The strain profile obtained for the wetted cantilever laminated with tape at the bottom is shown in Figure 16a, which provides information on bending strain (ϵ_b) in the B-MaC. The figure details that upon fluidic loading of the B-MaC, the paper undergoes hygroexpansion, to which the tape layer's mechanical response results in a bending state to obtain a sustainable deflection. The green and yellow sections in the plot represent the tension and compression loading respectively in the B-MaC layers. During fluidic loading, the paper layer ($h_t < z < h_p + h_t$) experiences tension loading due to the hygroexpansion of the constituent cellulose fiber of paper, and the tape layer ($0 < z < h_p$), as depicted, experiences compression loading to compensate for the actuation of the B-MaC. Figure 16b demonstrates the path traced for the zero-strain location in B-MaC for the varied thickness ratio h_r. The position of zero strains gives us information on the neutral axis. As the h_r is varied, the neutral axis moves up from the tape layer to the paper layer.

Comparative value plots for the hygroexpansion strain (ϵ_h) in W41 filter paper are presented for different lengths in Figure 17. All cases illustrate that irrespective of the length of the cantilever chosen, the hygroexpansion strain (ϵ_h) reaches a constant value after a given amount of time. The amount of time required by the cantilever to stabilize ϵ_h values indicate the time taken by the cantilever to fully saturate on fluidic loading. Figure 17 presents the value, ϵ_h, for 10 mm, 20 mm, and 3 mm cantilevers. The 10 mm filter paper, being the smallest in length, requires the least amount of time for 100% saturation on fluidic loading, evidently stabilizing the empirical value ϵ_h quickest, whereas 20 mm and 30 mm PBC take more time for saturation on fluidic loading and to reach the steady state value for the hygroexpansion strain (ϵ_h). At saturation, the ϵ_h for W41 filter paper is found to be $8 \times 10^{-3} \pm 0.5 \times 10^{-3}$.

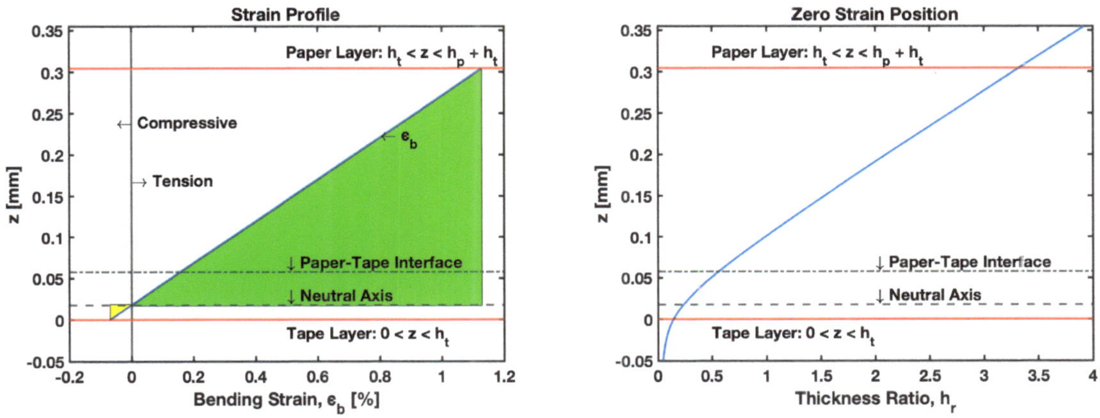

(a) (b)

Figure 16. Bending strain in B-MaC. (**a**) Strain profile in thickness of B-MaC; (**b**) zero strain position in B-MaC with varied thickness ratio h_r.

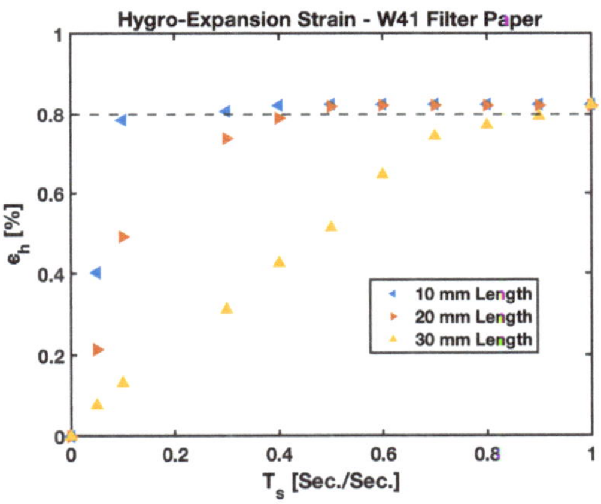

Figure 17. Hygroexpansion strain ϵ_h vs. characteristic time T_s for 10 mm, 20 mm, and 30 mm length PBC in Whatman Grade 41 filter paper.

Additionally, it is important to study the strain in thickness for the paper layer of the B-MaC. Upon wetting on the B-MaC, the volumetric change in configuration includes the deformation in thickness, along with the length (considered in previous sections). The strain in thickness (ϵ_t) is responsible for choosing the correct second moment of inertia and final thickness of B-MaC modeling. The thickness strain presented in Figure 18 follows a similar trend, with a sudden rise in the value on wetting and reaching a steady-state value of 0.12. As detailed in the previous sections, the fabrication of B-MaCs utilizes W-41; for PBC, the surface of the cantilever is not laminated with tape, encouraging loss of fluid from the surface due to evaporation. However, the B-MaC's bottom surface is laminated with tape not favoring evaporation from the laminated surface. This experiment is not suitable to capture the loss of source fluid from the specimen; hence, the results will not be discussed in detail due to unsuitable data for the scope of the experiment.

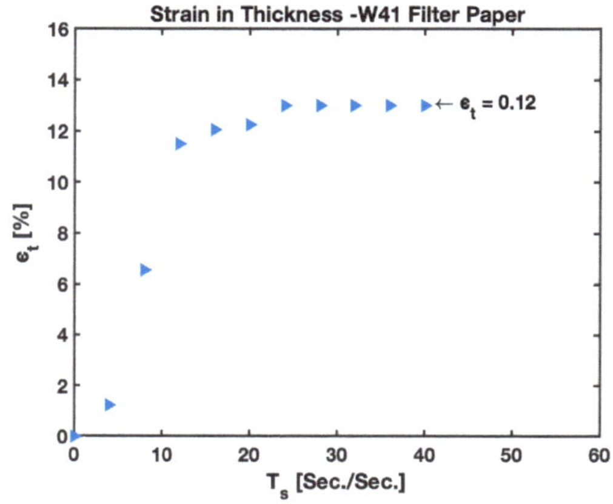

Figure 18. Plot for thickness strain ϵ_t vs. characteristic time T_s for Whatman Grade 41 filter paper.

3.4. Linear Coefficient of Hygroexpansion Paper

The coefficient of hygroexpansion is the ability of paper to change its shape, area, volume, and density in response to moisture change. This definition is tailored to study the change in the length of paper upon wetting, and the linear coefficient of hygroexpansion is introduced, as per Equation (7). This change in the length of the paper is also representative of hygroexpansive strain values at 100% saturation of the cantilever, providing information that can be relayed as the coefficient of hygroexpansion. This study presents the empirical value for the linear coefficient of hygroexpansion (β_h) for Whatman 41 filter paper. The value is found to be 0.008, based on the results obtained for the hygroexpansion strain upon the wetting of the paper. Figure 19 plots the change in the length of the paper to the original length vs. the change in moisture content for the W41 filter paper; empirical data were utilized for the study to obtain the slope of the line expressing the value of β_h for the W41 filter paper. The coefficient of hygroexpansion helps us understand the mechanical properties of paper used for the fabrication of cantilevers, which further assists in studying the behavior of B-MaC on fluidic loading.

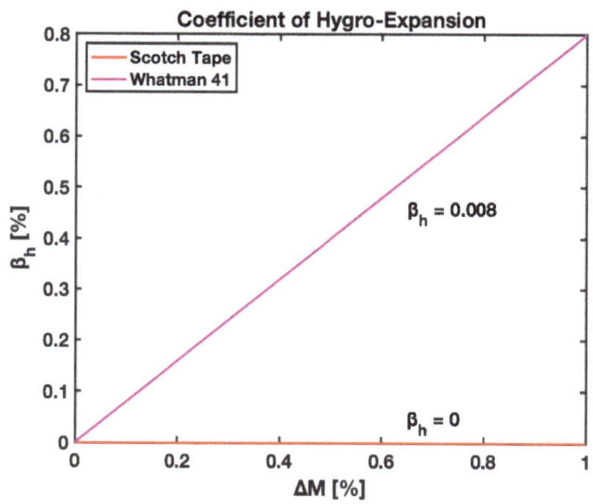

Figure 19. Linear coefficient of hygroexpansion β_h vs. moisture content ΔM for Whatman Grade 41 filter paper.

3.5. Deflection of B-MaC

To predict the sustainable deflection of B-MaCs on fluidic loading, theoretic analysis was carried out to obtain results for temporal shape evolutions. These experimental and numerical shape forms presented in Figure 20 illustrate that the observed behavior of B-MaCs can be precisely predicted from the model presented in this paper. The governing nondimensional Equation (25) was utilized to solve the deflection as a function of the wetted length of the cantilever. To obtain response deflection numerically, Neuman boundary conditions for deflection (w) and slope (w') of the cantilever at a fixed end were used. However, empirically, the fixed end of the cantilever upon fluidic loading experiences a sight deformation in the geometry of the B-MaC that constrains the use of the Neuman boundary condition. To obtain the best-fitting solution for the accuracy of bending strain, a polynomial fit was utilized in the plot. The result can also be utilized for the modeling of bimaterial cantilevers with the assumption of no stretching in the neutral axis. The modeling of the B-MaC is essential to understand the functionality of microfluidic paper-based analytical devices (μPADs), which are most suited for small–moderate deflection; for this purpose, this paper will focus on results for the deflection of B-MaCs.

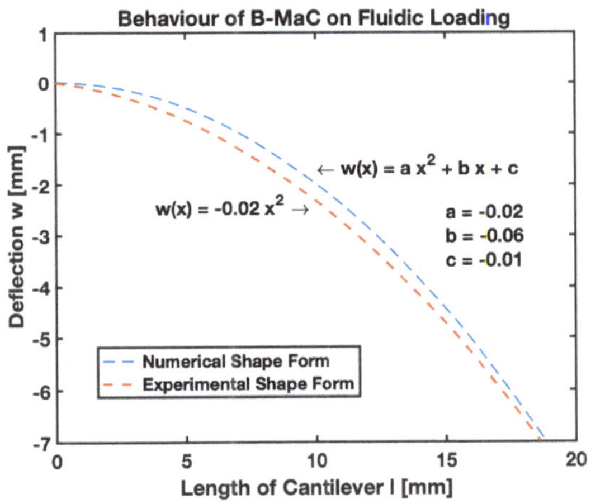

Figure 20. Deflection of B-MaC—numerical and experimental shape forms.

In order to validate the obtained results for the response deflection of B-MaCs, data for deflection were obtained experimentally. The numerical values predicted by the model were found to be within the predicted bound, shown in Figure 21. The empirical value for the maximum (tip) deflection of B-MaC is ~7 mm, which is almost three times the value obtained for PBC [1]. Tip deflection is the maximum deflection of the cantilever for any given wetted length. Figure 22a provides tip deflection for 0%, 20%, 40%, 60%, 80%, and 100% wetted lengths (l_w) of B-MaC. Other intermediate cases can be assessed by using Equation (1) for any given time period of fluidic loading, as shown in Figure 22b.

The response deflection of the B-MaC obtained in Equation (29) justifies that the curvature of the cantilever is constant over its length, and the deflection depends on the factors E_r, h_r, and ξ. These parameters help in predicting the behavior of B-MaCs on fluidic loading.

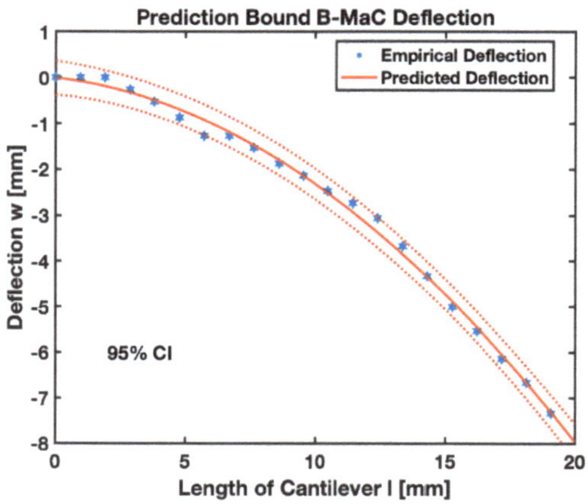

Figure 21. Predicted bound—numerical and experimental deflection of B-MaC.

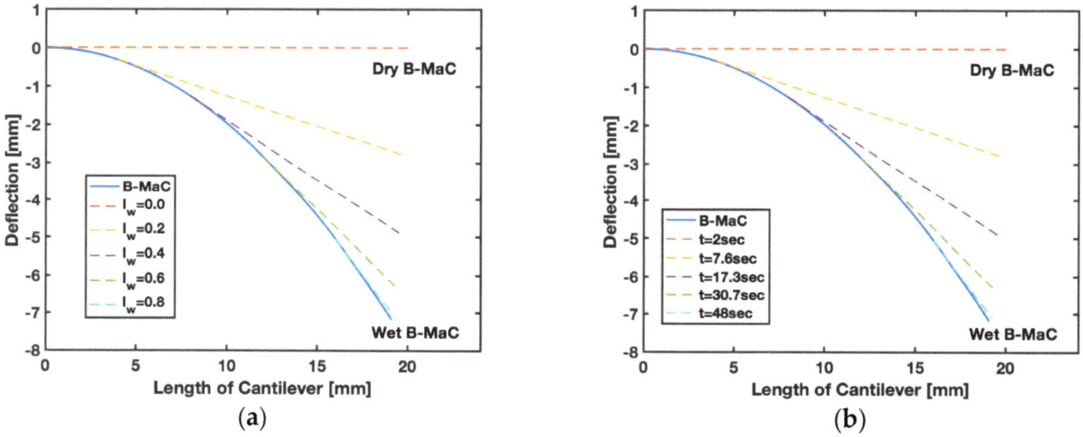

Figure 22. Response deflection of B-MaC. (**a**) For a given wetted length; (**b**) for a given time instance.

3.6. Parametric Models

The parametric model helps us understand the material, geometrical, and physical characteristics of B-MaCs under fluidic loading. The modeling of the B-MaC involves three vital parameters: E_r, h_r, and ξ, which seek to provide information on the choice of material and dimensional constraints for fabrication and the effect of fluid imbibition.

E_r is defined by Equation (20), as the ratio of the Young's modulus of tape to paper used to fabricate the B-MaC. This parameter provides the relative ability of the choice of material to withstand linear deformation of fluid imbibition. Figure 23, the parametric model for E_r–B-MaC, illustrates the behavior of fluidically loaded B-MaC deflection. The characteristic deflection is plotted against the characteristic length of the B-MaC for different E_r ratios of 15, 10, 05, 01, and 0. As the values of E_r start dropping, the response deflection of the B-MaC reduces until $E_r \rightarrow 0$, which is the case of no deflection. $E_r \rightarrow 0$ signifies a high Young's modulus of paper with respect to the tape. Cases where the paper is replaced with material of high E value on saturation will result in negligible deflection. $E_r = 0$ indicates the absence of tape in B-MaC, and the bilayer theory cannot be utilized to obtain a response deflection of B-MaC. Therefore, in such cases, the solution for a small deflection model for a monolayer paper-based cantilever [1] shall be utilized to obtain response deflection. Figure 24 indicates the case utilized for this study, which corresponds with the characteristic deflection of 0.36 for the tip on the B-MaC located at 0.95 characteristic lengths. Additionally, $E_r = 1$ represents the case for which the modulus of tape and paper are the same, resulting in maximum characteristic deflection ~0.2.

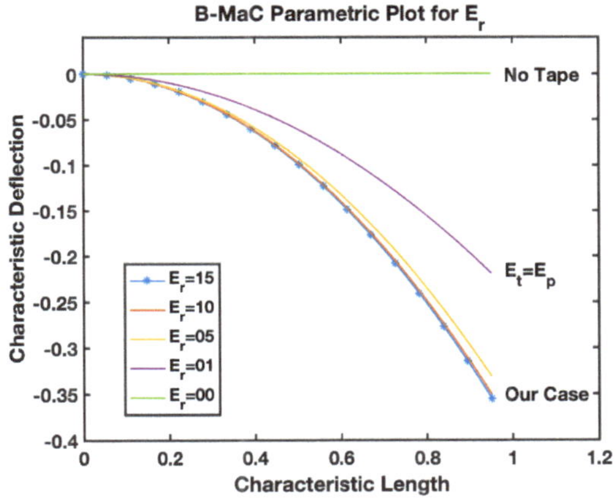

Figure 23. Parametric model for E_r–B-MaC.

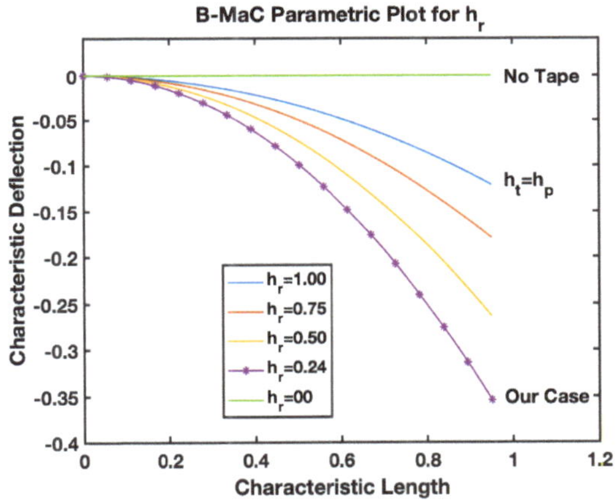

Figure 24. Parametric model for h_r–B-MaC.

The h_r is defined by Equation (20) as the ratio of the thickness of tape to the paper layer of the B-MaC. This parameter provides information on the geometrical parameters utilized for the fabrication of B-MaCs. Figure 24, the parametric model for h_r–B-MaC, plots the characteristic deflection of the B-MaC on fluidic loading for different h_r. As shown, for h_r from 0 to 1, the maximum characteristic response deflection reduces from 0.36 to 0.

This paper utilizes specific filter paper and tape types, described in the Materials and Methods section. The W-41 paper of thickness 220 μm is laminated by a layer of Scotch Tape of thickness 58 μm; this limits us to ignore the thickness of the tape. However, distinguishing characteristics of these B-MaCs may consist of a thin film layer; for such a case, where $h_r << 1$, Equation (24) can be simplified to obtain the deflection for the B-MaC.

The ξ is the ratio of change in length on wetting to the thickness of filter paper used for fabricating the B-MaC. This parameter provides information for the physical characteristics of filter paper. Figure 25, the parametric model for ξ–B-MaC, depicts the response deflection of B-MaC on wetting for different values of ξ. For this study, a ξ value of 0.0015 is obtained for the parameters, which corresponds to a maximum characteristic deflection of 0.36. The plot shows how reducing the value of ξ proportionally decreases the characteristic deflection of the B-MaC. $\xi = 0$, indicating the case of no fluidic loading, which is a result of a null value for the hygroexpansion strain of paper. When the paper is dry before the fluid is loaded onto the paper, the B-MaC remains at rest, corresponding to $\epsilon_h^p = 0$. Upon fluidic loading, the ϵ_h^p value increases until reaching a constant value. These increasing values of ϵ_h^p will give different ξ; some of these cases are presented in Figure 25.

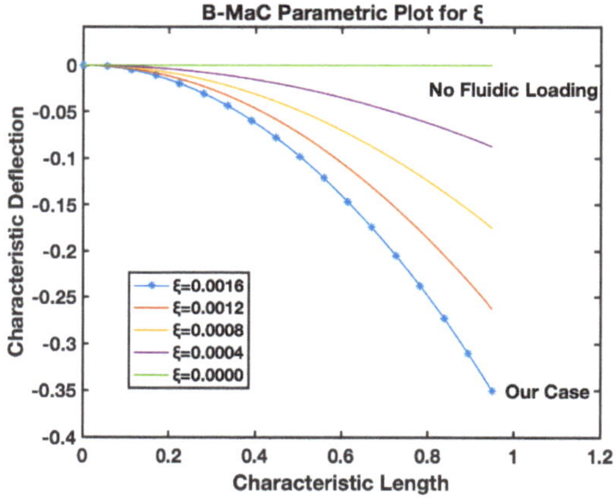

Figure 25. Parametric model for ξ–B-MaC.

It is interesting to notice the effect of the varied modulus ratio E_r, and the thickness ratio h_r together on the curvature of the B-MaC. In Figure 26, the Young's modulus and thickness ratio are varied from 0–15 to 0–10, respectively. This plot provides important results to select a geometrical parameter for B-MaC fabrication for the purpose of optimizing the curvature of the bilayer beam. A value for h_r of 0.2 and E_r of 15 will result in the maximum curvature of the B-MaC.

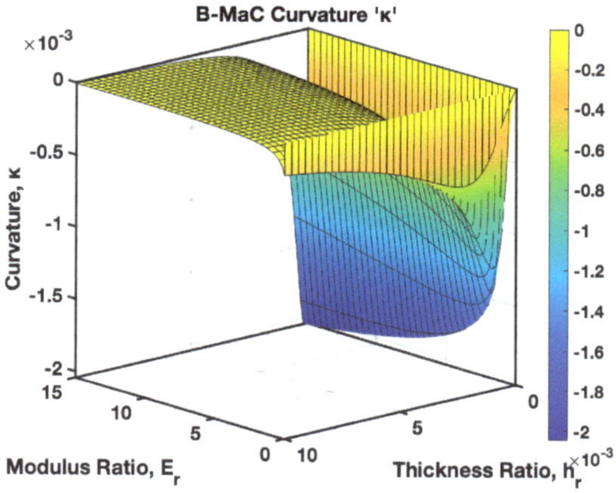

Figure 26. Curvature κ of B-MaC for varied modulus ratio E_r (0 to 15) and thickness ratio h_r (0 to 10).

The model presented in this paper considers small–moderate deflection of B-MaC, suitable for the application of NO-NC actuator/valves for µPADs and microfluidic assay. This model can be utilized for the large deflection of the cantilever, presented as a limited case for modeling. Figure 27 illustrates the curvature κ of the B-MaC for varied hygroexpansion/actuation strain ϵ_h; the curvature obtains its maximum value at $h_r \sim 0.2$ for different strains, and decreases in value upon further increasing h_r. The plot compares curvatures for 0.8%, 3.2%, 5.6%, and 8% of actuation strains. Moreover, the plot demonstrates that increasing the h_r beyond the value of 4 hardly affects the curvature of the B-MaC.

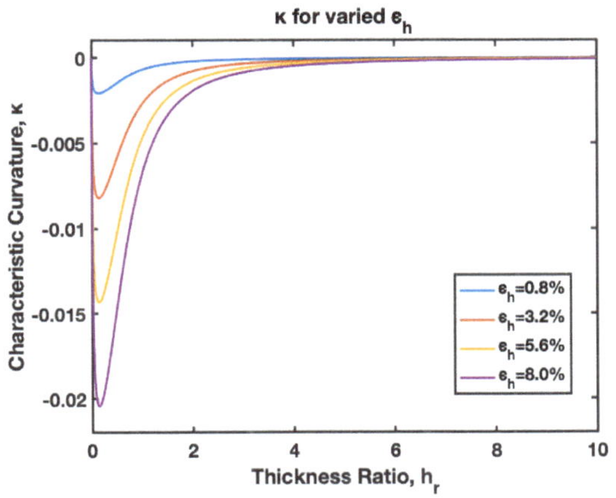

Figure 27. Curvature κ of B-MaC for varied hygroexpansion/actuation strain ϵ_h.

4. Discussion

This paper predicts the behavior of fluidically loaded B-MaCs and obtains response deflection for the purpose of modeling the moderate deflection of the cantilever. A capillary fluid flow model is considered for B-MaCs obeying the LW equation. Other important material properties, such as Young's modulus, hygroexpansive strain, and coefficient of hygroexpansion, were assessed to aid the modeling of the B-MaC. The B-MaC moderate deflection model was also assessed with different parameters to provide a full understanding of the change in behavior of response deflection of fluidic loading.

4.1. Fluid Flow in B-MaC

It is evident from the results obtained that fluid flow in B-MaC is governed by the LW equation with a good degree of reliability. The change in capillary height helps us understand the volumetric flux of fluid required for the saturation of B-MaC for this study. Upon wetting the filter paper layer of the B-MaC, it indulges in fluid imbibition, and the wetted length of the cantilever is obtained for the loading period. The results for the square of wetted length vs. time show a linear relation with a slope of 0.8344. The empirical data obtained for the plot closely follow this trend, asserting the choice of the capillary flow model for B-MaCs.

4.2. Young's Modulus of B-MaC

The material property helps us understand and predict the response deflection of B-MaCs. We are aware that B-MaCs consist of a layer of filter paper and laminated tape, and understanding the modulus of constituent material provides a complete picture of hygroexpansive strain and stress relationships. The tape layer provides additional rigidity to filter paper, resulting in strengthening the flexural rigidity of B-MaC. The fluidic loading drastically affects the flexural rigidity of paper, and the plots are evident of this reduced value, down to 7%, of dry paper vs. fully saturated paper. The higher values of the modulus for a fully saturated cantilever indicate better flexural rigidity, which leads to reduced response deflection. Similar trends are presented by Lee et al. for Whatman Grade 1 filter paper [45].

4.3. Stress and Strain in B-MaC

The stress–strain relationship helps us estimate the modulus of B-MaCs for different saturation levels/moisture content and predict the behavior of a fluidically loaded cantilever. We have learned that two kinds of strains, i.e., hygroexpansion and bending, are involved in a fluidically loaded B-MaC. The hygroexpansion strain helps us understand the change in length of paper upon wetting, and the bending strain provides details regarding the shape form of the cantilever. The hygroexpansion of W41 filter paper for different lengths of cantilever shows a similar response over time. The values of hygroexpansive strain increase until a stagnant value is achieved. For W41 filter paper, a value of 0.008 for hygroexpansive strain is attained. The bending strain is evident in the presence of laminated tape in the B-MaC. The bending strain results in 0 when no tape is present, or the hygroexpansive strain predominantly results in an effective strain in the absence of a tape layer. Lee et al. reports similar trends for Whatman Grade 1 filter paper.

4.4. Linear Coefficient of Hygroexpansion

As stated in this paper, the linear coefficient of hygroexpansion is the tendency of paper to change length upon wetting. The B-MaC is fabricated using W41 filter paper and a layer of Scotch Tape, and the linear coefficient of hygroexpansion is obtained for paper and tape upon wetting to understand the relative trends. $\beta_h = 0$ for tape shows its hydrophobic nature; therefore, it does not involve hygroexpansion. Comparatively, W41 filter paper has $\beta_h = 0.008$, reflecting the hydrophilic nature of paper.

4.5. Deflection of B-MaC

This paper brings important results in terms of the moderate response deflection of B-MaCs on fluidic loading. The experimental results obtained for the deflection of B-MaCs validate the predicted response of the B-MaC's mathematical model. The 2D quasi-static model provides a specific solution for the response deflection of the cantilever as a function of its characteristic length for given free fixed-end boundary conditions. The numerical and experimental shape forms were compared to understand the crucial details of inconsistent boundary conditions. The polynomial solutions for the response deflection assert the absence of the Neumann boundary condition of a slope at the fixed end; the mathematical model obtains the solution for the deflection with the boundary conditions of zero slopes and deflection at the fixed end. The filter paper, being docile on wetting, tends to soften and lose its rigidity, leading to a slight variation (nonzero boundary condition) at the fixed end of the cantilever. The moderate deflection model presented in this paper predicts the behavior of the B-MaC upon fluidic loading with a high degree of confidence. The empirical values obtained for the deflection fall under the 95% prediction bound, validating the adapted model for the B-MaC's deflection. The maximum (tip) deflection of the cantilever is a critical parameter that is obtained by using the interfacial boundary conditions for the wet and dry regions of B-MaC. The dry length of the cantilever remains unaffected by the fluidic loading and behaves as a straight line with rotation. The rotation is defined by the slope of the wetted length of the cantilever. The results for the tip deflection of dry length, corresponding to 0% to 100%, with an increment of 20% wetted length of the B-MaC, are presented to predict the behavior of the B-MaC upon wetting. Tip deflection is crucial to determine the design parameters for μPAD's assays. The B-MaCs are utilized as NO/NC valve actuators in microfluidic assays [3] for the sequential loading of reagents onto sample pads and detection zone. Along with the tip deflection, the rate of loading and unloading of reagents onto the reagent pads via B-MaCs is crucial to determine. The fluid flow models presented in this paper predict this with the wetted length on the cantilever, which is responsible for the loading time and the time to attain the response deflection for a given B-MaC.

4.6. Parametric Models

The behavior of the B-MaC is mainly asserted for three parameters in this paper, namely the material properties, geometrical properties, and physical properties of fabricated materials utilized for the cantilever. The material properties of B-MaCs in both layers help us predict the response of the cantilever, and the relative modulus value of the cantilever provides an understanding of their relative stiffnesses. The E_r as defined is the ratio of the Young's modulus of tape and paper, and B-MaC consists of a layer of paper laminated with tape; therefore, practically, $E_r = 1$ can be achieved when the values of the modulus of saturated paper will same as the tape modulus. In cases where $E_p >> E_t$, $E_r \rightarrow 0$, which will result in no response deflection of the B-MaC. Along with material properties, geometrical parameters are an important factor in accessing the response deflection of the B-MaC. h_r is given by the ratio of the thickness of paper to tape of the B-MaC and provides the relative sizes of the cantilever that come together in the formation of the B-MaC. The value h_r= 1 indicates identical paper and tape layer thickness, which results in ~0.1 maximum characteristic deflection. A thin film approximation is provided for applications where the substrate layer thickness is negligible; the simplified model for thin film approximation can be utilized to obtain the response deflection. Lastly the ξ parameter provides the physical aspect to understand the hygroexpansion of paper upon wetting to the thickness of paper utilized for the fabrication of B-MaCs. B-MaCs with papers with high expansion on fluidic loading and the tape layer will result in a higher response deflection; this result can be inferred from the increasing values of ξ. This study depicts the response deflection of the cantilever, with ξ varying from 0 to 0.0016. Ideally, a high B-MaC hygroexpansion will result in greater response deflection if the thickness of the paper layer is kept constant. This statement can be validated by comparing the deflection of the B-MaC in the machine and cross-machine direction [3].

4.7. Limitation of Model

In our study, the modeling was primarily focused on water-based solutions given their ubiquity in microfluidic biosensors. We did not consider other liquids such as oils due to their substantially different physical properties, including viscosity and surface tension, which would require considerable modifications to our current model to accurately depict their interaction with the Bi-Material cantilever (B-MaC). However, we acknowledge the potential of other fluids and foresee the expansion of our model to include such liquids in future work to broaden its applicability across various applications.

We opted for analytical modeling to comprehend the fundamental principles governing the behavior of Bi-Material cantilevers (B-MaCs) under fluidic loading. While this approach affords broad applicability and computational efficiency, we recognize that it might not capture all complexities and subtle effects inherent in bilayer systems, which may be better addressed through simulation software. Furthermore, the current scope of simulation tools for bilayer cantilevers remains somewhat limited. As such, future studies could substantially benefit from incorporating simulation methodologies. Building upon the groundwork of our current bilayer cantilever model under fluidic loading, simulation could serve as a logical next step for in-depth analysis, providing further validation and potentially extending the applicability of our research.

5. Conclusions

This research developed a comprehensive model to predict the behavior of a Bi-Material cantilever (B-MaC) under fluidic loading. The B-MaC, fabricated using Whatman 41 (W41) filter paper and Scotch Tape, demonstrated intriguing properties under fluidic loading. The interplay between these materials and fluidic loading became central to the understanding of the B-MaC's behavior. We significantly advanced our understanding of B-MaC behavior, as our model provided a specific solution for the response deflection of the cantilever. This was confirmed by experimental findings, with empirical values falling within the 95% prediction bound, asserting the credibility of our mathematical model. Furthermore, our investigation unveiled critical insights regarding the material properties of the constituent elements of B-MaCs. Specifically, the Young's modulus of W41 filter paper was found to decrease to approximately 20 MPa, which is about 7% of its dry state value, upon full saturation. This drastic reduction in flexural rigidity due to fluidic loading, along with the hygroexpansive behavior of W41 filter paper (which we empirically deduced to be 0.008), played a decisive role in determining the deflection of the B-MaC. We also examined different lengths of bilayer cantilever for strains in B-MaCs, including hygroexpansion and bending strains. These investigations were crucial for predicting the B-MaC's behavior under different saturation levels/moisture contents. Moreover, we conducted an in-depth analysis of the parameters affecting the response deflection. By altering these parameters, such as the ratio of the Young's modulus of tape to the paper layer (E_r) and the ratio of tape to the paper layer height (h_r), we can optimize the curvature of B-MaCs for specific applications. This study's significance is multifold. It opens new possibilities for the development of compact, low-cost, and biodegradable microfluidic devices. Our ability to predict and manipulate B-MaC behavior allows for the design and optimization of these devices for various applications, such as environmental monitoring and point-of-care diagnostics.

For future research, we propose investigating alternative materials for B-MaCs, improving the model, and considering more complex interactions. These steps could lead to a more comprehensive understanding of B-MaC behavior under different conditions and further enhance microfluidic device versatility and effectiveness. In conclusion, our work has elucidated the potential of B-MaCs for microfluidic biosensors. The findings of this study, coupled with our developed model, are poised to pave the way for future advancements in the field of microfluidics.

Author Contributions: Conceptualization, A.K. and H.H.-B.; methodology, A.K. and H.H.-B.; software, A.K.; validation, A.K. and H.H.-B.; formal analysis, A.K.; investigation, A.K. and H.H.-B.; resources, N.R., C.A. and M.F.; writing—original draft preparation, A.K. and H.H.-B.; writing—review and editing, A.K., H.H.-B., N.R., C.A. and M.F.; visualization, A.K. and H.H.-B.; supervision, N.R., C.A. and M.F.; project administration, C.A. and M.F.; funding acquisition, N.R., C.A. and M.F. All authors have read and agreed to the published version of the manuscript.

Funding: This research received no external funding.

Institutional Review Board Statement: Not applicable.

Informed Consent Statement: Not applicable.

Data Availability Statement: Data are contained within the article. Additional data not presented in this article are available on request from the corresponding author.

Acknowledgments: The authors would like to acknowledge the students, research scientists, and visiting scholars at the Microfluidics Laboratory at the University of Rhode Island for their help and support.

Conflicts of Interest: The authors declare no conflict of interest.

Appendix A

From Figure 10:

$$ds = R \cdot d\theta \tag{A1}$$

Radius of curvature:

$$\kappa = \frac{1}{R} = \frac{d\theta}{ds} \tag{A2}$$

$$ds = \frac{dx}{\cos\theta} \tag{A3}$$

Substituting Equation (A3) in (A2), we obtain

$$\kappa = \frac{1}{R} = \frac{d\theta}{dz}\cos\theta \tag{A4}$$

For small θ, $\cos\theta \approx 1$:

$$\kappa = \frac{d\theta}{dx} = \frac{d\left(\frac{dw}{dx}\right)}{dx} = \frac{d^2w}{dx^2} \quad \text{(A5)}$$

References

1. Kumar, A.; Heidari-Bafroui, H.; Charbaji, A.; Rahmani, N.; Anagnostopoulos, C.; Faghri, M. Numerical and Experimental Modeling of Paper-Based Actuators. *Chem. Proc.* **2021**, *5*, 15. [CrossRef]
2. Binnig, G.; Quate, C.F.; Gerber, C. Atomic force Microscope. *Phys. Rev. Lett.* **1986**, *56*, 930–933. [CrossRef] [PubMed]
3. Heidari-Bafroui, H.; Kumar, A.; Charbaji, A.; Smith, W.; Rahmani, N.; Anagnostopoulos, C.; Faghri, M. A Parametric Study on a Paper-Based Bi-Material Cantilever Valve. *Micromachines* **2022**, *13*, 1502. [CrossRef] [PubMed]
4. Calleja, M.; Kosaka, P.M.; Paulo, A.S.; Tamayo, J. Challenges for nanomechanical sensors in biological detection. *Nanoscale* **2012**, *4*, 4925–4938. [CrossRef] [PubMed]
5. Xu, R.; Lee, J.W.; Pan, T.; Ma, S.; Wang, J.; Han, J.H.; Ma, Y.; Rogers, J.A.; Huang, Y. Designing Thin, Ultrastretchable Electronics with Stacked Circuits and Elastomeric Encapsulation Materials. *Adv. Funct. Mater.* **2017**, *27*, 1604545. [CrossRef]
6. Quirion, D.; Manna, M.; Hidalgo, S.; Pellegrini, G. Manufacturability and Stress Issues in 3D Silicon Detector Technology at IMB-CNM. *Micromachines* **2020**, *11*, 1126. [CrossRef]
7. Zhang, L.; Pan, J.; Liu, Y.; Xu, Y.; Zhang, A. NIR–UV Responsive Actuator with Graphene Oxide/Microchannel-Induced Liquid Crystal Bilayer Structure for Biomimetic Devices. *ACS Appl. Mater. Interfaces* **2020**, *12*, 6727–6735. [CrossRef]
8. Wang, L.; Wang, D.; Huang, S.; Guo, X.; Wan, G.; Fan, J.; Chen, Z. Controllable Shape Changing and Tristability of Bilayer Composite. *ACS Appl. Mater. Interfaces* **2019**, *11*, 16881–16887. [CrossRef]
9. Morteza, A. Functional Nanomaterial Composites for Soft Sensing and Actuation. Ph.D. Thesis, ETH Zurich, Zurich, Switzerland, 2019. [CrossRef]
10. Liu, F.; Alici, G.; Zhang, B.; Beirne, S.; Li, W. Fabrication and characterization of a magnetic micro-actuator based on deformable Fe-doped PDMS artificial cilium using 3D printing. *Smart Mater. Struct.* **2015**, *24*, 035015. [CrossRef]
11. Makino, E.; Mineta, T.; Mitsunaga, T.; Kawashima, T.; Shibata, T. Sphincter actuator fabricated with PDMS/SMA bimorph cantilevers. *Microelectron. Eng.* **2011**, *88*, 2662–2665. [CrossRef]
12. Maleki, T.; Chitnis, G.; Ziaie, B. A batch-fabricated laser-micromachined PDMS actuator with stamped carbon grease electrodes. *J. Micromech. Microeng.* **2011**, *21*, 027002. [CrossRef]
13. Bhattacharjee, N.; Urrios, A.; Kanga, S.; Folch, A. The upcoming 3D- printing revolution in microfluidics. *Lab Chip* **2016**, *16*, 1720–1742. [CrossRef] [PubMed]
14. Mirzaee, I.; Song, M.; Charmchi, M.; Sun, H. A microfluidics-based on-chip impinger for airborne particle collection. *Lab Chip* **2016**, *16*, 2254–2264. [CrossRef] [PubMed]
15. Pandya, H.J.; Park, K.; Desai, J.P. Design, and fabrication of a flexible MEMS-based electro-mechanical sensor array for breast cancer diagnosis. *J. Micromech. Microeng.* **2015**, *25*, 13. [CrossRef]
16. Gaitas, A.; Malhotra, R.; Pienta, K. A method to measure cellular adhesion utilizing a polymer micro-cantilever. *Appl. Phys. Lett.* **2013**, *103*, 4. [CrossRef] [PubMed]
17. Kim, J.Y.H.; Nandra, M.; Tai, Y.C. IEEE, Cantilever Actuated by Piezoelectric Parylene-C. In Proceedings of the 25th IEEE International Conference on Micro Electromechanical Systems (MEMS), Paris, France, 29 January–2 February 2012. [CrossRef]
18. Shang, Y.F.; Ye, X.Y.; Feng, J.Y.; Zhou, H.Y.; Wang, Y. Fabrication and Characterization of a Polymer/Metal Bimorph Microcantilever for Ultrasensitive Thermal Sensing. *IEEE Sens. J.* **2014**, *14*, 1304–1312. [CrossRef]
19. Li, X.; Tian, J.F.; Nguyen, T.; Shen, W. Paper-Based Microfluidic Devices by Plasma Treatment. *Anal. Chem.* **2008**, *80*, 9131–9134. [CrossRef]
20. Jahanshahi-Anbuhi, S.; Chavan, P.; Sicard, C.; Leung, V.; Hossain, S.M.Z.; Pelton, R.; Brennan, J.D.; Filipe, C.D. Creating fast flow channels in paper fluidic devices to control the timing of sequential reactions. *Lab Chip* **2012**, *12*, 5079–5085. [CrossRef]
21. Han, K.N.; Choi, J.-S.; Kwon, J. Three-dimensional paper-based slip device for one-step point-of-care testing. *Sci. Rep.* **2016**, *6*, 25710. [CrossRef]
22. Martinez, A.W.; Phillips, S.T.; Nie, Z.; Cheng, C.-M.; Carrilho, E.; Wiley, B.; Whitesides, G.M. Programmable diagnostic devices made from paper and tape. *Lab Chip* **2010**, *10*, 2499–2504. [CrossRef]
23. Rodriguez, N.M.; Wong, W.S.; Liu, L.; Dewar, R.; Klapperich, C.M. A fully integrated paper fluidic molecular diagnostic chip for the extraction, amplification, and detection of nucleic acids from clinical samples. *Lab Chip* **2016**, *16*, 753–763. [CrossRef] [PubMed]
24. Jayawardane, B.M.; Wei, S.; McKelvie, I.D.; Kolev, S. Microfluidic Paper-Based Analytical Device for the Determination of Nitrite and Nitrate. *Anal. Chem.* **2014**, *86*, 7274–7279. [CrossRef] [PubMed]
25. Fu, H.; Song, P.; Wu, Q.; Zhao, C.; Pan, P.; Li, X.; Li-Jessen, N.Y.K.; Liu, X. A paper-based microfluidic platform with shape-memory-polymer-actuated fluid valves for automated multi-step immunoassays. *Microsyst. Nanoeng.* **2019**, *5*, 50. [CrossRef] [PubMed]
26. Bowden, N.; Brittain, S.; Evans, A.G.; Hutchinson, J.W.; Whitesides, G.M. Spontaneous formation of ordered structures in thin films of metals supported on an elastomeric polymer. *Nature* **1998**, *393*, 146–149. [CrossRef]

27. Efimenko, K.; Rackaitis, M.; Manias, E.; Vaziri, A.; Mahadevan, L.; Genzer, J. Nested self-similar wrinkling patterns in skins. *Nat. Mater.* **2005**, *4*, 293–297. [CrossRef]
28. Huang, J.; Juszkiewicz, M.; de Jeu, W.H.; Cerda, E.; Emrick, T.; Menon, N.; Russell, T.P. Capillary wrinkling of floating thin polymer films. *Science* **2007**, *317*, 650–653. [CrossRef]
29. Holmes, D.P.; Roche, M.; Sinha, T.; Stone, H.A. Bending and twisting of soft materials by non-homogenous swelling. *Soft Matter* **2011**, *7*, 5188–5193. [CrossRef]
30. Siddique, J.I.; Anderson, D.M.; Bondarev, A. Capillary rise of a liquid into a deformable porous material. *Phys. Fluids* **2009**, *21*, 013106. [CrossRef]
31. Ma, J.; Zhang, C.; Xi, F.; Chen, W.; Jiao, K.; Du, Q.; Bai, F.; Liu, Z. Experimental Study on the Influence of Environment Conditions on the Performance of Paper-Based Microfluidic Fuel Cell. *SSRN Electron. J.* **2022**, *219*, 119487. [CrossRef]
32. Liu, Z.; Hu, J.; Zhao, Y.; Qu, Z.; Xu, F. Experimental and numerical studies on liquid wicking into filter papers for paper-based diagnostics. *Appl. Therm. Eng.* **2015**, *88*, 280–287. [CrossRef]
33. Cheng, H.; Liu, J.; Zhao, Y.; Hu, C.; Zhang, Z.; Chen, N.; Jiang, L.; Qu, L. Graphene Fibers with Predetermined Deformation as Moisture-Triggered Actuators and Robots. *Angew. Chem. Int. Ed.* **2013**, *52*, 10482–10486. [CrossRef] [PubMed]
34. Leppa, T.; Sorvari, J.; Erkkila, A.; Ha, J. Mathematical modeling of moisture induced out-of-plane deformation of a paper sheet. *Model. Simul. Mater. Sci. Eng.* **2005**, *13*, 841. [CrossRef]
35. Madsen, B.; Hoffmeyer, P.; Lilholt, H. Hemp yarn reinforced composites-III. Moisture content and dimensional changes. *Compos. Part A-Appl. Sci. Manuf.* **2012**, *43*, 2151–2160. [CrossRef]
36. Yao, Y.; Chen, X.; Guo, H.; Wu, Z.; Li, X. Humidity sensing behaviors of graphene oxide-silicon bi-layer flexible structure. *Sens. Actuators B Chem.* **2012**, *161*, 1053–1058. [CrossRef]
37. Duigou, L.; Castro, M. Moisture-induced self-shaping flax-reinforced polypropylene biocomposite actuator. *Ind. Crops Prod.* **2015**, *71*, 1–6. [CrossRef]
38. Mao, Z.; Yoshida, K.; Kim, J.W. A micro vertically-allocated SU-8 check valve and its characteristics. *Microsyst. Technol.* **2019**, *25*, 245–255. [CrossRef]
39. Chen, G.; Kumar, A.; Heidari-Bafroui, H.; Smith, W.; Charbaji, A.; Rahmani, N.; Anagnostopoulos, C.; Faghri, M. Paper-Based Bi-Material Cantilever Actuator Bending Behavior and Modeling. *Micromachines* **2023**, *14*, 924. [CrossRef]
40. Gendron, G.; Dano, M.L.; Cloutier, A. A numerical study of the hygro- mechanical deformation of two cardboard layups. *Compos. Sci. Technol.* **2004**, *64*, 619–627. [CrossRef]
41. Dano, M.-L.; Bourque, J.-P. Deformation behavior of paper and board subjected to moisture diffusion. *Int. J. Solids Struct.* **2009**, *46*, 1305–1316. [CrossRef]
42. Heidari-Bafroui, H.; Kumar, A.; Hahn, C.; Scholz, N.; Charbaji, A.; Rahmani, N.; Anagnostopoulos, C.; Faghri, M. Development of a New Lab-on-Paper Microfluidics Platform Using Bi-Material Cantilever Actuators for ELISA on Paper. *Biosensors* **2023**, *13*, 310. [CrossRef]
43. Mäkelä, P.; Östlund, S. Orthotropic elastic-plastic material model for paper materials. *Int. J. Solids Struct.* **2003**, *40*, 5599–5620. [CrossRef]
44. Washburn, E.W. The dynamics of capillary flow. *Phys. Rev.* **1921**, *17*, 273. [CrossRef]
45. Lee, M.; Kim, S.; Kim, H.-Y.; Mahadevan, L. Bending and buckling of wet paper. *Phys. Fluids* **2016**, *28*, 042101. [CrossRef]

biosensors

MDPI

Review

Revolutionizing Drug Discovery: The Impact of Distinct Designs and Biosensor Integration in Microfluidics-Based Organ-on-a-Chip Technology

Sheng Yuan [1], Huipu Yuan [2], David C. Hay [3], Huan Hu [4,*] and Chaochen Wang [1,5,*]

1 Centre of Biomedical Systems and Informatics, Zhejiang University-University of Edinburgh Institute (ZJU-UoE Institute), School of Medicine, International Campus, Zhejiang University, Haining 314400, China
2 Sir Run Run Shaw Hospital, School of Medicine, Zhejiang University, Hangzhou 310020, China
3 Centre for Regenerative Medicine, Institute for Regeneration and Repair, The University of Edinburgh, Edinburgh EH16 4UU, UK; david.hay@ed.ac.uk
4 Zhejiang University-University of Illinois Urbana-Champaign Institute (ZJU-UIUC Institute), International Campus, Zhejiang University, Haining 314400, China
5 Department of Gynecology, The Second Affiliated Hospital, School of Medicine, Zhejiang University, Hangzhou 310020, China
* Correspondence: huanhu@intl.zju.edu.cn (H.H.); chaochenwang@intl.zju.edu.cn (C.W.)

Abstract: Traditional drug development is a long and expensive process with high rates of failure. This has prompted the pharmaceutical industry to seek more efficient drug development frameworks, driving the emergence of organ-on-a-chip (OOC) based on microfluidic technologies. Unlike traditional animal experiments, OOC systems provide a more accurate simulation of human organ microenvironments and physiological responses, therefore offering a cost-effective and efficient platform for biomedical research, particularly in the development of new medicines. Additionally, OOC systems enable quick and real-time analysis, high-throughput experimentation, and automation. These advantages have shown significant promise in enhancing the drug development process. The success of an OOC system hinges on the integration of specific designs, manufacturing techniques, and biosensors to meet the need for integrated multiparameter datasets. This review focuses on the manufacturing, design, sensing systems, and applications of OOC systems, highlighting their design and sensing capabilities, as well as the technical challenges they currently face.

Keywords: organ-on-a-chip; drug development; biomedicine

Citation: Yuan, S.; Yuan, H.; Hay, D.C.; Hu, H.; Wang, C. Revolutionizing Drug Discovery: The Impact of Distinct Designs and Biosensor Integration in Microfluidics-Based Organ-on-a-Chip Technology. *Biosensors* **2024**, *14*, 425. https://doi.org/10.3390/bios14090425

Received: 24 July 2024
Revised: 20 August 2024
Accepted: 30 August 2024
Published: 3 September 2024

1. Introduction

The process of developing new drugs is a complex, costly, and time-consuming process with a high attrition rate [1]. Specifically, a new drug must undergo laboratory experiments, preclinical studies, and clinical trials (phases I, II, and III) before it can be approved for marketing. Following market entry, it must also undergo phase IV clinical trials and post-marketing approval [2]. Many candidate drugs are eliminated at different stages throughout the development process due to poor efficacy, safety issues, high production costs, and other factors [3]. Consequently, the average development cycle of a new drug takes 10–15 years and costs over 2 billion US dollars [4].

Given these challenges, the pharmaceutical industry has sought more sophisticated drug development models that are better at screening out compounds with serious off-target effects. This urgent need has spurred the development of organ-on-a-chip (OOC) technology based on microfluidics [5]. In OOC systems, microfluidic cell cultures are integrated with circuits that precisely manipulate the cells' microenvironment to simulate the activities and physiological responses of various human organs, thereby providing an efficient research model for biomedical study [6]. OOC systems have many advantages compared to traditional research methods. They can replicate more complex in vivo environments, including interactions among cells, tissues, and blood vessels. Additionally, OOCs require fewer reagents, cells, and space while facilitating rapid analysis, high-throughput experimentation, and automation [7,8]. Overall, this technology is expected to mitigate the risks and costs associated with new drug development while enhancing efficiency and output quality.

The global demand for OOC technology is rapidly increasing due to its enormous potential in new drug development [5]. It is projected to grow from 131.11 million US dollars in 2024 to 1.3883 billion US dollars by 2032, with a compound annual growth rate of 34.3% during the forecast period [9]. This significant growth trend reflects the market expansion and widespread acceptance and promotion of OOC technology from academic research to commercial applications. As more biotechnology and pharmaceutical companies invest in the research and development of OOC technology, this field is expected to maintain rapid development momentum in the coming years.

This review summarizes the technical basis of OOCs based on microfluidic technology, existing designs, integrated sensors, and their applications in drug discovery and preclinical screening. Additionally, we analyze the main challenges and recent breakthroughs of the technology, highlight future research directions, and discuss the broad application prospects of this technology in drug development, personalized medicine, and disease treatment. Overall, OOC technology is advancing swiftly, especially with respect to enhanced design and the integration of multi-functional sensors. Despite existing challenges, OOC holds significant and expansive potential for the foreseeable future.

2. Fabrication and Sensors of Organ-on-a-Chip

2.1. OOC Fabrication

The major framework of OOCs is generally based on microfluidic chips, which leverage microfabrication technology or micro-electro-mechanical systems (MEMS) technology to fabricate micrometer-scale channels, reservoirs, valves, etc [10]. The most widely used microfluidic chip technology employs soft lithography [11]. As shown in Figure 1, a mold is first fabricated using ultra-violet (UV) lithography to define microstructures. The typical mold material is photoresist, such as SU-8 (an epoxy-based negative photoresist), or it can be silicon (Figure 1A,B). After the mold is ready, the liquid PDMS is typically mixed with its curing agent in a 10:1 ratio then poured on top of the mold. After curing, the PDMS mold is detached. The final step of the process is to bond the PDMS to a glass cover after plasma treatment. After bonding, the PDMS and the glass substrate can be post-baked for 30 min at 70 degrees C to further improve the strength of the bonding. Microchannels of other materials such as thermoplastic are fabricated using a similar process but with hot embossing as the patterning step instead of material curing. After microchannel patterning, the bonding between thermoplastic can be thermal bonding, adhesive bonding, or ultrasonic bonding [12].

Since OOCs normally require fluorescence imaging measurements, even though the PDMS is almost transparent in the visible light range, it emits a certain amount of fluorescence that contributes to the background noise of the fluorescence signal [13]. In addition to glass, which offers much better optical transparency and minimal fluorescence background, researchers have also explored plastic materials such as polystyrene (PS), poly (methyl methacrylate) PMMA, and polycarbonate (PC), which are usually patterned through hot embossing and injection molding [14].

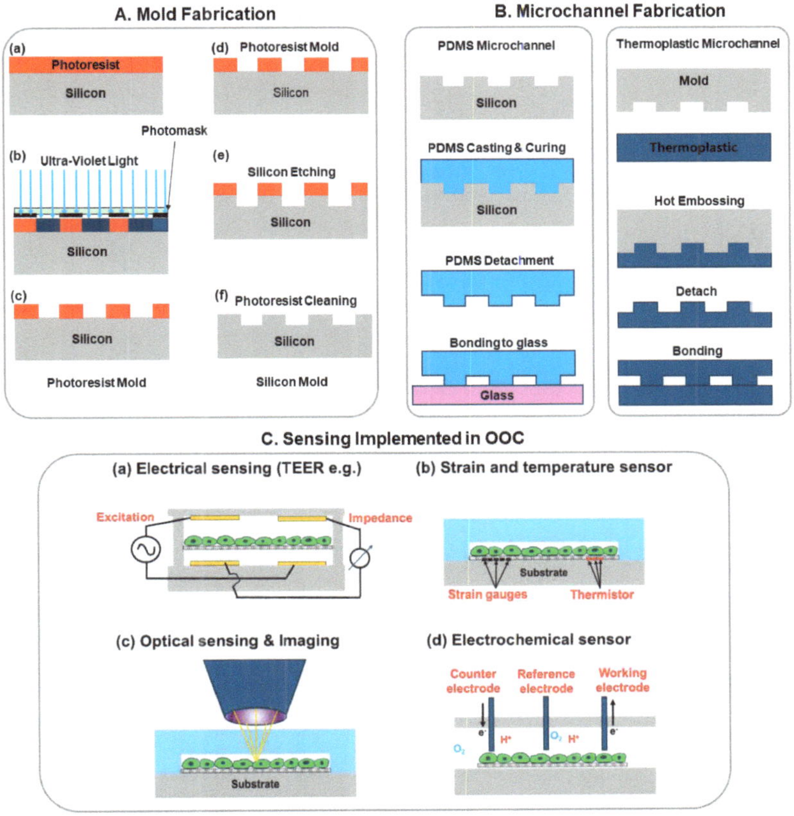

Figure 1. Illustration of OOC chip fabrication: mold fabrication and microchannel fabrication. (**A**) Mold fabrication using conventional UV lithography and etching: (a) Photoresist spincoating and prebake; (b) UV-lithography exposure with a photomask; (c) photoresist development; (d) in cases where photoresist mold is not sufficient, a more reliable silicon mold is fabricated starting with a photoresist mold on silicon as step (c); (e) Reactive ion etching of silicon with photoresist as etching mask; (f) Removing photoresist and rendering a pristine silicon mold; (**B**) Microchannel fabrication using molding for PDMS or hot embossing for thermoplastics; (**C**) Various sensing implemented in OOCs: (a) Electrical sensing mostly impedance sensing such as trans-epithelial electrical resistance (TEER); (b) Strain and temperature sensors mostly made of strain gauges and thermistors; (c) Optical sensing and imaging to investigate the cell morphology and photoluminescence of molecules; (d) Electrochemical sensors detect the presence and concentration of ions, gas molecules, glucose, etc.

2.2. Sensing Systems Implemented in OOC

Sensors can detect and analyze various parameters in biological systems, such as physiological signals, biochemical reactions, and physical changes, providing real-time detection of health or disease status [15]. Within OOC systems, sensors are crucial [16]. Integrated sensors allow for the real-time monitoring and analysis of cell behavior, tissue function, and drug effects, thereby enhancing the OOC system's capability to simulate and predict biological responses accurately [17]. Figure 1C shows four representative types of sensing including electrical impedance, strain, temperature, and optical and electrochemical sensing.

Various types of sensors have been developed and integrated into OOCs to measure distinct biological parameters, as illustrated in Figure 1C. For example, Trans-epithelial electrical resistance (TEER) sensors are used to assess the barrier functions of epithelial and endothelial cells in OOC models simulating the intestine, lung, and blood–brain barrier [18–20]. Electric cell-substrate impedance sensing (ECIS) sensors provide real-time tracking of cell proliferation, migration, and differentiation, contributing significantly to the study of diverse cell types, including cancer and endothelial cells under various culture conditions [21–23].

In addition to these, microelectrode arrays (MEAs) are employed to record the electrical activity of heart and neuronal cells, providing insights into their electrophysiological properties. These sensors have been implemented in heart-on-chip and brain-on-chip models to study the effects of drugs and other interventions [24–26]. Strain gauges are used to measure the mechanical forces generated by cells, particularly cardiomyocytes, in heart-on-a-chip models, thereby helping us to understand the impact of drugs and diseases on heart function [24,25].

Environmental factors such as temperature, pH, and oxygen levels are crucial to driving specific cell phenotypes and are monitored using various types of sensors. For instance, a general temperature sensor can monitor temperature in real-time by measuring the electrical resistance change of temperature sensors [26]. Different methods are used for detecting pH and oxygen levels [27]. A microfluidic optical platform constructed by Mousavi Shaegh et al. allows for real-time monitoring of pH and oxygen in OOC systems using low-cost electro-optical devices [24].

Mechanical force, including shear stress, is another key parameter that sensors need to detect. For instance, heartbeat dynamics, a primary indicator for assessing heart health, are studied using various methods [23]. Lind et al. integrated piezoresistive sensors into a multi-layered cantilever beam to guide the growth of cardiac tissue and measure the tissue's contraction force [25]. In another approach, Aung et al. proposed a detection method that is highly compatible with the OOC design, where the contraction of cardiac tissue generates mechanical forces transmitted to the surrounding hydrogel, resulting in measurable deformation [28].

Electrochemical sensors monitor the metabolic activity of cells by detecting the release of metabolites such as glucose, lactate, and oxygen. For example, amperometric sensors have been used to measure glucose consumption and lactate production in liver-on-chip models [29–31]. Optical sensors, particularly those based on photoluminescence, monitor parameters like oxygen levels and pH, providing insights into cellular respiration and metabolic activity. These sensors have been integrated into various OOC models, including liver-on-chip and lung-on-chip, to study cellular responses to hypoxia, drugs, and other stimuli [32–35]. Sensors for real-time monitoring of specific proteins are also available. For instance, Li et al. designed a novel label-free optofluidic nanosensor for real-time analysis of single-cell cytokine secretion [36]. This sensor monitors the dynamic secretion of cytokines without molecular markers, which can interfere with cell integrity and time resolution.

In most instances, OOC systems incorporate multiple sensor types to simultaneously monitor various parameters. An example of this is a liver-on-a-chip model that combines optical oxygen sensors with electrochemical glucose and lactate sensors. This approach has been used to investigate the metabolic response of liver cells to drugs, offering a more comprehensive understanding of cell function and toxicity [29].

In summary, the integration of diverse sensors into OOC systems underscores the versatility of these platforms and their potential to revolutionize our understanding of cellular behavior and response to stimuli. Examples of sensors implemented in different types of OOC systems are summarized in Table 1.

Table 1. Sensors implemented in different OOC platforms.

OOC Platform	Measurements	Sensors	Applications	Reference
Lung	The barrier integrity of the cells, the secretion of inflammatory markers, Mechanical stress, and changes in cell mechanics	Trans epithelial electrical resistance (TEER) measurement, sodium fluorescein permeability test, ELISA and ATP luminescence assay, and special material that changes color in sync with air pressure	Lung disease models, drug evaluation, mechanical stretching effect	[37–39]
Heart	Electrophysiological signals and mechanical contractions of cardiac tissue, and dynamic tissue beating pulse	Microelectrode arrays (MEAs), piezoresistive sensors, calcium transient dye, optical sensing technology, and nanowire probe	Drug evaluation, cardiotoxicity detection	[24,25,40,41]
Liver	Oxygen concentration, cell growth population	Electrochemical dissolved oxygen sensors produced by inkjet printing technology, electrochemical impedance spectroscopy, and amperometric sensors	Metabolic activity monitoring, hepatotoxicity tests	[29–31,42]
Intestine	pH, oxygen, temperature, barrier integrity, ion flow resistance, sequential impedance measurement and cell migration	Fluorescent probes, TEER sensors, electrochemical sensors, electrical cell-impedance sensors, monitoring sensors	Barrier function test, ion transport monitoring, anti-inflammation test, human disease models	[43–51]
Brain	pH, oxygen, temperature, shear stress, secreted molecules (e.g., cytokines, insulin), blood flow, cell viability, cell-cell interactions, and BBB crossing of drugs and nanoparticles	MEA, External sensor-integrated BOC (TEER measurement and multi-parameter measurement), and internal sensor-integrated BOC (microelectrode arrays and multi-sensor integration platform)	Real-time brain activity monitoring, neurodegenerative disease model, drug development and screening, pre-clinical test of novel therapies	[32,52–61]
Skin	pH, oxygen, temperature, tight junction formation	Optical pH, oxygen and temperature monitors, TEER sensors, electrochemically activated immune biosensors attached to physical microelectrodes	Skin barrier function test, drug evaluation, toxicity test, biomimetic artificial skin model	[62–66]

3. Design of Organ-on-a-Chip

The design of the OOC is crucial to its functionality. The complexity and diversity of these designs determine the breadth and depth of the chip's application in drug development (Figure 2). In general, microfluidics-based OOCs can be divided into two categories: single-OOCs and multi-OOCs. Single-OOCs simulate the function of a single organ, while multi-OOCs integrate the functions of multiple organs to simulate complex physiological systems.

3.1. *Single-Organ-on-a-Chip Systems*

Single-OOCs are microfluidic devices that mimic the microenvironment and function of a single organ, providing an ideal in vitro model for studying a specific organ's physiological and pathological processes. Researchers can use these devices to analyze drug interactions with specific organs and assess how organs respond to various physiological conditions [67]. Consequently, these chips are widely used in drug screening, toxicity testing, and disease modeling, identifying key biological mechanisms and thus providing reliable references for clinical trials [5,68]. In this section, we mainly discuss several prevalent types of Single-OOC models and elucidate how their unique designs enhance the in vitro simulation of respective organs. We also briefly summarize the other types of OOCs established in recent studies (Table 2).

Table 2. Designs of different Single- Organ-on-a-Chips.

Organ Type	Special Structure	Morphological Simulation	Environmental Simulation	Special Indicator Tests	General Indicator Tests	References
Lung	Alveoli	Dynamic deformation and gas exchange between alveoli and capillaries	Simulating the gas exchange environment during respiration	Gas exchange efficiency	Temperature, pH, oxygen concentration, cell viability, etc.	[69,70]
Heart	Myocardial tissue	Periodic mechanical contraction of heart tissue	Simulating the electrophysiological environment and mechanical stress during heartbeats	Contraction stress of heart tissue, electrophysiological parameters		[24]
Intestine	Intestinal epithelial cells	Periodic mechanical contraction of the intestine; interaction among intestinal epithelial cells, vascular endothelial cells, and microbiome	Simulating the chemical environment inside the intestine, including pH and microbial communities	Barrier function, microbiome balance, inflammation markers		[71,72]
Kidney	Glomerulus	Imitates the filtering action of the kidney glomerulus	Simulating fluid flow, electrolyte concentration gradient, and pressure changes	Glomerular filtration rate, metabolite concentration, renal tubule reabsorption function		[73,74]
	Renal tubules	Imitating the reabsorption function of the nephron				
Liver	Liver lobule	Imitating the special shape of the liver lobules and the multiple blood vessels through the liver lobules	Simulating the liver's metabolic environment, including oxygen concentration, nutrient, and metabolite concentrations	Metabolic activity, toxicity response, liver enzyme activity		[75,76]
	Hepatic Sinus	Cultivation of endothelial cells from perforated, discontinuous hepatic sinusoids and associated macrophages				
Spleen	Spleen red pulp	Imitates the red bone marrow, stores red blood cells and white blood cells, and screens for healthy red blood cells	Simulating the closed-fast and open-slow microcirculation in the spleen	Mechanical and physiological responses of red blood cells		[77]
Bone	Bone marrow	Three-dimensional bone tissue and bone marrow cavities to mimic the spatial layout of bone	The hematopoietic microenvironment includes stromal cells that support hematopoietic stem cells, vascular networks, signaling molecules and cytokines that regulate cell activity, marrow signaling molecules, and cytokines that regulate cell activity.	Hematopoietic function, cell type, cytokine level		[68]
	Osteoblasts, osteocytes, and osteoclasts	By adjusting the ratio of osteoblasts, osteocytes, and osteoclasts, different bone conditions can be simulated.	The permeability of the vascular system under different bone conditions is simulated through a simulated vascular channel lined with endothelial cells.	Cell co-culture ratio, vascular permeability, tissue mineralization level		[78]
Brain	Blood–brain barrier	Cultured human brain microvascular endothelial cells, human brain astrocytes, and pericytes formed a blood–brain barrier	Simulates the hypoxic microenvironment with less oxygen that the blood–brain barrier is exposed to during development.	TEER, apparent permeability, tight junction protein expression, efflux pump function,		[79]
Lymphatic system	Lymphoid follicle	Using B and T cells, ectopic lymphoid follicles were simulated in 3D extracellular matrix gel	3D extracellular matrix gel as a platform for the spontaneous assembly of ectopic lymphoid follicles.	Lymphoid follicle formation and number, B cell activation status, cytokine secretion		[80]

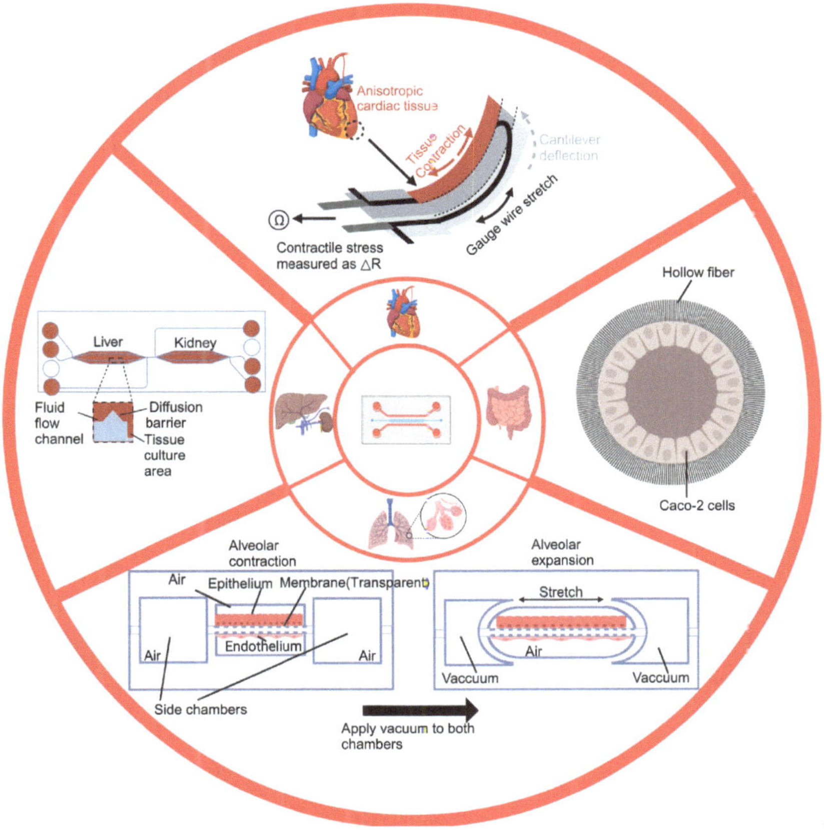

Figure 2. Overview of distinct Organ-on-a-chip designs. The figure shows four chips and their corresponding designs. The lung-on-a-chip simulates the dynamic deformation and gas exchange between the alveoli and capillaries. The heart-on-a-chip simulates the periodic mechanical contraction of heart tissue. The intestinal-on-a-chip simulates the intestinal environment. The liver-kidney chip simulates the interaction between the liver and kidneys in the human body. Created with BioRender.com.

3.1.1. Lung-on-a-Chip

The lungs are the central organs of the respiratory system in humans and most other vertebrates, playing a crucial role in maintaining cellular respiration by regulating blood oxygen and carbon dioxide levels [81]. Microscopically, gas exchange occurs within millions of alveolar units, the basic functional units of the lung. The alveolar walls comprise a thin epithelial cell layer and a rich capillary network, facilitating efficient oxygen and carbon dioxide exchange [82]. To replicate the biological functions of the lungs effectively, an in vitro lung model must be designed with appropriate cellular components and a structure that supports gas exchange. Microfluidic technology offers precise fluid flow and constant gas exchange, creating a three-dimensional microstructure and microenvironment that mimics the human lung [69,83].

Huh et al. produced the first biomimetic microfluidic lung model using classic soft lithography [69] (Figure 3). This model consisted of an upper and lower PDMS (polydimethylsiloxane) frame with a PDMS porous membrane in the middle. The upper and lower PDMS frames were seeded with epithelial and endothelial cells to mimic the microchannels of the airways and blood vessels, respectively. The PDMS membrane, coated with extracellular matrix (ECM) proteins, separated the two chambers to simulate the alveolar–capillary barrier, facilitating gas and nutrient exchange. Applying a vacuum to the two chambers reduces the pressure inside the microchambers, causing the PDMS membrane to deform elastically. This deformation simulates the expansion of the alveoli during inhalation. When the vacuum is released, the PDMS membrane returns to its original state due to its elastic properties, causing the membrane and attached cells to relax, simulating alveolar contraction during exhalation. This elastic recoil effect is crucial for mimicking the dynamic deformation and interaction between alveoli and capillaries.

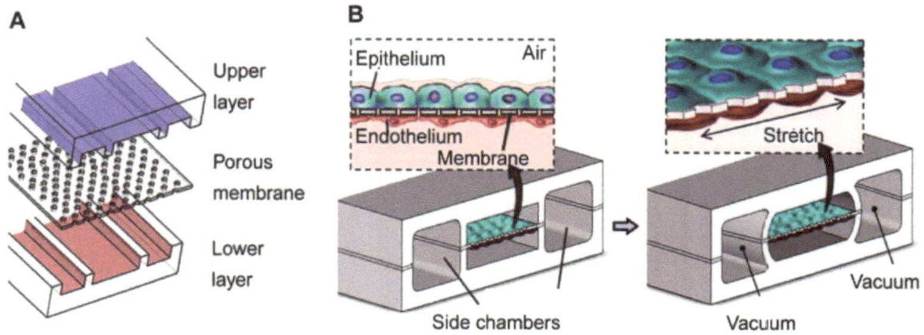

Figure 3. Illustration of the lung model produced by Huh et al. (**A**) The model consists of a PDMS frame and a PDMS porous membrane coated with alveolar epithelial cells in the upper part of the membrane (mimicking air channels) and endothelial cells in the lower part (mimicking microvascular channels). (**B**) The left part simulates the contraction of the alveoli during expiration. The right part simulates the application of vacuum to the lateral lumen, which produces a cyclic stretch that simulates the expansion of the alveoli during inspiration. Reproduced with permission from [69].

In terms of applications, the research team has successfully simulated the inflammatory process in the lungs, including the production of early response cytokines by epithelial cells, the activation of the vascular endothelium, and the adhesion and penetration of white blood cells. Furthermore, the team has explored the pulmonary toxicity of nanoparticles. Findings indicate that respiratory motions amplify the inflammation triggered by nanoparticles, evidenced by heightened ICAM-1 expression in endothelial cells and the increased adhesion and infiltration of neutrophils. The lung-on-a-chip device thus shows great potential for studying the pathological mechanisms of lung diseases, toxicology, and drug development [69,83].

In addition, based on this design, Dasgupta et al. conducted a study on radiation-induced lung injury (RILI), including the effects of radiation on cell structure and function and the potential therapeutic effects of drugs [70]. The study showed that within 6 hours of radiation exposure, both pulmonary epithelial and endothelial cells exhibited DNA damage, increased cell proliferation, upregulation of inflammatory factors, and loss of barrier function; the test drugs (lovastatin and prednisolone) showed potential effects in inhibiting acute RILI. These results support the use of organ-on-a-chip models as a novel method for studying the molecular mechanisms of acute RILI and evaluating new radiation protection therapies.

3.1.2. Heart-on-a-Chip

The heart is the central organ in the circulatory system, primarily responsible for generating pneumatic pressure and driving blood circulation. The myocardium constitutes the main body of the heart. The periodic contraction and relaxation of the uniformly arranged cardiomyocytes in each layer provides the heart's pumping action [84]. In order to effectively mimic the heart's biological functions, in vitro heart models must simulate the periodic mechanical contraction of cardiac tissue and allow for real-time monitoring of contraction stress.

Cardiotoxicity is one main reason for drug recalls. Drugs, if cardiotoxicity levels are unidentified in preclinical studies, may cause lethal arrhythmias and death. Lind et al. developed a 3D-printed heart-on-a-chip (HOC) system to study how cardiac tissue responds to drugs [24] (Figure 4). This system was constructed by sequentially printing multiple materials using direct ink writing (DIW). It mainly comprises three layers: a base layer (dextran film, thermoplastic polyurethane (TPU)), a sensor layer (carbon black-doped TPU ink (CB: TPU)), and a tissue-guiding layer (PDMS). These layers provide structural support for the device, placement of sensors, and guidance for cell alignment and assembly. Cardiac muscle cells are seeded onto the tissue guide layer, where they self-assemble into a layered structure that simulates the natural arrangement of cardiac tissue. Furthermore, the integrated sensors can non-invasively detect the contraction stress of the heart tissue and send out the data in real time [85,86]. The researchers conducted a drug dose–response study using this model to investigate the effects of the drug on the contractility and rhythm of the heart tissue. The results showed that a drug (verapamil) produced a negative inotropic effect (attenuated contractility) on cardiac tissue after administration, which is consistent with previous studies. Similarly, a drug (isoprenaline) exhibited a positive inotropic effect (enhanced contractility), which is also consistent with previous studies [87,88]. This model construction method enhances the efficiency and accuracy of heart disease research, allowing for long-term studies of heart tissue function. It provides new tools for heart disease models and drug screening.

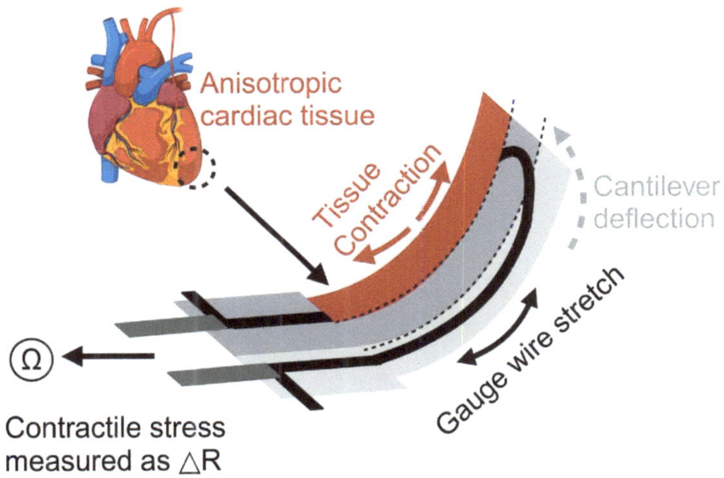

Figure 4. Illustration of the heart model produced by Lind et al. The device is constructed with a U-shaped cantilever that is connected to electrodes for measuring resistance changes. Contraction of cardiac tissue causes deformation of the cantilever, which in turn changes the electrical resistance, and the change in resistance can be used to quantify the contractile stresses exerted on the cardiac tissue. Created with BioRender.com.

It is worth noting that classic HOC models lack the study of epicardial cells. Bannerman et al. constructed a heart tissue model containing the outer layer of the heart, the epicardium, based on the traditional HOC model containing the inner layer of the heart muscle structure, and tracked the migration of epicardial cells in the experiment [89]. The results showed that under conditions simulating ischemia-reperfusion injury, the epicardial heart tissue had less cell death and less impact on functional characteristics than the control group without epicardial tissue, which laid a foundation for future applications in the study of heart disease and the testing of treatment methods.

3.1.3. Liver-on-a-Chip

The liver, an organ unique to vertebrates, serves as the principal site for critical physiological processes including metabolism, detoxification, bile secretion, immune responses, and the synthesis of various biochemical substances [90]. The structural and functional cornerstone of the liver is the liver lobule, typically hexagonal, comprising millions of hepatocytes. This structure includes hepatocyte plates, hepatic sinusoids, and portal areas, orchestrating the primary physiological functions of the liver through their complex structural and functional integration [91].

Liver-on-a-chip (LOC) is extensively employed to explore drug metabolism and liver diseases, leveraging designs that emulate the architecture of liver lobules and sinusoids [92]. For instance, lobule chips designed by Ya et al. feature a hexagonal configuration mirroring the liver lobule, incorporating a six-layer assembly with perfusion inlets and outlets at the top, a central liquid distribution system, and a basal cell culture area [75] (Figure 5). The chip simulates the hexagonal structure of the liver lobule, with the portal vein (PV) at each vertex, the hepatic artery (HA) at the midpoint of each hexagonal edge, and the prominent central vein (CV) at the center of the top surface. This design reproduces the blood flow path of the liver lobule, with the cell culture medium entering from the PV and HA inlets and finally converging to the CV through the flow channel system.

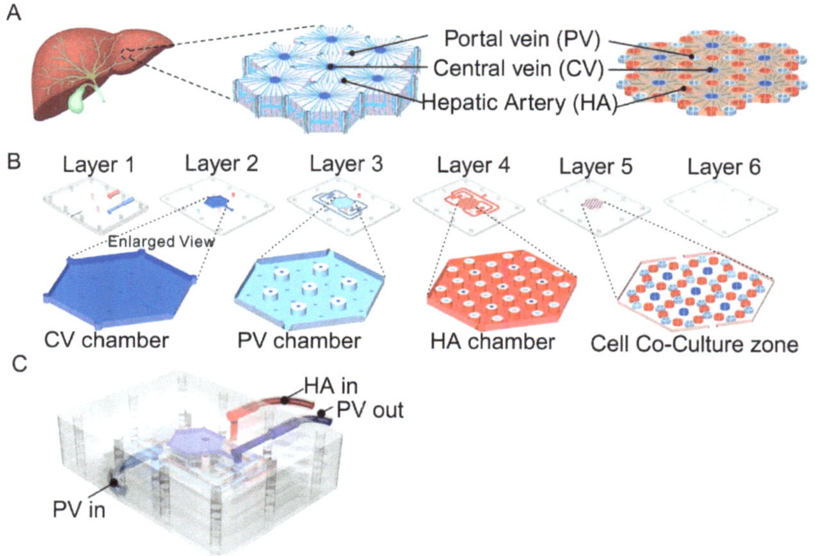

Figure 5. Illustration of the lobules model produced by Ya et al. (**A**) The chip mimics the hexagonal structure of a liver lobule, with the portal vein (PV) at each vertex, the hepatic artery (HA) at the midpoint of each hexagonal edge, and the prominent central vein (CV) at the center of the top surface. (**B**) The chip has a six-layer structure: the first and second layers contain fluid collection chambers for the CV and PV. The third and fourth layers are designed with a raised platform to separate the flow of the PV, HA, and CV. The fifth layer is the co-culture area, which is designed with micro-columns to support cell growth and guide cell alignment to form hepatic sinuses. The sixth layer is the sealing layer, which keeps the entire chip sealed and structurally stable. (**C**) Schematic diagram of the overall structure of the chip and the inlet (PV/HA) and outlet (PV) of the culture medium. Reproduced with permission from [75].

The Ya et al. study showed that hepatocytes cultured on a chip exhibited longer-term biochemical functions (far exceeding conventional two-dimensional culture systems), including protein synthesis, urea, and bile acid production. In addition, the researchers also conducted drug toxicity tests and tumor growth simulations. For the toxicity test, the long-term effects of the drug (acetaminophen) on liver function were observed, including changes in enzyme activity and a sustained decrease in cellular metabolism. For the tumor growth simulation, the chip simulated the growth of HepG2 liver cancer cells in liver tissue.

In the experiment, HepG2 cells formed tumor spheres on the chip and simulated the tumor microenvironment, including vascular networks and hypoxic areas.

Similarly, the sinus chip developed by Jang et al., akin to the lung-on-a-chip by Huh et al., includes dual microchannels: an upper hepatocyte-seeded channel and a lower vascular channel lined with sinusoidal endothelial cells [76] (Figure 6). This design may also incorporate Kupffer and stellate cells to stimulate immune responses and fibrotic processes within the liver. The continuous fluid dynamics of the upper and lower channels are controlled by microfluidic channels connected to an external pump system, simulating liver blood flow and enhancing cell–cell interactions and drug metabolism.

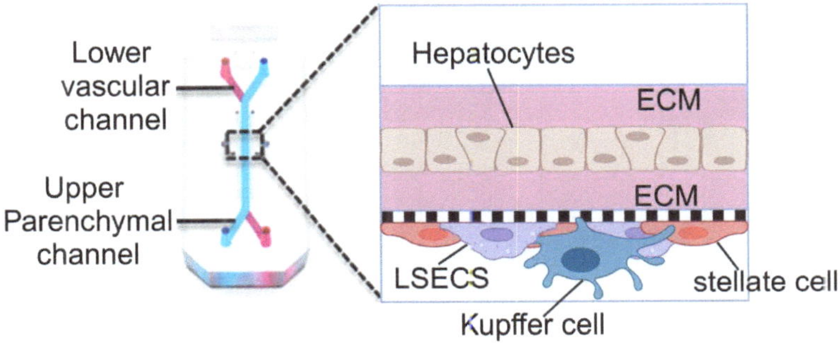

Figure 6. Illustration of the sinus model produced by Jang et al. The chip contains two parallel microchannels, the upper one for hepatocytes and the lower one for blood vessels, separated by a porous membrane. The chip is coated with an extracellular matrix (ECM) to support cell attachment and growth. The upper channel is seeded with primary hepatocytes and the lower channel with liver sinusoidal endothelial cells (LSECs), Kupffer cells (macrophages), and stellate cells. Reproduced with permission from [76].

Additionally, the research team employed TAK-875, an experimental drug intended for type 2 diabetes management, to investigate the risk of drug-induced liver injury. The results showed that the potential accumulation of TAK-875 and its metabolites in liver cells may lead to cell dysfunction, including inhibition of bile excretion and the activation of cellular stress and inflammatory responses.

Based on the classic LOC, Jiao, Xie, and Xing built a gravity-driven perfusion model of a pump-free LOC [93]. This design avoids the need for an external pump and complex tubing connections, simplifies the construction and operation of experimental equipment, and is more suitable for large-scale drug screening or multi-sample comparison studies. In addition, because there are no external pumps or tubing restrictions, the pump-free chip is easier to integrate with other experimental systems or equipment for a variety of research scenarios. The results found that the chip exhibited similar hepatotoxicity responses to traditional two-dimensional models in drug toxicity testing (aristolochic acid I and its analog aristolochic acid II) but was more sensitive in detecting toxic substances. Overall, this study demonstrates the potential application of this system in drug hepatotoxicity research, especially in high-throughput drug screening.

3.1.4. Other Single-Organ-on-a-Chip Systems

In addition to lung-on-a-chip and heart-on-a-chip, developing other single-OOCs is also crucial for biomedical research and drug development. For example, kidney-on-a-chip (KOC), intestine-on-a-chip (IOC), and spleen-on-a-chip (SOC) are indispensable tools for studying drug metabolism, toxicology assessment, and disease mechanisms.

KOCs replicate the kidney's filtration, absorption, and secretion functions. These devices simulate the glomerulus or renal tubule and facilitate the study of acute kidney injury, chronic kidney disease, and drug effects on the kidney [94]. Like lung-on-a-chip, glomerulus chips feature an upper channel cultured with glomerular endothelial cells to mimic capillaries and a lower channel cultured with podocytes to replicate Bowman's capsule. An injection pump adjusts flow rates to simulate physiological and pathological hemodynamic microenvironments [73]. Similarly, renal tubule chips have upper and lower channels cultured with renal tubular epithelial cells to simulate luminal and interstitial regions, respectively, with fluid shear stress applied by an injection pump [74].

IOCs simulate the intestinal environment of the human digestive system and contribute to the study of microbial interaction, intestinal barrier function, and inflammatory bowel diseases [47]. Typically, an IOC comprises intestinal epithelial cells, vascular endothelial cells, and a representative microbial community [47]. In the design by Jalili-Firoozinezhad et al., the upper channel is used to culture intestinal epithelial cells to simulate the intestinal lumen environment, while the lower channel cultures vascular endothelial cells to mimic the vascular environment [71]. Additionally, an anaerobic chamber with an oxygen sensor facilitates microbial culture.

The SOC replicates the spleen's function of filtering blood, which helps to study the function of the spleen in blood diseases such as malaria [77]. At the core of the SOC developed by Rigat-Brugarolas et al. are two major microfluidic channels that simulate the closed-fast microcirculation (the fast path through the spleen without filtration, which accounts for about 90% of the total blood flow) and the open-slow microcirculation (the blood passes through the reticular structure in the red pulp of the spleen and is filtered, accounting for about 10% of the total blood flow). To mimic the filtration of blood by the red pulp, the open-slow microcirculation is equipped with a columnar matrix area and micro-contractile structures, which are responsible for increasing the residence time of red blood cells and screening functional red blood cells, respectively.

In addition to the above-mentioned OOCs, which are widely used, more specialized OOCs have been developed recently, including bone marrow-on-a-chip, bone-on-a-chip, blood–brain barrier chips, and lymphoid follicle chips [68,78–80]. Bone marrow chips, lymphoid follicle chips, and blood–brain barrier chips all adopt the classic lung-on-a-chip structure, with a porous membrane separating the tissue culture chamber from the blood flow channel [68,79,80]. A micropump continuously perfuses oxygen and nutrients while removing waste products. The blood–brain barrier chip can change the oxygen concentration to induce cells to differentiate into different blood–brain barrier phenotypes [79]. A similar design was used for the bone chips [78]. However, instead of a porous membrane, it uses five PDMS triangular pillars to separate the two vascular channels and the bone microenvironment channel. The bone microenvironment channel can simulate different bone conditions by changing the proportion of three types of bone cells (osteoblasts, osteocytes, and osteoclasts).

3.2. *Multi-Organ-on-a-Chip Systems*

Multi-OOCs integrate various single OOCs to simulate complex interactions between different organs, providing a model that closely replicates physiology [95]. Although still in the developmental stages, significant advancements have been made with systems incorporating two to ten organ models [67]. This section elaborates on two prominent design approaches and their applications in biomedical research.

3.2.1. Horizontal Design

The horizontal design of a multi-OOC involves the integration of multiple organ models that interact functionally with each other. The chambers containing different organs are positioned horizontally and interconnected by microfluidic channels. This configuration is instrumental for studying systemic diseases, multi-organ responses to pharmacological interventions, and the metabolic pathways of drugs in vivo [96]. A typical example is the liver-kidney chip, which combines these two critical organs to assess drug metabolism [97]. The liver, a central metabolic organ, and the kidney, a primary excretory organ, are pivotal in evaluating the pharmacokinetics and potential toxicity of drugs, which are crucial for clinical trial outcomes and market approval [98,99].

This system is constructed with two interconnected bioreactors, each with independent fluid input and output ports, facilitating isolated or combined organ studies (Figure 7). Each bioreactor comprises fluidic channels, diffusion barriers, and cell culture areas. The fluidic channels ensure precise control of nutrient and waste flow, which is essential for maintaining physiologically relevant conditions. The diffusion barriers minimize convective flow, protecting the cells from shear stress and air bubble formation. The liver and kidney cells are co-cultured in the culture areas under microfluidic conditions simulated by a low-pressure syringe pump system. This setup not only mimics the blood flow-induced shear stress on cells but also enhances the metabolic and absorptive functionalities of the cultured cells, offering insights into cellular responses to various drugs and their metabolites [100].

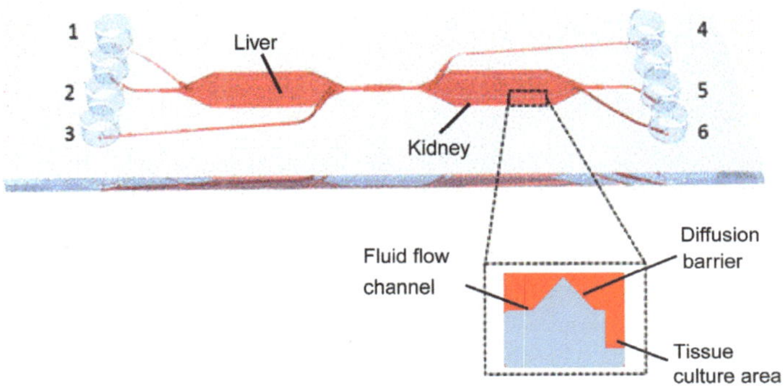

Figure 7. Illustration of the liver-kidney chip produced by Theobald et al. This chip has a total of six inlets/outlets, three on each side (denoted by the number 1–6). And two bioreactors, the liver and kidney, are connected by microfluidic channels that mimic the interaction between the liver and kidney in the human body. Reproduced with permission from [97].

At the applied level, the system was also used to study the biotransformation and toxicity of aflatoxin B1 (AFB1) and benz[a]pyrene (BαP). In the liver-kidney chip system, the toxins were first exposed to the liver cells, and then their metabolites were transported to the kidney cells via a simulated blood flow. The results showed that this metabolic activity of the liver cells significantly increased the toxic burden on the kidney cells.

The results show that the system can be used to evaluate the toxicity and metabolic response of drugs in a flow-dependent manner, demonstrating the great potential of the horizontally designed multi-OOCs for studying the response of multiple organs to drug intervention and the metabolic pathways of drugs in the body.

3.2.2. Vertical Design

Contrasting with the horizontal arrangement, the vertical design incorporates additional organ models to expand the scope of drug delivery studies and evaluate the effects of different administration routes on drug toxicity and efficacy. Vertical designs are critical in simulating body systems that involve tissue barriers, transcellular transport, and absorption. For instance, integrating skin, intestine, and bone marrow modules facilitates the exploration of transdermal, oral, and intravenous drug deliveries [101].

A notable implementation of such a design is the intestinal-liver-cancer chip, developed by Jie et al. [102] (Figure 8). This chip integrates a Caco-2 cell-lined hollow fiber simulating the intestine atop chambers containing HepG2 and U251 cells, representing the liver and glioblastoma, respectively. Drugs administered into the hollow fiber undergo simulated intestinal absorption, subsequent hepatic metabolism, and finally interact with cancer cells, providing a comprehensive model for evaluating the efficacy and toxicity of orally administered glioblastoma therapeutics. The design ensures that all drugs and molecules must undergo cellular processing in the top layer of the intestine chamber before reaching the organ chambers in the bottom layer. This functionality cannot be achieved with a standard horizontal design.

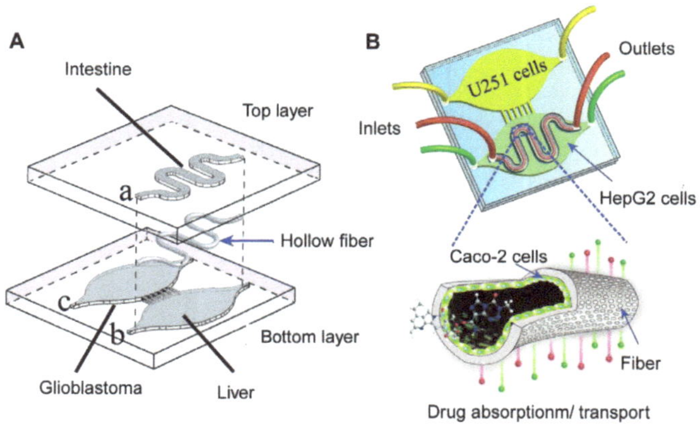

Figure 8. Illustration of the intestinal-liver-cancer chip produced by Jie et al. (**A**) The intestinal bioreactor is embedded in the upper layer (hollow fiber) of the chip; the liver and cancer bioreactors are embedded in the lower layer of the chip. (**B**) The chip has three inlets and outlets for introducing and draining medium or drug solutions, respectively, to maintain the cell growth environment and experimental conditions. U251 cells were cultured in the cancer bioreactor; HepG2 cells were cultured in the liver bioreactor; and Caco-2 cells were cultured in the intestinal bioreactor. Reproduced with permission from [102].

In practical applications, the research team used the model to evaluate the therapeutic effect of a drug combination (irinotecan, temozolomide, and cyclophosphamide) on glioblastoma. The results showed that the drug combination of irinotecan and temozolomide exhibited stronger anti-cancer activity than single drugs and other drug combinations. Interestingly, the authors have demonstrated that metabolites from compounds processed by HepG2 cells were detectable in the extracellular medium, influencing the combined efficacy of the drugs [49]. This provides an alternative to traditional animal models for studying the effect of drug metabolites in cancer therapy. In addition, the multi-OOC model, which employs human cells, might better represent human drug metabolism and efficacy compared to animal models. However, it's worth noting that the use of an immortalized hepatocyte cell line in the OOC may not fully mimic the body's drug metabolism in this study. For instance, the metabolite concentrations measured in the current OOC may not be suitable as a reference for further pharmacokinetic studies on the effects of these metabolites. As liver organoid culture techniques mature [75,103], the system could be optimized further by incorporating liver organoids, thereby better replicating the conditions in the human body. Nevertheless, these findings highlight vertical design in multi-OOCs for investigating the impacts of specific or varied drug delivery methods on drug toxicity and effectiveness.

Another implementation of this design is the heart-liver-skin chip, which assesses the kinetics of drug absorption through the skin and its systemic effects [95]. The chip has three bioreactors: skin, heart, and liver. The vertical design of the multi-OOC system simulates transdermal administration (local administration). Cardiomyocytes cultured on microelectrode arrays and MEMS cantilevers within the heart bioreactor at the bottom provide real-time data on mechanical and electrical activities. In addition, this approach mimics body-wide fluid dynamics through a rocking mechanism, ensuring effective medium circulation and simulating physiological interactions among the organs.

At the application level, the research team used the chip to study the different toxicities of drugs (diclofenac, paracetamol, hydrocortisone, and ketoconazole) between transdermal and systemic administration. The results showed that the effects of all drugs on cardiac and liver function were lower in the transdermal administration mode than in systemic administration. When administered systemically, high concentrations of diclofenac and acetaminophen had a significant effect on cardiac function, such as altering cardiac electrical activity and contractility, while hydrocortisone and ketoconazole mainly affected liver function.

4. Applications of OOC in Biomedicine and Clinics

The successful development of OOCs of various organs has significantly benefited biomedical research and clinics. In response to the diverse and complex demands of biomedical research and clinical therapy, a burgeoning body of research has been dedicated to the precise design and development of OOC platforms. These innovative models showcase a paradigm shift in the way biological systems are studied and offer immense potential across various domains. Within the realms of biomedicine and clinical practice, OOC technology holds promise in simulating organ-level functions, disease modeling, drug screening, and personalized medicine approaches.

Specifically, OOC systems offer a dynamic and physiologically relevant microenvironment that closely mimics the intricate structures and functions of human organs, providing researchers and clinicians with a sophisticated tool for studying disease mechanisms and drug responses. In comparison to traditional methodologies, OOC models offer advantages such as enhanced physiological relevance, scalability, cost-effectiveness, and the ability to perform high-throughput experiments with reduced reliance on animal models These attributes collectively underscore the transformative impact of OOC technology in advancing biomedical research and clinical applications.

4.1. Disease Modeling and Drug Evaluation

Drug discovery is fundamentally an innovative process that relies on an in-depth understanding of biological targets to design or screen potential drug candidates from vast chemical compound libraries [104]. This process is based on extensive research into the pathological mechanisms of specific diseases to identify molecular targets for drug intervention. By utilizing disease-specific OOC systems, researchers can study the pathological mechanisms of diseases. For instance, OOC technology is a critical method for exploring cancer development, including cancer cell phenotype, growth, migration, and invasion [105–108]. Specifically, Toh et al. created a system that mimics the tumor microenvironment, representing cancer cell migration dynamics in a microfluidic environment and providing an effective platform for evaluating anti-cancer drugs. Similarly, Marturano-Kruik et al. developed a perivascular niche chip for studying breast cancer cell metastasis and drug resistance in bone [105]. Beyond solid tumors, OOCs have also been instrumental in studying hematological cancers [5,109]. A chip-based leukemia model revealed potential mechanisms of chemoresistance in the bone marrow microenvironment of B-cell acute lymphoblastic leukemia (B-ALL), where B-ALL cells utilize factors from the surrounding vasculature, endosteal, and hematopoietic microenvironment (e.g., CXCL12 cytokine signaling, VCAM-1/OPN adhesion signaling, and leukemia-specific NF-κB pathways) to maintain survival and quiescence [5]. These studies suggest that OOCs can potentially replace or complement existing animal models in exploring disease pathomechanisms. One of the primary advantages of OOC models is the utilization of human cells or organoids, or even patient-derived cells or organoids. This approach potentially provides a more accurate reflection of human pathologic mechanisms. However, current OOCs primarily mimic the micro or local environment of targeted organs, encompassing a limited variety of cells or organs. In the human body, disease development and drug responses often involve more complex systemic metabolisms and interactions between organs. For instance, tumor development is not only influenced by the interactions between cancer cells and neighboring fibroblasts, adipocytes, endothelial cells, and macrophages, but also by the infiltration of neutrophils, T cells, and other immune cells from the bloodstream. Additionally, hormones from the endocrine system can also significantly impact tumor progression.

These interactions between tumor cells and other cells and organs also play crucial roles in the tumor's response to drugs. Therefore, there is a need for more complex OOC models that incorporate as many cell types as possible.

4.2. Drug Screening and Discovery

Drug screening and development require the identification of targeted and effective compounds from chemical libraries [110]. An efficient and precise screening platform is crucial for successful drug development. Traditional drug screening often relies on the use of common or engineered cell lines. While these approaches offer advantages such as easy access to cell lines and simple platform design, they come with inherent limitations. A major shortcoming is that 2D cultured cells lack complex structure, composition, and intercellular communications found in vivo, leading to differences in drug metabolism and responses. In contrast, OOC technology offers a more sophisticated solution by better mimicking the in vivo environment and addressing these issues. OOC platforms replicate the physiological conditions of organs more accurately, enabling the study of complex cell interactions, tissue responses, and drug effects in a more physiologically relevant manner. By providing a more representative model of human biology, OOC systems have the potential to revolutionize drug screening processes and enhance the efficiency and accuracy of drug development efforts [5]. For example, Jalili-Firoozinezhad et al. investigated radiation-induced cell death and drug responses using a Gut-on-a-Chip platform. By simulating gamma radiation exposure, they observed increases in the generation of reactive oxygen species, cytotoxicity, apoptosis, and DNA fragmentation. Furthermore, pretreatment with the radioprotective drug dimethyloxalylglycine (DMOG) significantly inhibited these damage responses [111]. In addition, Liu et al. developed a vascular-cancer chip model to evaluate the efficacy and impact of drugs on microvascular networks and tumor cells. In this model, human endothelial cells and fibroblasts are co-cultured to form a microvascular network, and colorectal cancer cells are introduced simultaneously. The research team tested the effects of the anti-cancer drugs fluorouracil, vincristine, and sorafenib at different concentrations. The results showed that the effects of these drugs on endothelial cells and tumor cells at different concentrations were dose- and time-dependent [112].

4.3. Preclinical Studies

Preclinical studies serve as a critical stage in the drug development process following the initial identification of a drug candidate. The goal is to collect essential feasibility data, conduct iterative testing, and assess the drug's safety and efficacy prior to its progression to clinical trials [2]. Traditionally, preclinical studies have been primarily conducted using animal studies [113,114]. However, animal studies are inherently low-throughput and cannot accurately predict drug effects in humans due to differences in pharmacodynamic (PD) and/or pharmacokinetic (PK) responses [115–117]. In June 2022, the U.S. House of Representatives passed the Food and Drug Amendments of 2022, which included OOC and micro-physiological systems as separate evaluation systems for non-clinical trials of drugs. In August, the FDA approved the first new drug (NCT04658472) to enter clinical trials based solely on preclinical efficacy data obtained from OOC studies.

Different types of drugs require various preclinical studies, including PD, PK, and toxicology testing. PK studies focus on the drug's concentration in different organ sites during its metabolism, encompassing absorption, distribution, metabolism, and elimination (ADME). PD studies examine the drug's biological effects on the target organ or tissue and its mechanism of action. Toxicology testing evaluates the potentially harmful effects of the drug to ensure that it does not pose unacceptable risks to humans and the environment. These preclinical indicators can be detected by the OCC system, which offers a powerful tool to optimize preclinical research [118].

4.3.1. PD-PK Testing

Single-OOCs, such as lung-on-a-chip models, play a pivotal role in PK–PD testing for specific organs with unique drug delivery methods. For instance, pulmonary drug delivery is a prominent approach for treating respiratory conditions like asthma and cystic fibrosis [119]. Drugs administered via inhalation traverse bronchial and alveolar tissues, ensuring high bioavailability and therapeutic effects in the lung. In the study by Barros, Costa, and Sarmento, a lung-on-a-chip model was employed to replicate the alveolar–capillary interface, providing a sophisticated platform for predicting PK–PD parameters during drug screening processes. This innovative approach allows researchers to evaluate how drugs interact with lung tissues at the cellular level, offering insights into drug absorption, distribution, metabolism, and excretion within the pulmonary system [120].

Moreover, the advent of multi-organ OOC systems has significantly advanced PK–PD studies by enabling comprehensive investigations under various drug administration conditions [121]. As vital organs for drug metabolism and detoxification, a multi-organ OOC system with liver and kidney chips is particularly important for systematic in situ PK–PD study [122]. Researchers such as Sung, Kam, and Shuler have pioneered the development of multi-OOC systems, including the tumor-liver-bone marrow model, to explore PK–PD relationships [123]. This device comprises multiple cell culture chambers with fluid systems representing different organs. The toxicity and mechanism of action of the anti-cancer drug 5-fluorouracil were evaluated by adding the chemotherapeutic drug and its modulator uracil to the cell culture medium. Herland et al. developed an integrated intestinal-hepatic-renal and bone marrow-hepatic-renal multi-OOC system for PK–PD studies under diverse dosing conditions [113]. This system, utilizing OOC lined with blood vessels to mimic human organ functions, facilitates drug transfer between organs via a microfluidic network containing a blood substitute. By simulating different drug administration routes (e.g., oral and intravenous), researchers were able to accurately predict drug absorption, metabolism, and excretion pathways, in line with clinical data. The intestinal-hepatic-renal chip effectively simulated cisplatin's metabolism and excretion, aligning with clinical data. The bone marrow-hepatic-renal chip also assessed cisplatin's pharmacodynamic effects, particularly its toxicity to hematopoietic cells, consistent with clinical observations. This pioneering work highlights the development of OOC platforms in integrating multiple metabolic organs into a single chip, a feature that was lacking in previous studies. On one hand, this system facilitates the discovery and evaluation of the PK and PD parameters of drugs in a systemic metabolic manner, closely mimicking the process in the human body. On the other hand, it potentially enables simultaneous monitoring of therapeutic effects of the metabolites processed by the liver and the potential toxicity of these liver-processed metabolites on the kidney. This multi-organ integration on a single chip offers a more comprehensive and realistic model for studying drug metabolism, efficacy, and toxicity, paving the way for more accurate predictions of drug behavior in the human body.

4.3.2. Toxicology Testing

As noted, preclinical studies traditionally involve laboratory animals. However, many drugs may not exhibit adverse effects in animals during preclinical stages but can cause liver, heart, or kidney damage in patients during clinical trials [124]. This underscores the importance of toxicology testing, where OOCs can serve as more effective and accurate systems than animal models [125]. Ma et al. developed a LOC featuring three-dimensional liver lobular microtissues [126]. These biomimetic microtissues maintain high basal CYP-1A1/2 and UGT enzyme activities, dynamically responding to drug induction and inhibition, effectively simulating toxicology studies in vitro, and providing a screening platform for drug toxicity during combination therapy.

Moreover, multi-OOCs can facilitate the study of drug metabolism between organs. Zhang et al. developed a multi-OOC platform integrating physical, biochemical, and optical sensors for non-invasive and continuous monitoring of microenvironment parameters and dynamic organ responses to drug compounds. The platform assessed drug toxicity in liver and heart chips separately, verifying that anti-cancer drugs like capecitabine exhibit liver toxicity when metabolized by liver cells and cardiotoxicity [110]. In an impressive study by Novak et al., eight vascularized organs, including the intestine, heart, lung, liver, kidney, skin, brain, and blood–brain barrier, were unified [117]. This comprehensive multi-OOC platform allows for systematic toxicity testing across all major organs simultaneously. More importantly, such a platform provides a superior simulation of drug and metabolite circulation within the body. This is crucial because it takes into account potential synergy, feedback, or consequential reactions of different organs to a drug. This kind of holistic approach could revolutionize drug testing, potentially leading to more accurate predictions of drug responses and side effects in humans.

4.4. Precision Medicine in Clinics

Precision medicine customizes healthcare choices for individual patients based on their anticipated response or disease risk. Due to genetic and microenvironment heterogeneity, patients often respond differently to drugs, necessitating accurate assessment and optimization of individual efficacy. OOCs can be used to construct patient-specific models for evaluating drug efficacy and safety in specific patient groups, aiding in developing precise treatment plans.

By integrating primary cells from patient donors into an OOC system, it is possible to evaluate the patient-specific drug response [127]. For instance, using epithelial cells from patients with chronic obstructive pulmonary disease allows the OOC system to reproduce the characteristics of the disease [127]. Similarly, by differentiating patient-derived induced pluripotent stem cells (iPSC) into cells that differentiate into blood–brain barrier (BBB) cells and using this as the basis for establishing a BBB chip, researchers can create personalized models that reflect the genetic background and pathological characteristics of specific patients [128].

Despite promising results of cancer immunotherapies in clinical trials, most patients' responses remain suboptimal, emphasizing the need for precision medicine to optimize treatment [129]. For example, PD-1 checkpoint immunotherapy has shown potential in clinical trials. However, its effectiveness varies due to the genetic heterogeneity of glioblastoma (GBM) in patients and the immunosuppressive tumor microenvironment [130]. Researchers used a GBM-on-a-Chip system to explore immunosuppressive tumor microenvironment heterogeneity and optimized anti-PD-1 immunotherapy for different GBM subtypes, providing a personalized screening approach [131]. In addition, researchers compared CAR T cells from different donors and constructs using micro-patterned tumor arrays (MiTA), finding significant differences in migration, aggregation, and tumor cell killing efficiency, highlighting individual treatment differences [132].

5. Technical Challenges and Future Prospects

As OOC and microfluidic technologies rapidly develop, they show significant potential in drug discovery and development but also face numerous technical challenges.

5.1. Cost and Manufacturing

First, the high cost of developing and producing OOCs is a primary barrier to their widespread use. Currently, most OOCs are manually fabricated in research laboratories using soft lithography and PDMS, which leads to high costs and design limitations. High-precision microfabrication techniques and expensive materials, such as PDMS and microfluidic devices, are required for precision processing. PDMS also has other disadvantages, such as adsorption for specific drugs and poor light transmission for imaging. Additionally, specialized technicians are needed to maintain and operate these complex instruments. To achieve large-scale applications, it is necessary to standardize the manufacturing process, develop more cost-effective manufacturing methods and materials, and reduce operational complexity and costs. Three possible approaches include current standard manufacturing materials and techniques, advanced additive manufacturing methods, and modular design. For standard manufacturing materials and processes, injection molding and laser cutting can replace PDMS with plastics [97]. Advanced additive manufacturing methods like 3D printing can pre-program and automatically print high-fidelity and controllable tissue structures, complex scaffolds, or device templates [24]. This technology offers a one-step tissue reconstruction and culture platform, potentially revolutionizing OOC manufacturing [133]. Modular design involves breaking OOCs into multiple independent functional modules, which can be flexibly combined to simulate different physiological systems and pathological states. This design enhances system flexibility and applicability while reducing development and production costs.

Integration of sensors also poses a challenge in OOC manufacturing, as they require precise microfabrication processes, including thin-film deposition and photolithography. These processes must be meticulously controlled to achieve the desired sensor geometry and placement within the OOC system. Sensors should be placed in positions that provide the most relevant and least perturbed measurements, which can be influenced by factors like cell distribution and flow dynamics within the device. Optical access is a concern for sensors placed above or below cells, as it may obstruct the view for microscopy-based cell characterization. The use of transparent electrode materials or strategic positioning of the electrodes can help mitigate this issue. Standardization of sensor integration methods and OOC designs is another ongoing challenge. Sterilization of OOC devices is essential for maintaining sterile conditions, but standard sterilization techniques like autoclaving or gamma irradiation can be harmful to some sensor materials. After all, the integration of sensors significantly adds complexity to an OOC system, increasing manufacturing costs and the potential for variability between devices.

5.2. Sensing Systems in OOC

In the development of OOCs, the miniaturization and complexity of the OOC systems necessitate precise and continuous monitoring of various parameters such as oxygen levels, pH, temperature, and metabolite concentrations. However, the integration of real-time sensors also presents a multitude of challenges that span various scientific disciplines. First, material selection is a fundamental aspect of sensor integration. The chosen materials must exhibit biocompatibility and non-toxicity to avoid interference with cell growth or function. Additionally, the materials need to possess the necessary electrical, electrochemical, or optical properties suitable for the various sensing applications simultaneously. Therefore, innovative polymer materials other than PDMS are needed. Second, the design of the sensors is another crucial factor. For electrical sensors like TEER and ECIS, electrode configuration can significantly affect the measurement outcome. The design must ensure uniform electric fields and minimize contact impedance. Third, the long-term stability of the sensors is a critical aspect. Sensors must maintain their performance over the duration of the experiment, which can span several weeks. Issues like sensor drift, degradation of the sensing element, and biofouling, where biological materials adhere to the sensor surface, can affect long-term stability. In addition, the biocompatibility of sensor materials and fabrication processes is paramount to avoid adverse effects on cell viability and function in OOCs. To prevent biofouling, which can alter sensor responses, antifouling coatings may be applied to the sensor surfaces. This is particularly important for sensors that are used over extended periods or for multiple experiments. When multiple sensors are integrated into a single OOC system, there is a risk of crosstalk or interference between the sensors, which can complicate data interpretation. Lastly, the integration of sensors must be compatible with mass production processes, including the use of materials and fabrication techniques that can be scaled up for commercial applications. Addressing these challenges requires a deep understanding of materials science, microfabrication, cell biology, and sensor technology. Advances in these areas are essential for the development of reliable and effective sensor-integrated OOC systems. Future research should focus on overcoming these obstacles to fully exploit the potential of OOC technology in drug discovery and personalized medicine.

5.3. Cell Source and Variability

Second, the scarcity and variability of patient-derived cells hinder the development of precision medicine based on OOC systems. OOCs developed from patient-derived materials can play a crucial role in precision medicine. However, the invasive collection of specific samples is impossible in some special conditions. For example, obtaining brain tissue samples from patients with neurological diseases may be too risky and unacceptable The limited number of patient tissue cells and low proliferation potential can also be problematic. However, iPSCs are cells that have been genetically reprogrammed to an embryonic stem cell-like state, which means they have the potential to differentiate into any cell type in the body [134]. This makes them an excellent source for generating various cell types required in OOC systems. The controlled differentiation of iPSCs could provide a continuous, patient-specific source of cells for OOCs, overcoming the limitations of donor availability and ethical concerns related to primary human cells.

For instance, iPSCs derived from skin fibroblasts can be an alternative source of unlimited cells to generate autologous target organs or tissues [135,136]. This enables the construction of patient-specific organ OOCs for personalized disease modeling and drug screening [137]. Despite their potential, iPSCs face challenges in OOC systems, including optimizing differentiation efficiency and consistency to ensure stable expression of target cell characteristics. Cells derived from iPSCs may undergo phenotypic drift during long-term culture, necessitating the development of refined culture and differentiation strategies to maintain cell stability [138].

5.4. Integration of the Immune System

Third, most OOC systems currently do not integrate immune cells or the immune system. Incorporating the immune system into OOCs represents a significant, yet necessary, frontier in the advancement of this technology. The immune system's integral role in numerous biological processes and pathologies underscores the current limitations of OOC models that do not include this complex network. The development of an immune system OOC, integrating a diverse array of immune cell types such as T cells, B cells, and macrophages, presents an ambitious research direction. The challenge lies in accurately emulating the intricate interactions between immune cells and other cell types, which encompasses mechanisms like cell signaling and cytotoxicity. The task is further complicated by the organ-specific functionality of the immune system, making the integration of immune system OOCs with existing single-organ or multi-organ OOCs a daunting endeavor. Incorporating a specific sensing system into OOC models to monitor and evaluate immune responses presents another notable challenge. Immune responses, including inflammation, are typically characterized by altered levels of representative cytokines, as well as the proliferation and migration of immune cells. The development and integration of sensors that can detect these changes would significantly enhance the utility and functionality of immune system-integrated OOCs. For instance, electrochemical sensors might be employed to measure cytokine levels, while optical sensors could track cell migration and proliferation [139]. The development of such sensors, however, is not without its challenges. These sensors must exhibit high sensitivity and specificity to accurately detect and quantify the biological markers of interest. Additionally, they must be biocompatible to avoid interfering with the biological processes they aim to monitor. Moreover, they should be capable of continuous, real-time monitoring to track dynamic changes in immune response.

Despite these hurdles, the potential applications are vast. Immune system OOCs could revolutionize drug discovery and clinical trials by facilitating the identification and screening of potential anti-inflammatory or anti-tumor drugs and the evaluation of novel immunomodulatory drugs. Furthermore, when these models utilize cells derived from specific patients, they could enable precision medicine by predicting individual immune responses to treatments, thereby optimizing therapeutic strategies. Despite the challenges, the integration of the immune system into OOC models is pivotal for creating more accurate human biological models, with the potential to drastically enhance our understanding of disease mechanisms and therapeutic development.

5.5. Influence of OOC Nanostructures

Fourth, there is currently no comprehensive study on the influence of OOC nanostructures on the growth and metabolism of organoids. Some studies indicate that nanostructures can significantly affect cell behavior, including adhesion and proliferation. Research shows that moderate surface energy and roughness provide optimal cell adhesion and growth conditions, with higher proliferation and differentiation under these conditions [140]. Additionally, pore nanostructures significantly affect cell behavior. For example, electrospun scaffolds with pore sizes ranging from 50 to 400 microns can influence cell adhesion and proliferation, with specific pore sizes having different optimal effects on various cell types, such as bone marrow mesenchymal stem cells, chondrocytes, and tendon cells [141]. Future research should further explore the impact of different nanostructures on multiple cell types and organoids, providing more theoretical and practical guidance for OOC design and application.

5.6. Prospects for the Commercialization of OOC Technology

As technology matures and applications expand, the commercialization of OOC technology will accelerate. Despite the challenges of commercialization, such as limited venture capital in the field, publications in the field of OOC are booming worldwide, and government agencies and universities in developed economies are working with companies to promote the widespread use of OOC technology [142–144]. Among them, the EU has adopted Directive 2010/63/EU and established a regulatory framework to support the development of new microphysiological systems and other bioengineering alternatives to animal research, which is of great significance for the promotion of OOC systems [145]. As the use of OOC technology in drug development, disease research, and personalized medicine increases, market demand is expected to grow rapidly, potentially reaching billions of dollars [142]. This growth will drive more investment and research to improve OOCs' performance and application diversity [142]. As standardized production processes and regulatory frameworks are established, the accessibility and reliability of OOCs will significantly improve, providing more effective and accurate tools for medical research and clinical applications [142]. In the future, OOCs are expected to become a vital pillar of biomedical research, driving innovation and progress in medical technology [144].

6. Conclusions

This review article focuses on the role that microfluidic-based OOC technologies play in transforming drug discovery. By replicating human physiology, OOCs offer a developing platform for biomedical research, particularly in drug discovery. This technology can potentially reduce the risks and costs associated with drug development while significantly enhancing the efficiency and safety of new drug post-market entry. From early drug discovery to preclinical trials, OOC combined with sensors constitutes a system that can precisely control experimental conditions and deliver real-time data, optimizing the drug development pipeline. Additionally, the various OOC designs, and multi-OOCs, enable the simulation of complex biological processes, organ crosstalk, and pathological states, which is crucial for understanding disease mechanisms and evaluating drug efficacy. Despite the considerable potential of OOC technology in advancing drug development and personalized medicine, it is important to acknowledge the technical challenges and commercialization barriers it faces. Future research must focus on improving manufacturing efficiency, reducing costs, and continually exploring and optimizing OOC designs, as well as developing and integrating novel sensors to support a broader range of biomedical applications. In conclusion, although challenges remain, the future of OOC technology is highly promising. With ongoing technological advancements and enhanced interdisciplinary collaboration, OOC is poised to play a pivotal role in future biomedical research, bringing revolutionary changes in drug development and disease treatment.

Author Contributions: Conceptualization, H.H. and C.W.; writing—original draft preparation, S.Y.; writing—review and editing, H.Y., D.C.H., H.H. and C.W.; visualization, S.Y., H.H. and H.Y.; supervision, H.H. and C.W.; project administration, H.H. and C.W.; funding acquisition, H.Y., H.H. and C.W. All authors have read and agreed to the published version of the manuscript.

Funding: This research was funded by the Natural Science Foundation of Zhejiang Province, China, grant number LQ22C070002, the National Natural Science Foundation of China, grant number 32300674 and 32370872. The APC was funded by 2024 International Campus Talent Special Plan Fund and Center of Pathogen Detection in the Dynamic Research Enterprise for Multidisciplinary Engineering Sciences (DREMES).

Institutional Review Board Statement: Not applicable.

Informed Consent Statement: Not applicable.

Data Availability Statement: Not applicable.

Conflicts of Interest: The authors declare no conflicts of interest.

References

1. Low, L.A.; Sutherland, M.; Lumelsky, N.; Selimovic, S.; Lundberg, M.S.; Tagle, D.A. Organs-on-a-Chip. In *Biomaterials- and Microfluidics-Based Tissue Engineered 3D Models*; Oliveira, J.M., Reis, R.L., Eds.; Springer International Publishing: Cham, Switzerland, 2020; pp. 27–42, ISBN 978-3-030-36588-2.
2. Kaitin, K. Deconstructing the Drug Development Process: The New Face of Innovation. *Clin. Pharmacol. Ther.* **2010**, *87*, 356–361. [CrossRef]
3. Singh, N.; Vayer, P.; Tanwar, S.; Poyet, J.-L.; Tsaioun, K.; Villoutreix, B.O. Drug Discovery and Development: Introduction to the General Public and Patient Groups. *Front. Drug Discov.* **2023**, *3*, 1201419. [CrossRef]
4. Berdigaliyev, N.; Aljofan, M. An Overview of Drug Discovery and Development. *Future Med. Chem.* **2020**, *12*, 939–947. [CrossRef]
5. Ma, C.; Peng, Y.; Li, H.; Chen, W. Organ-on-a-Chip: A New Paradigm for Drug Development. *Trends Pharmacol. Sci.* **2021**, *42*, 119–133. [CrossRef] [PubMed]
6. Bhatia, S.N.; Ingber, D.E. Microfluidic Organs-on-Chips. *Nat. Biotechnol.* **2014**, *32*, 760–772. [CrossRef]
7. Hamon, M.; Hong, J.W. New Tools and New Biology: Recent Miniaturized Systems for Molecular and Cellular Biology. *Mol. Cells* **2013**, *36*, 485–506. [CrossRef]
8. Yoon, S.; Kilicarslan You, D.; Jeong, U.; Lee, M.; Kim, E.; Jeon, T.-J.; Kim, S.M. Microfluidics in High-Throughput Drug Screening: Organ-on-a-Chip and C. Elegans-Based Innovations. *Biosensors* **2024**, *14*, 55. [CrossRef] [PubMed]
9. Polaris. Organ-On-Chip Market Size, Growth & Revenue Report, 2024–2032. Available online: https://www.polarismarketresearch.com/industry-analysis/organ-on-chip-market (accessed on 4 June 2024).
10. Leung, C.M.; de Haan, P.; Ronaldson-Bouchard, K.; Kim, G.-A.; Ko, J.; Rho, H.S.; Chen, Z.; Habibovic, P.; Jeon, N.L.; Takayama, S.; et al. A Guide to the Organ-on-a-Chip. *Nat. Rev. Methods Primer* **2022**, *2*, 33. [CrossRef]
11. Xia, Y.; Whitesides, G.M. Soft Lithography. *Angew. Chem. Int. Ed.* **1998**, *37*, 550–575. [CrossRef]
12. Chen, X.; Shen, J.; Zhou, M. Rapid Fabrication of a Four-Layer PMMA-Based Microfluidic Chip Using CO_2-Laser Micromachining and Thermal Bonding. *J. Micromech. Microeng.* **2016**, *26*, 107001. [CrossRef]
13. When PDMS Isn't the Best. *Anal. Chem.* **2007**, *79*, 3248–3253. [CrossRef] [PubMed]
14. Liu, S.; Kumari, S.; He, H.; Mishra, P.; Singh, B.N.; Singh, D.; Liu, S.; Srivastava, P.; Li, C. Biosensors Integrated 3D Organoid/Organ-on-a-Chip System: A Real-Time Biomechanical, Biophysical, and Biochemical Monitoring and Characterization. *Biosens. Bioelectron.* **2023**, *231*, 115285. [CrossRef] [PubMed]
15. Oleaga, C.; Riu, A.; Rothemund, S.; Lavado, A.; McAleer, C.W.; Long, C.J.; Persaud, K.; Narasimhan, N.S.; Tran, M.; Roles, J.; et al. Investigation of the Effect of Hepatic Metabolism on Off-Target Cardiotoxicity in a Multi-Organ Human-on-a-Chip System. *Biomaterials* **2018**, *182*, 176–190. [CrossRef] [PubMed]
16. Sabaté del Río, J.; Ro, J.; Yoon, H.; Park, T.-E.; Cho, Y.-K. Integrated Technologies for Continuous Monitoring of Organs-on-Chips: Current Challenges and Potential Solutions. *Biosens. Bioelectron.* **2023**, *224*, 115057. [CrossRef] [PubMed]
17. Sung, J.H.; Wang, Y.I.; Sriram, N.N.; Jackson, M.; Long, C.; Hickman, J.J.; Shuler, M.L. Recent Advances in Body-on-a-Chip Systems. *Anal. Chem.* **2019**, *91*, 330–351. [CrossRef] [PubMed]
18. Ferrell, N.; Desai, R.R.; Fleischman, A.J.; Roy, S.; Humes, H.D.; Fissell, W.H. A Microfluidic Bioreactor with Integrated Transepithelial Electrical Resistance (TEER) Measurement Electrodes for Evaluation of Renal Epithelial Cells. *Biotechnol. Bioeng.* **2010**, *107*, 707–716. [CrossRef] [PubMed]

19. Douville, N.J.; Tung, Y.-C.; Li, R.; Wang, J.D.; El-Sayed, M.E.H.; Takayama, S. Fabrication of Two-Layered Channel System with Embedded Electrodes to Measure Resistance Across Epithelial and Endothelial Barriers. *Anal. Chem.* **2010**, *82*, 2505–2511. [CrossRef]
20. Odijk, M.; van der Meer, A.D.; Levner, D.; Kim, H.J.; van der Helm, M.W.; Segerink, L.I.; Frimat, J.-P.; Hamilton, G.A.; Ingber, D.E.; Berg, A. van den Measuring Direct Current Trans-Epithelial Electrical Resistance in Organ-on-a-Chip Microsystems. *Lab Chip* **2015**, *15*, 745–752. [CrossRef]
21. Wu, Q.; Wei, X.; Pan, Y.; Zou, Y.; Hu, N.; Wang, P. Bionic 3D Spheroids Biosensor Chips for High-Throughput and Dynamic Drug Screening. *Biomed. Microdevices* **2018**, *20*, 82. [CrossRef]
22. An, Y.; Jin, T.; Zhang, F.; He, P. Electric Cell-Substrate Impedance Sensing (ECIS) for Profiling Cytotoxicity of Cigarette Smoke. *J. Electroanal. Chem.* **2019**, *834*, 180–186. [CrossRef]
23. Giaever, I.; Keese, C.R. A Morphological Biosensor for Mammalian Cells. *Nature* **1993**, *366*, 591–592. [CrossRef] [PubMed]
24. Lind, J.U.; Busbee, T.A.; Valentine, A.D.; Pasqualini, F.S.; Yuan, H.; Yadid, M.; Park, S.-J.; Kotikian, A.; Nesmith, A.P.; Campbell, P.H.; et al. Instrumented Cardiac Microphysiological Devices via Multimaterial Three-Dimensional Printing. *Nat. Mater.* **2017**, *16*, 303–308. [CrossRef]
25. Lind, J.U.; Yadid, M.; Perkins, I.; O'Connor, B.B.; Eweje, F.; Chantre, C.O.; Hemphill, M.A.; Yuan, H.; Campbell, P.H.; Vlassak, J.J.; et al. Cardiac Microphysiological Devices with Flexible Thin-Film Sensors for Higher-Throughput Drug Screening. *Lab Chip* **2017**, *17*, 3692–3703. [CrossRef] [PubMed]
26. Koutsouras, D.A.; Perrier, R.; Villarroel Marquez, A.; Pirog, A.; Pedraza, E.; Cloutet, E.; Renaud, S.; Raoux, M.; Malliaras, G.G.; Lang, J. Simultaneous Monitoring of Single Cell and of Micro-Organ Activity by PEDOT:PSS Covered Multi-Electrode Arrays. *Mater. Sci. Eng. C* **2017**, *81*, 84–89. [CrossRef]
27. Maoz, B.M.; Herland, A.; Henry, O.Y.F.; Leineweber, W.D.; Yadid, M.; Doyle, J.; Mannix, R.; Kujala, V.J.; FitzGerald, E A.; Parker, K.K.; et al. Organs-on-Chips with Combined Multi-Electrode Array and Transepithelial Electrical Resistance Measurement Capabilities. *Lab Chip* **2017**, *17*, 2294–2302. [CrossRef] [PubMed]
28. Ştefănescu, D.M. Strain Gauges and Wheatstone Bridges—Basic Instrumentation and New Applications for Electrical Measurement of Non-Electrical Quantities. In Proceedings of the Eighth International Multi-Conference on Systems, Signals & Devices, Sousse, Tunisia, 22–25 March 2011; pp. 1–5.
29. Bavli, D.; Prill, S.; Ezra, E.; Levy, G.; Cohen, M.; Vinken, M.; Vanfleteren, J.; Jaeger, M.; Nahmias, Y. Real-Time Monitoring of Metabolic Function in Liver-on-Chip Microdevices Tracks the Dynamics of Mitochondrial Dysfunction. *Proc. Natl. Acad. Sci. USA* **2016**, *113*, E2231–E2240. [CrossRef] [PubMed]
30. Misun, P.M.; Rothe, J.; Schmid, Y.R.F.; Hierlemann, A.; Frey, O. Multi-Analyte Biosensor Interface for Real-Time Monitoring of 3D Microtissue Spheroids in Hanging-Drop Networks. *Microsyst. Nanoeng.* **2016**, *2*, 16022. [CrossRef]
31. Moya, A.; Ortega-Ribera, M.; Guimerà, X.; Sowade, E.; Zea, M.; Illa, X.; Ramon, E.; Villa, R.; Gracia-Sancho, J.; Gabriel, G. Online Oxygen Monitoring Using Integrated Inkjet-Printed Sensors in a Liver-on-a-Chip System. *Lab Chip* **2018**, *18*, 2023–2035. [CrossRef]
32. Zirath, H.; Rothbauer, M.; Spitz, S.; Bachmann, B.; Jordan, C.; Müller, B.; Ehgartner, J.; Priglinger, E.; Mühleder, S.; Redl, H.; et al. Every Breath You Take: Non-Invasive Real-Time Oxygen Biosensing in Two- and Three-Dimensional Microfluidic Cell Models. *Front. Physiol.* **2018**, *9*, 815. [CrossRef]
33. Wang, X.; Wolfbeis, O.S. Optical Methods for Sensing and Imaging Oxygen: Materials, Spectroscopies and Applications. *Chem. Soc. Rev.* **2014**, *43*, 3666–3761. [CrossRef]
34. Prill, S.; Bavli, D.; Levy, G.; Ezra, E.; Schmälzlin, E.; Jaeger, M.S.; Schwarz, M.; Duschl, C.; Cohen, M.; Nahmias, Y. Real-Time Monitoring of Oxygen Uptake in Hepatic Bioreactor Shows CYP450-Independent Mitochondrial Toxicity of Acetaminophen and Amiodarone. *Arch. Toxicol.* **2016**, *90*, 1181–1191. [CrossRef] [PubMed]
35. Rennert, K.; Steinborn, S.; Gröger, M.; Ungerböck, B.; Jank, A.-M.; Ehgartner, J.; Nietzsche, S.; Dinger, J.; Kiehntopf, M.; Funke, H.; et al. A Microfluidically Perfused Three Dimensional Human Liver Model. *Biomaterials* **2015**, *71*, 119–131. [CrossRef] [PubMed]
36. Hulanicki, A.; Glab, S.; Ingman, F. Chemical Sensors: Definitions and Classification. *Pure Appl. Chem.* **1991**, *63*, 1247–1250. [CrossRef]
37. Shrestha, J.; Ryan, S.T.; Mills, O.; Zhand, S.; Bazaz, S.R.; Hansbro, P.M.; Ghadiri, M.; Warkiani, M.E. A 3D-Printed Microfluidic Platform for Simulating the Effects of CPAP on the Nasal Epithelium. *Biofabrication* **2021**, *13*, 035028. [CrossRef] [PubMed]
38. Zhu, Y.; Sun, L.; Wang, Y.; Cai, L.; Zhang, Z.; Shang, Y.; Zhao, Y. A Biomimetic Human Lung-on-a-Chip with Colorful Display of Microphysiological Breath. *Adv. Mater.* **2022**, *34*, 2108972. [CrossRef]
39. Najjar, D.; Rainbow, J.; Sharma Timilsina, S.; Jolly, P.; de Puig, H.; Yafia, M.; Durr, N.; Sallum, H.; Alter, G.; Li, J.Z.; et al. A Lab-on-a-Chip for the Concurrent Electrochemical Detection of SARS-CoV-2 RNA and Anti-SARS-CoV-2 Antibodies in Saliva and Plasma. *Nat. Biomed. Eng.* **2022**, *6*, 968–978. [CrossRef] [PubMed]
40. Li, X.; Soler, M.; Szydzik, C.; Khoshmanesh, K.; Schmidt, J.; Coukos, G.; Mitchell, A.; Altug, H. Label-Free Optofluidic Nanobiosensor Enables Real-Time Analysis of Single-Cell Cytokine Secretion. *Small* **2018**, *14*, 1800698. [CrossRef]
41. Zhang, F.; Qu, K.-Y.; Zhou, B.; Luo, Y.; Zhu, Z.; Pan, D.-J.; Cui, C.; Zhu, Y.; Chen, M.-L.; Huang, N.-P. Design and Fabrication of an Integrated Heart-on-a-Chip Platform for Construction of Cardiac Tissue from Human iPSC-Derived Cardiomyocytes and in Situ Evaluation of Physiological Function. *Biosens. Bioelectron.* **2021**, *179*, 113080. [CrossRef]
42. Meissner, R.; Eker, B.; Kasi, H.; Bertsch, A.; Renaud, P. Distinguishing Drug-Induced Minor Morphological Changes from Major Cellular Damage via Label-Free Impedimetric Toxicity Screening. *Lab Chip* **2011**, *11*, 2352–2361. [CrossRef] [PubMed]

43. Wang, L.; Wu, J.; Chen, J.; Dou, W.; Zhao, Q.; Han, J.; Liu, J.; Su, W.; Li, A.; Liu, P.; et al. Advances in Reconstructing Intestinal Functionalities *in Vitro*: From Two/Three Dimensional-Cell Culture Platforms to Human Intestine-on-a-Chip. *Talanta* **2021**, *226*, 122097. [CrossRef] [PubMed]
44. Liang, D.; Su, W.; Tan, M. Advances of Microfluidic Intestine-on-a-Chip for Analyzing Anti-Inflammation of Food. *Crit. Rev. Food Sci. Nutr.* **2022**, *62*, 4418–4434. [CrossRef] [PubMed]
45. Zirath, H.; Spitz, S.; Roth, D.; Schellhorn, T.; Rothbauer, M.; Müller, B.; Walch, M.; Kaur, J.; Wörle, A.; Kohl, Y.; et al. Bridging the Academic–Industrial Gap: Application of an Oxygen and pH Sensor-Integrated Lab-on-a-Chip in Nanotoxicology. *Lab Chip* **2021** *21*, 4237–4248. [CrossRef] [PubMed]
46. Wang, L.; Han, J.; Su, W.; Li, A.; Zhang, W.; Li, H.; Hu, H.; Song, W.; Xu, C.; Chen, J. Gut-on-a-Chip for Exploring the Transport Mechanism of Hg(II). *Microsyst. Nanoeng.* **2023**, *9*, 2. [CrossRef]
47. Marrero, D.; Pujol-Vila, F.; Vera, D.; Gabriel, G.; Illa, X.; Elizalde-Torrent, A.; Alvarez, M.; Villa, R. Gut-on-a-Chip: Mimicking and Monitoring the Human Intestine. *Biosens. Bioelectron.* **2021**, *181*, 113156. [CrossRef] [PubMed]
48. Li, X.-G.; Chen, M.; Zhao, S.; Wang, X. Intestinal Models for Personalized Medicine: From Conventional Models to Microfluidic Primary Intestine-on-a-Chip. *Stem Cell Rev. Rep.* **2022**, *18*, 2137–2151. [CrossRef] [PubMed]
49. Trietsch, S.J.; Naumovska, E.; Kurek, D.; Setyawati, M.C.; Vormann, M.K.; Wilschut, K.J.; Lanz, H.L.; Nicolas, A.; Ng, C.P.; Joore, J. et al. Membrane-Free Culture and Real-Time Barrier Integrity Assessment of Perfused Intestinal Epithelium Tubes. *Nat. Commun* **2017**, *8*, 262. [CrossRef] [PubMed]
50. Bein, A.; Shin, W.; Jalili-Firoozinezhad, S.; Park, M.H.; Sontheimer-Phelps, A.; Tovaglieri, A.; Chalkiadaki, A.; Kim, H.J.; Ingber D.E. Microfluidic Organ-on-a-Chip Models of Human Intestine. *Cell. Mol. Gastroenterol. Hepatol.* **2018**, *5*, 659–668. [CrossRef]
51. Sciurti, E.; Blasi, L.; Prontera, C.T.; Barca, A.; Giampetruzzi, L.; Verri, T.; Siciliano, P.A.; Francioso, L. TEER and Ion Selective Transwell-Integrated Sensors System for Caco-2 Cell Model. *Micromachines* **2023**, *14*, 496. [CrossRef]
52. Soscia, D.A.; Lam, D.; Tooker, A.C.; Enright, H.A.; Triplett, M.; Karande, P.; Peters, S.K.G.; Sales, A.P.; Wheeler, E.K.; Fischer, N.O A Flexible 3-Dimensional Microelectrode Array for in Vitro Brain Models. *Lab Chip* **2020**, *20*, 901–911. [CrossRef]
53. Cecen, B.; Saygili, E.; Zare, I.; Nejati, O.; Khorsandi, D.; Zarepour, A.; Alarcin, E.; Zarrabi, A.; Topkaya, S.N.; Yesil-Celiktas, O. et al. Biosensor Integrated Brain-on-a-Chip Platforms: Progress and Prospects in Clinical Translation. *Biosens. Bioelectron.* **2023**, *225*, 115100. [CrossRef]
54. Weltin, A.; Slotwinski, K.; Kieninger, J.; Moser, I.; Jobst, G.; Wego, M.; Ehret, R.; Urban, G.A. Cell Culture Monitoring for Drug Screening and Cancer Research: A Transparent, Microfluidic, Multi-Sensor Microsystem. *Lab Chip* **2013**, *14*, 138–146. [CrossRef] [PubMed]
55. Huang, S.-H.; Lin, S.-P.; Liang, C.-K.; Chen, J.-J.J. Impedimetric Monitoring of IGF-1 Protection of in Vitro Cortical Neurons under Ischemic Conditions. *Biomed. Microdevices* **2013**, *15*, 135–143. [CrossRef]
56. Wang, Y.I.; Abaci, H.E.; Shuler, M.L. Microfluidic Blood-Brain Barrier Model Provides In Vivo-Like Barrier Properties for Drug Permeability Screening. *Biotechnol. Bioeng.* **2017**, *114*, 184–194. [CrossRef]
57. Gehre, C.; Flechner, M.; Kammerer, S.; Küpper, J.-H.; Coleman, C.D.; Püschel, G.P.; Uhlig, K.; Duschl, C. Real Time Monitoring of Oxygen Uptake of Hepatocytes in a Microreactor Using Optical Microsensors. *Sci. Rep.* **2020**, *10*, 13700. [CrossRef]
58. Zhao, C.; Wang, Z.; Tang, X.; Qin, J.; Jiang, Z. Recent Advances in Sensor-Integrated Brain-on-a-Chip Devices for Real-Time Brain Monitoring. *Colloids Surf. B Biointerfaces* **2023**, *229*, 113431. [CrossRef] [PubMed]
59. Spitz, S.; Schobesberger, S.; Brandauer, K.; Ertl, P. Sensor-integrated Brain-on-a-chip Platforms: Improving the Predictive Validity in Neurodegenerative Research. *Bioeng. Transl. Med.* **2023**, *9*, e10604. [CrossRef]
60. Marino, A.; Battaglini, M.; Lefevre, M.C.; Ceccarelli, M.C.; Ziaja, K.; Ciofani, G. Sensorization of Microfluidic Brain-on-a-Chip Devices: Towards a New Generation of Integrated Drug Screening Systems. *Trends Anal. Chem. TRAC* **2023**, *168*, 117319. [CrossRef] [PubMed]
61. Chen, H.; Luo, Z.; Lin, X.; Zhu, Y.; Zhao, Y. Sensors-Integrated Organ-on-a-Chip for Biomedical Applications. *Nano Res.* **2023**, *16*, 10072–10099. [CrossRef]
62. Sukhodub, L.F. Interactions between nucleotide bases in coplanar and stacking dimers under vacuum. Mass spectrometric study. *Biofizika* **1987**, *32*, 994–1005.
63. Biswas, C.; Nugent, M.A. Membrane Association of Collagenase Stimulatory Factor(s) from B-16 Melanoma Cells. *J. Cell. Biochem.* **1987**, *35*, 247–258. [CrossRef]
64. Sutterby, E.; Thurgood, P.; Baratchi, S.; Khoshmanesh, K.; Pirogova, E. Microfluidic Skin-on-a-Chip Models: Toward Biomimetic Artificial Skin. *Small* **2020**, *16*, 2002515. [CrossRef]
65. Risueño, I.; Valencia, L.; Jorcano, J.L.; Velasco, D. Skin-on-a-Chip Models: General Overview and Future Perspectives. *APL Bioeng.* **2021**, *5*, 030901. [CrossRef] [PubMed]
66. Zoio, P.; Oliva, A. Skin-on-a-Chip Technology: Microengineering Physiologically Relevant In Vitro Skin Models. *Pharmaceutics* **2022**, *14*, 682. [CrossRef] [PubMed]
67. Kumar, D.; Nadda, R.; Repaka, R. Advances and Challenges in Organ-on-Chip Technology: Toward Mimicking Human Physiology and Disease in Vitro. *Med. Biol. Eng. Comput.* **2024**, *62*, 1925–1957. [CrossRef]
68. Torisawa, Y.; Mammoto, T.; Jiang, E.; Jiang, A.; Mammoto, A.; Watters, A.L.; Bahinski, A.; Ingber, D.E. Modeling Hematopoiesis and Responses to Radiation Countermeasures in a Bone Marrow-on-a-Chip. *Tissue Eng. Part C Methods* **2016**, *22*, 509–515. [CrossRef]

69. Huh, D.; Matthews, B.D.; Mammoto, A.; Montoya-Zavala, M.; Yuan Hsin, H.; Ingber, D.E. Reconstituting Organ-Level Lung Functions on a Chip. *Science* **2010**, *328*, 1662–1668. [CrossRef] [PubMed]
70. Dasgupta, Q.; Jiang, A.; Wen, A.M.; Mannix, R.J.; Man, Y.; Hall, S.; Javorsky, E.; Ingber, D.E. A Human Lung Alveolus-on-a-Chip Model of Acute Radiation-Induced Lung Injury. *Nat. Commun.* **2023**, *14*, 6506. [CrossRef]
71. Jalili-Firoozinezhad, S.; Gazzaniga, F.S.; Calamari, E.L.; Camacho, D.M.; Fadel, C.W.; Bein, A.; Swenor, B.; Nestor, B.; Cronce, M.J.; Tovaglieri, A.; et al. A Complex Human Gut Microbiome Cultured in an Anaerobic Intestine-on-a-Chip. *Nat. Biomed. Eng.* **2019**, *3*, 520–531. [CrossRef]
72. Bein, A.; Fadel, C.W.; Swenor, B.; Cao, W.; Powers, R.K.; Camacho, D.M.; Naziripour, A.; Parsons, A.; LoGrande, N.; Sharma, S.; et al. Nutritional Deficiency in an Intestine-on-a-Chip Recapitulates Injury Hallmarks Associated with Environmental Enteric Dysfunction. *Nat. Biomed. Eng.* **2022**, *6*, 1236–1247. [CrossRef]
73. Zhou, M.; Zhang, X.; Wen, X.; Wu, T.; Wang, W.; Yang, M.; Wang, J.; Fang, M.; Lin, B.; Lin, H. Development of a Functional Glomerulus at the Organ Level on a Chip to Mimic Hypertensive Nephropathy. *Sci. Rep.* **2016**, *6*, 31771. [CrossRef]
74. Jang, K.-J.; Suh, K.-Y. A Multi-Layer Microfluidic Device for Efficient Culture and Analysis of Renal Tubular Cells. *Lab Chip* **2010**, *10*, 36–42. [CrossRef] [PubMed]
75. Ya, S.; Ding, W.; Li, S.; Du, K.; Zhang, Y.; Li, C.; Liu, J.; Li, F.; Li, P.; Luo, T.; et al. On-Chip Construction of Liver Lobules with Self-Assembled Perfusable Hepatic Sinusoid Networks. *ACS Appl. Mater. Interfaces* **2021**, *13*, 32640–32652. [CrossRef]
76. Jang, K.-J.; Otieno, M.A.; Ronxhi, J.; Lim, H.-K.; Ewart, L.; Kodella, K.R.; Petropolis, D.B.; Kulkarni, G.; Rubins, J.E.; Conegliano, D.; et al. Reproducing Human and Cross-Species Drug Toxicities Using a Liver-Chip. *Sci. Transl. Med.* **2019**, *11*, eaax5516 [CrossRef] [PubMed]
77. Rigat-Brugarolas, L.G.; Elizalde-Torrent, A.; Bernabeu, M.; Niz, M.D.; Martin-Jaular, L.; Fernandez-Becerra, C.; Homs-Corbera, A.; Samitier, J.; Portillo, H.A. del A Functional Microengineered Model of the Human Splenon-on-a-Chip. *Lab Chip* **2014**, *14*, 1715–1724. [CrossRef] [PubMed]
78. Lee, S.; Kim, Y.G.; Jung, H.-I.; Lim, J.S.; Nam, K.C.; Choi, H.S.; Kwak, B.S. Bone-on-a-Chip Simulating Bone Metastasis in Osteoporosis. *Biofabrication* **2024**, *16*, 045025. [CrossRef] [PubMed]
79. Park, T.-E.; Mustafaoglu, N.; Herland, A.; Hasselkus, R.; Mannix, R.; FitzGerald, E.A.; Prantil-Baun, R.; Watters, A.; Henry, O.; Benz, M.; et al. Hypoxia-Enhanced Blood-Brain Barrier Chip Recapitulates Human Barrier Function and Shuttling of Drugs and Antibodies. *Nat. Commun.* **2019**, *10*, 2621. [CrossRef]
80. Goyal, G.; Prabhala, P.; Mahajan, G.; Bausk, B.; Gilboa, T.; Xie, L.; Zhai, Y.; Lazarovits, R.; Mansour, A.; Kim, M.S.; et al. Ectopic Lymphoid Follicle Formation and Human Seasonal Influenza Vaccination Responses Recapitulated in an Organ-on-a-Chip. *Adv. Sci.* **2022**, *9*, 2103241. [CrossRef]
81. Hansen, J.E.; Ampaya, E.P.; Bryant, G.H.; Navin, J.J. Branching Pattern of Airways and Air Spaces of a Single Human Terminal Bronchiole. *J. Appl. Physiol.* **1975**, *38*, 983–989. [CrossRef]
82. Manickavel, S. Pathophysiology of Respiratory Failure and Physiology of Gas Exchange during ECMO. *Indian J. Thorac. Cardiovasc. Surg.* **2021**, *37*, 203–209. [CrossRef]
83. Shrestha, J.; Razavi Bazaz, S.; Aboulkheyr Es, H.; Yaghobian Azari, D.; Thierry, B.; Ebrahimi Warkiani, M.; Ghadiri, M. Lung-on-a-Chip: The Future of Respiratory Disease Models and Pharmacological Studies. *Crit. Rev. Biotechnol.* **2020**, *40*, 213–230. [CrossRef]
84. Chen, X.; Liu, S.; Han, M.; Long, M.; Li, T.; Hu, L.; Wang, L.; Huang, W.; Wu, Y. Engineering Cardiac Tissue for Advanced Heart-on-a-Chip Platforms. *Adv. Healthc. Mater.* **2024**, *13*, 2301338. [CrossRef] [PubMed]
85. Cho, K.W.; Lee, W.H.; Kim, B.-S.; Kim, D.-H. Sensors in Heart-on-a-Chip: A Review on Recent Progress. *Talanta* **2020**, *219*, 121269. [CrossRef]
86. Criscione, J.; Rezaei, Z.; Hernandez Cantu, C.M.; Murphy, S.; Shin, S.R.; Kim, D.-H. Heart-on-a-Chip Platforms and Biosensor Integration for Disease Modeling and Phenotypic Drug Screening. *Biosens. Bioelectron.* **2023**, *220*, 114840. [CrossRef]
87. Zimmermann, W.-H.; Schneiderbanger, K.; Schubert, P.; Didié, M.; Münzel, F.; Heubach, J.F.; Kostin, S.; Neuhuber, W.L.; Eschenhagen, T. Tissue Engineering of a Differentiated Cardiac Muscle Construct. *Circ. Res.* **2002**, *90*, 223–230. [CrossRef]
88. Oštádalová, I.; Kolář, F.; Oštádal, B.; Rohlíček, V.; Rohlíček, J.; Procházka, J. Early Postnatal Development of Contractile Performance and Responsiveness to Ca^{2+}, Verapamil and Ryanodine in the Isolated Rat Heart. *J. Mol. Cell. Cardiol.* **1993**, *25*, 733–740. [CrossRef] [PubMed]
89. Bannerman, D.; Pascual-Gil, S.; Wu, Q.; Fernandes, I.; Zhao, Y.; Wagner, K.T.; Okhovatian, S.; Landau, S.; Rafatian, N.; Bodenstein, D.F.; et al. Heart-on-a-Chip Model of Epicardial–Myocardial Interaction in Ischemia Reperfusion Injury. *Adv. Healthc. Mater.* **2024**, *13*, e2302642. [CrossRef]
90. Abdel-Misih, S.R.Z.; Bloomston, M. Liver Anatomy. *Surg. Clin. N. Am.* **2010**, *90*, 643–653. [CrossRef] [PubMed]
91. Kruepunga, N.; Hakvoort, T.B.M.; Hikspoors, J.P.J.M.; Köhler, S.E.; Lamers, W.H. Anatomy of Rodent and Human Livers: What Are the Differences? *Biochim. Biophys. Acta BBA-Mol. Basis Dis.* **2019**, *1865*, 869–878. [CrossRef]
92. Liu, M.; Xiang, Y.; Yang, Y.; Long, X.; Xiao, Z.; Nan, Y.; Jiang, Y.; Qiu, Y.; Huang, Q.; Ai, K. State-of-the-Art Advancements in Liver-on-a-Chip (LOC): Integrated Biosensors for LOC. *Biosens. Bioelectron.* **2022**, *218*, 114758. [CrossRef] [PubMed]
93. Jiao, D.; Xie, L.; Xing, W. A Pumpless Liver-on-a-Chip for Drug Hepatotoxicity Analysis. *Analyst* 2024, *advance article*. [CrossRef]
94. Wang, D.; Gust, M.; Ferrell, N. Kidney-on-a-Chip: Mechanical Stimulation and Sensor Integration. *Sensors* **2022**, *22*, 6889. [CrossRef] [PubMed]

95. de Mello, C.P.P.; Carmona-Moran, C.; McAleer, C.W.; Perez, J.; Coln, E.A.; Long, C.J.; Oleaga, C.; Riu, A.; Note, R.; Teissier, S.; et al Microphysiological Heart-Liver Body-on-a-Chip System with Skin Mimic for Evaluating Topical Drug Delivery. *Lab Chip* **2020**, *20*, 749–759. [CrossRef]
96. Maschmeyer, I.; Lorenz, A.K.; Schimek, K.; Hasenberg, T.; Ramme, A.P.; Hübner, J.; Lindner, M.; Drewell, C.; Bauer, S.; Thomas, A.; et al. A Four-Organ-Chip for Interconnected Long-Term Co-Culture of Human Intestine, Liver, Skin and Kidney Equivalents. *Lab Chip* **2015**, *15*, 2688–2699. [CrossRef]
97. Theobald, J.; Ghanem, A.; Wallisch, P.; Banaeiyan, A.A.; Andrade-Navarro, M.A.; Taškova, K.; Haltmeier, M.; Kurtz, A.; Becker, H.; Reuter, S.; et al. Liver-Kidney-on-Chip To Study Toxicity of Drug Metabolites. *ACS Biomater. Sci. Eng.* **2018**, *4*, 78–89. [CrossRef] [PubMed]
98. Beger, R.D.; Sun, J.; Schnackenberg, L.K. Metabolomics Approaches for Discovering Biomarkers of Drug-Induced Hepatotoxicity and Nephrotoxicity. *Toxicol. Appl. Pharmacol.* **2010**, *243*, 154–166. [CrossRef] [PubMed]
99. Trefts, E.; Gannon, M.; Wasserman, D.H. The Liver. *Curr. Biol. CB* **2017**, *27*, R1147–R1151. [CrossRef]
100. Huang, Q.; Yang, T.; Song, Y.; Sun, W.; Xu, J.; Cheng, Y.; Yin, R.; Zhu, L.; Zhang, M.; Ma, L.; et al. A Three-Dimensional (3D) Liver–Kidney on a Chip with a Biomimicking Circulating System for Drug Safety Evaluation. *Lab Chip* **2024**, *24*, 1715–1726. [CrossRef] [PubMed]
101. Monteduro, A.G.; Rizzato, S.; Caragnano, G.; Trapani, A.; Giannelli, G.; Maruccio, G. Organs-on-Chips Technologies—A Guide from Disease Models to Opportunities for Drug Development. *Biosens. Bioelectron.* **2023**, *231*, 115271. [CrossRef]
102. Jie, M.; Mao, S.; Liu, H.; He, Z.; Li, H.-F.; Lin, J.-M. Evaluation of Drug Combination for Glioblastoma Based on an Intestine–Liver Metabolic Model on Microchip. *Analyst* **2017**, *142*, 3629–3638. [CrossRef]
103. Tanimizu, N.; Ichinohe, N.; Sasaki, Y.; Itoh, T.; Sudo, R.; Yamaguchi, T.; Katsuda, T.; Ninomiya, T.; Tokino, T.; Ochiya, T.; et al. Generation of Functional Liver Organoids on Combining Hepatocytes and Cholangiocytes with Hepatobiliary Connections Ex Vivo. *Nat. Commun.* **2021**, *12*, 3390. [CrossRef]
104. Zhou, S.-F.; Zhong, W.-Z. Drug Design and Discovery: Principles and Applications. *Mol. J. Synth. Chem. Nat. Prod. Chem.* **2017**, *22*, 279. [CrossRef] [PubMed]
105. Marturano-Kruik, A.; Nava, M.M.; Yeager, K.; Chramiec, A.; Hao, L.; Robinson, S.; Guo, E.; Raimondi, M.T.; Vunjak-Novakovic, G. Human Bone Perivascular Niche-on-a-Chip for Studying Metastatic Colonization. *Proc. Natl. Acad. Sci. USA* **2018**, *115*, 1256–1261. [CrossRef]
106. Toh, Y.-C.; Raja, A.; Yu, H.; van Noort, D. A 3D Microfluidic Model to Recapitulate Cancer Cell Migration and Invasion. *Bioengineering* **2018**, *5*, 29. [CrossRef] [PubMed]
107. Imparato, G.; Urciuolo, F.; Netti, P.A. Organ on Chip Technology to Model Cancer Growth and Metastasis. *Bioengineering* **2022**, *9*, 28. [CrossRef]
108. Regmi, S.; Poudel, C.; Adhikari, R.; Luo, K.Q. Applications of Microfluidics and Organ-on-a-Chip in Cancer Research. *Biosensors* **2022**, *12*, 459. [CrossRef] [PubMed]
109. Witkowski, M.T.; Dolgalev, I.; Evensen, N.A.; Ma, C.; Chambers, T.; Roberts, K.G.; Sreeram, S.; Dai, Y.; Tikhonova, A.N.; Lasry, A.; et al. Extensive Remodeling of the Immune Microenvironment in B-Cell Acute Lymphoblastic Leukemia. *Cancer Cell* **2020**, *37*, 867–882.e12. [CrossRef] [PubMed]
110. Zhang, Y.S.; Aleman, J.; Shin, S.R.; Kilic, T.; Kim, D.; Mousavi Shaegh, S.A.; Massa, S.; Riahi, R.; Chae, S.; Hu, N.; et al. Multisensor-Integrated Organs-on-Chips Platform for Automated and Continual in Situ Monitoring of Organoid Behaviors. *Proc. Natl. Acad. Sci. USA* **2017**, *114*, E2293–E2302. [CrossRef]
111. Jalili-Firoozinezhad, S.; Prantil-Baun, R.; Jiang, A.; Potla, R.; Mammoto, T.; Weaver, J.C.; Ferrante, T.C.; Kim, H.J.; Cabral, J.M.S.; Levy, O.; et al. Modeling Radiation Injury-Induced Cell Death and Countermeasure Drug Responses in a Human Gut-on-a-Chip. *Cell Death Dis.* **2018**, *9*, 223. [CrossRef]
112. Liu, Y.; Sakolish, C.; Chen, Z.; Phan, D.T.T.; Bender, R.H.F.; Hughes, C.C.W.; Rusyn, I. Human In Vitro Vascularized Micro-Organ and Micro-Tumor Models Are Reproducible Organ-on-a-Chip Platforms for Studies of Anticancer Drugs. *Toxicology* **2020**, *445*, 152601. [CrossRef]
113. Herland, A.; Maoz, B.M.; Das, D.; Somayaji, M.R.; Prantil-Baun, R.; Novak, R.; Cronce, M.; Huffstater, T.; Jeanty, S.S.F.; Ingram, M.; et al. Quantitative Prediction of Human Pharmacokinetic Responses to Drugs via Fluidically Coupled Vascularized Organ Chips. *Nat. Biomed. Eng.* **2020**, *4*, 421–436. [CrossRef]
114. Yan, J.; Li, Z.; Guo, J.; Liu, S.; Guo, J. Organ-on-a-Chip: A New Tool for in Vitro Research. *Biosens. Bioelectron.* **2022**, *216*, 114626. [CrossRef] [PubMed]
115. Ajalik, R.E.; Alenchery, R.G.; Cognetti, J.S.; Zhang, V.Z.; McGrath, J.L.; Miller, B.L.; Awad, H.A. Human Organ-on-a-Chip Microphysiological Systems to Model Musculoskeletal Pathologies and Accelerate Therapeutic Discovery. *Front. Bioeng. Biotechnol.* **2022**, *10*, 846230. [CrossRef] [PubMed]
116. Singh, G.; Mishra, A.; Mathur, A.; Shastri, S.; Nizam, A.; Rizwan, A.; Dadial, A.S.; Firdous, A.; Hassan, H. Advancement of Organ-on-Chip towards next Generation Medical Technology. *Biosens. Bioelectron. X* **2024**, *18*, 100480. [CrossRef]
117. Novak, R.; Ingram, M.; Marquez, S.; Das, D.; Delahanty, A.; Herland, A.; Maoz, B.M.; Jeanty, S.S.F.; Somayaji, M.R.; Burt, M.; et al. Robotic Fluidic Coupling and Interrogation of Multiple Vascularized Organ Chips. *Nat. Biomed. Eng.* **2020**, *4*, 407–420. [CrossRef]
118. Kratz, S.R.A.; Höll, G.; Schuller, P.; Ertl, P.; Rothbauer, M. Latest Trends in Biosensing for Microphysiological Organs-on-a-Chip and Body-on-a-Chip Systems. *Biosensors* **2019**, *9*, 110. [CrossRef]

MDPI AG
Grosspeteranlage 5
4052 Basel
Switzerland
Tel.: +41 61 683 77 34

Biosensors Editorial Office
E-mail: biosensors@mdpi.com
www.mdpi.com/journal/biosensors

www.ingramcontent.com/pod-product-compliance
Lightning Source LLC
Chambersburg PA
CBHW041411210926
51850CB00013B/3242

* 9 7 8 3 7 2 5 8 2 8 2 1 0 *